ENCYCLOPAEDIA OF

FOOD TECHNOLOGY

ENCYCLOPAEDIA OF
FOOD TECHNOLOGY

Sandhya Trivedi

ANMOL PUBLICATIONS PVT. LTD.
NEW DELHI - 110 002 (INDIA)

ANMOL PUBLICATIONS PVT. LTD.
H.O.: 4374/4B, Ansari Road, Darya Ganj,
New Delhi-110 002 (India)
Ph.: 23278000, 23261597
B.O.: No. 1015, Ist Main Road, BSK IIIrd Stage
IIIrd Phase, IIIrd Block
Bangalore - 560 085 (India)
Visit us at: www.anmolpublications.com

Encyclopaedia of Food Technology

First Published, 2006
ISBN 81-261-3034-2

PRINTED IN INDIA

Printed at Mehra Offset Press, Delhi.

Preface

This Encyclopaedia is the perfect guide for those studying Food technology, either at university level or anyone interested in Food technology, such as consumers, cooks, food manufacturers, salespeople or students, who must understand, interpret, and communicate information to others. It provides a comprehensive coverage of Food technology. The aim of this encyclopaedia is to provide teachers, students and laymen interested in Food technology with clear, concise, and correct definitions and descriptions of the Food technology terms, throughout.

This volume provides clear and lucid definitions of terms associated with all aspects of food technology and various aspects of food science such as nutrition, diets, food values, different types of foods and various processing techniques. In this volume all the branches of Food technology are represented. This volume covers whole realm of Food technology.

The aim of this Encyclopaedia is to satisfy the immediate technical needs of the user. Concepts are presented in a clear and lucid manner. In terms of its broader objectives, the present book may be viewed as an outline or compendium of Food technology ideas correlated with a large collection of specialized items.

The aim of this volume is to provide a concise repository of number of major concepts in this field. In preparation of this book, the author has freely consulted large number of books and journals so no authenticity is claimed. Author is especially thankful to Anmol Publications Pvt. Ltd., New Delhi for shaping this book in its final form. Suggestions for further improvement of this book are not only welcome but also greatly appreciated.

Author

A

A Vitamin A is a vitamin with antioxidant properties, important for eye protection and bone growth; protein and hormone synthesis; supports tissue maintenance. Helps reduce susceptibility to infection. Essential for healthy skin, good blood, strong bones and teeth, kidneys, bladder, lungs and membranes.

Abalone A shellfish, Haliotus splendens, H. rufescens, H. cracherodii, also sometimes called ormer, or sea ear. Found especially in waters around Australia, and also California, Japan, the Channel Islands, and France. A rich source of protein and niacin; a source of iron and vitamin B1.

Absinthe A herb liqueur flavoured with wormwood; it is toxic, and banned in many countries.

Absolute alcohol Pure ethyl alcohol.

Absorb The process by which nutrients are taken into the bloodstream from the body's intestinal tract after food digestion.

Acarbose The name of a group of complex carbohydrates which inhibit the enzymes of starch and disaccharide digestion; used experimentally to reduce the digestion of starch, and so slow the rate of absorption of carbohydrates. It has been marketed for use in association with weight-reducing diet regimes as a 'starch blocker', but there is no evidence that it is of any use whatsoever in weight reduction.

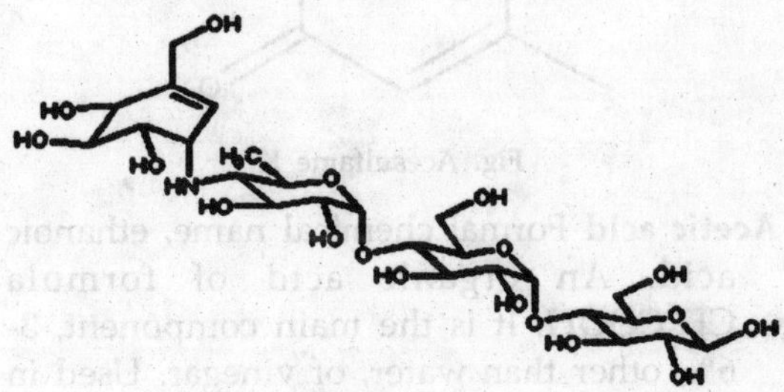

Fig. Acarbose

Acaricides Pesticides used to kill mites and ticks which cause animal diseases and the spoilage of flour and other foods in storage.

Accelase A mixture of enzymes which hydrolyse proteins, including an exopeptidase from the bacterium Streptococcus lactis, which is one of the starter organisms in dairy processing. The mixed enzymes are used to shorten the maturation time of cheeses and intensify the flavour of processed cheese.

Acceptable daily intake The amount of a food additive that could be taken daily for an entire life-span without appreciable risk. Determined by measuring the highest dose of the substance that has no effect on experimental animals, then dividing by a safety factor of 100. Substances that are not given an ADI are regarded as having no adverse effect at any level of intake.

Acesulfame K An artificial sweetener. Approved in 1988, this sweetener is 200 times sweeter than sugar, yet has a bitter taste. The 'K' in its name stands or potassium. Acesulfame-K is not metabolized by the body. You can cook and bake with it. There are no reported side effects with this sweetener although it may increase insulun secretion, which can cause concern for diabetics.

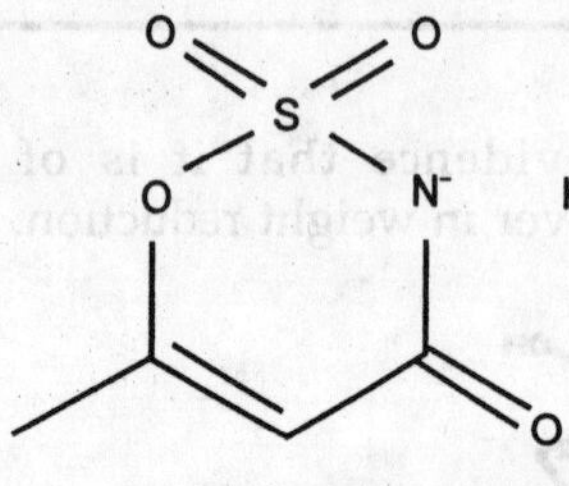

Fig. Acesulfame K

Acetic acid Formal chemical name, ethanoic acid. An organic acid of formula CH_3COOH. It is the main component, 3-6%, other than water, of vinegar. Used in fish technology in preparation of marinades.

Acetobacter A genus of bacteria which oxidize ethyl alcohol to acetic acid, used in the manufacture of vinegar. They also grow as a film on the surface of beer wort, pickle brine, and fruit juices, when they are commonly known as 'mother of vinegar'.

Acetomenaphthone Synthetic compound with vitamin K activity; vitamin K3, also known as menaquinone-0.

Acetone One of the ketone bodies formed in the body in fasting. It is a metabolically useless side-product of fat metabolism, but detection of acetone in blood, urine, or breath may be clinically useful in cases of diabetes, as a means of detecting ketosis. Also used as a solvent, e.g. in varnishes and lacquer. Chemically dimethyl ketone or propan-2-one.

Acetyl L carnitine The acetyl ester of carnitine, ALC acts as an antioxidant, has protective effects in the brain, and stimulates hormone release.

Acetylcholine The acetyl derivative of choline, produced at some nerve endings both in the brain, where it acts as a chemical transmitter, and at the junctions between nerves and muscles, where it stimulates muscle contraction.

Achene Botanical term for small, dry, one-seeded fruit which does not open to liberate the seed, e.g. a nut.

Achromotricia Loss of the pigment of hair. One of the signs of pantothenic acid deficiency in animals, but there is no evidence that pantothenic acid affects loss of hair colour in human beings.

Acid Chemically, compounds that dissociate in water to give rise to hydrogen ions; they taste sour. Mineral acids such as hydrochloric, sulphuric, and nitric are more or less completely dissociated and so are strong acids. Organic acids are generally weak since they are not completely dissociated.

Acid drops Boiled sweets with sharp flavour from tartaric acid; known as sourballs in USA.

Acidophilus milk Milk seeded with living Lactobacillus acidophilus bacteria used to improve the balance of beneficial bacteria in the intestines of the person who drinks the milk.

Acidosis An increase in the acidity of blood plasma to below the normal range of pH 7.3-7.45, resulting from a loss of the buffering capacity of the plasma, alteration in the excretion of carbon dioxide, excessive loss of base from the body or metabolic overproduction of acids.

Acorn Fruit of the oak tree, used both for animal feed and to make a flour for baking. Roasted acorns have been used as a coffee substitute.

Acorn sugar A sweet compound extracted from acorns.

Acrolein An aldehyde formed when glycerol is heated to a high temperature. It is responsible for the acrid odour and lachrymatory vapour produced when fats are overheated. Chemically CH 2 =CH.CHO.

Acronize Trade name for the antibiotic chlortetracycline; 'acronized' is used to describe products that have been treated with chlortetracycline, as, e.g. acronized ice.

Activators Compounds that increase the activity of enzymes.

Active oxygen method A method of measuring the stability of fats and oils to oxidative damage by bubbling air through the heated material and following the formation of peroxides. Also known as the Swift stability test.

Actomyosin A combination of the two main proteins, actin and myosin, present in all muscle tissues.

Additive A chemical added to a food to affect its properties. Objectives of including additives in a product include increased stability during storage; inhibition of growth of microorganisms or production of microbial toxins; prevention or reduction of formation of off-flavours; improved sensory properties, particularly colours and appearance, affecting acceptability to the consumer; improved properties related to preparation and processing of food, for example, ability to create stable foams or emulsions, or to stabilise or thicken sauces. Additives permitted for use in foods are regulated by food authorities. Typically, there are two lists additives that may be used without limit in any food; and additives that may be used within maximum limits and/or only in specified foods. Some permitted additives are chemicals naturally present in foods such as ascorbic acid and sodium chloride.

ADI The Acceptable Daily Intake is defined as an estimate of the amount of a food additive, expressed on a bodyweight basis, that can be ingested on a daily basis in the diet over a lifetime without appreciable risk to health. "Without appreciable risk" means the practical, in view of the actual level of knowledge, certainty that no harm will result, even after a lifetime of exposure to the chemical additive concerned. The ADI is usually given as a range of 0–x milligrams per kilogram of bodyweight per day.

Adipectomy Surgical removal of subcutaneous fat.

Adipose tissue Body fat—the cells that synthesize and store fat, releasing it for metabolism in fasting. Also known as white adipose tissue, to distinguish it from the metabolically more active brown adipose tissue, which is involved in heat production to maintain body temperature. Much of the body fat reserve is subcutaneous; in addition there is adipose tissue around the organs, which serves to protect them from physical damage. In lean people, 20-25% of body weight is adipose tissue, increasing with age; the proportion is greater in people who are overweight or obese. Adipose tissue contains 82-88% fat, 2-2.6% protein and 10-14% water.

Adirondack bread American baked product made from ground maize, butter, wheat flour, eggs, and sugar.

Adlay The seeds of a wild grass botanically related to maize, growing wild in parts of Africa and Asia and eaten especially in the south-east Pacific region.

Adrenocorticotrophic hormone A hormone secreted by the anterior part of the pituitary gland which stimulates the adrenal gland to secrete corticosteroids.

Adulteration The adding of cheaper, inferior or less desirable materials to a food.

Adverse reactions to foods Food aversion, unpleasant reactions caused by emotional responses to certain foods rather than to the foods themselves, which are unlikely to occur in blind testing when the foods are disguised. Food allergy, physiological reactions to specific foods or ingredients due to an immunological response. Antibodies to the allergen are formed as a result of previous exposure or sensitization, and cause a variety of symptoms when the food is eaten, including gastro-intestinal disturbances, skin rashes, asthma, and, in severe cases, anaphylactic shock, which may be fatal. Food intolerance, physiological reactions to specific foods or ingredients which are not due to immunological responses, but may result from the irritant action of spices, pharmacological actions of some naturally occurring compounds, or an inability to metabolize a component of the food as a result of an enzyme defect.

Aerate To add a gas to, or to let the air act on.

In bread making, adding air to the mixture makes the bread light and fluffy.

Aerobic Requires oxygen.

Aerosol cream Cream sterilized and packaged in aerosol canisters with a propellant gas to expel it from the container, giving conveniently available whipped cream. Gelling agents and stabilizers may also be added.

Aesculln A glucoside of dihydroxycoumarin found in the leaves and bark of the horse chestnut tree which has an effect on capillary fragility.

Affective test A class of experimental procedures used in sensory assessment of foods to determine an assessor's personal response the food. Also referred to as a subjective test. Examples of affective tests are pair comparison tests in which the assessor is asked to express their preference between a pair of products, and hedonic tests in which the assessor expresses their degree of liking/disliking or acceptability/unacceptability of a product.

Aflatoxins Aflatoxins are toxic metabolites produced by Aspergillus flavus or A.

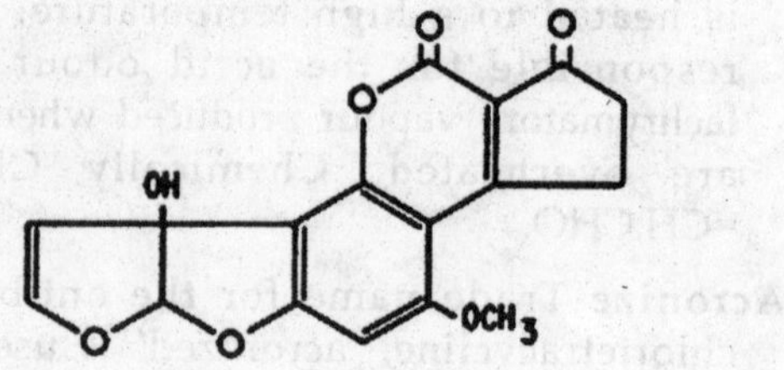

M1

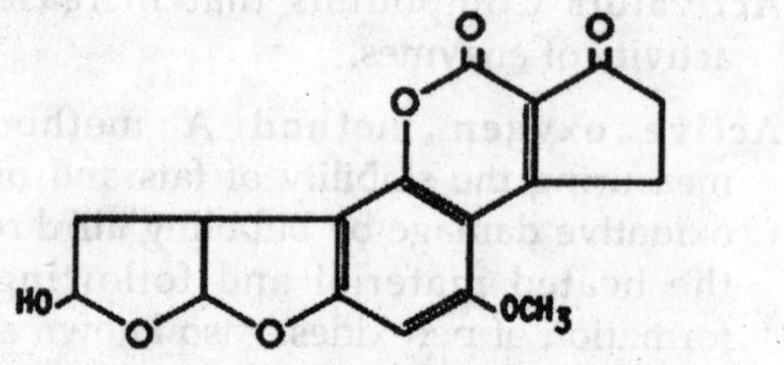

B2A

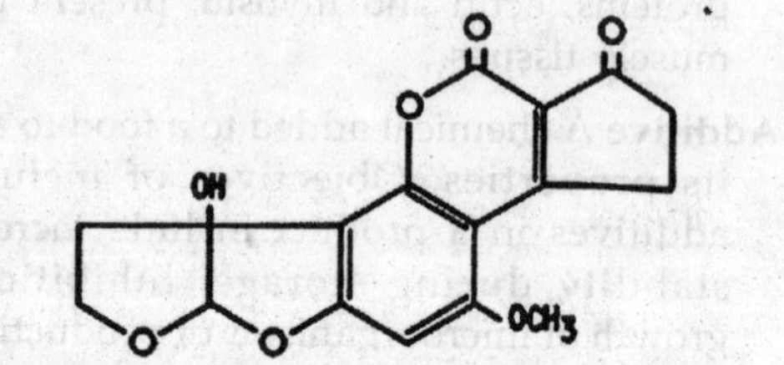

M2

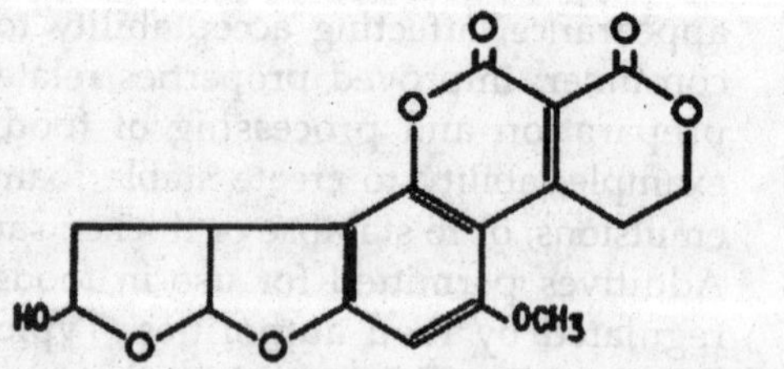

G2A

Fig. Aflatoxins

parasiticus in/on ioods and feeds. They are probably the best known and most intensively researched mycotoxins in the world. Aflatoxins have been associated with various diseases, such as aflatoxicosis, in livestock, domestic animals and humans throughout the world. The occurrence of aflatoxins is influenced by certain environmental factors; hence the extent of contamination will vary with geographic location, agricultural and agronomic practices, and the susceptibility of commodities to fungal invasion during preharvest, storage, and/or processing periods. Aflatoxins have received greater attention than any other mycotoxins because of their demonstrated potent carcinogenic effect in susceptible laboratory animals and their acute toxicological effects in humans. As it is realized that absolute safety is never achieved, many countries have attempted to limit exposure to aflatoxins by imposing regulatory limits on commodities intended for use as food and feed.

In the 1960 more than 100,000 young turkeys on poultry farms in England died in the course of a few months from an apparently new disease that was termed "Turkey X disease" . It was soon found that the difficulty was not limited to turkeys. Ducklings and young pheasants were also affected and heavy mortality was experienced.

A careful survey of the early outbreaks showed that they were all associated with feeds, namely Brazilian peanut meal. An intensive investigation of the suspect peanut meal was undertaken and it was quickly found that this peanut meal was highly toxic to poultry and ducklings with symptoms typical of Turkey X disease.

Speculations made during 1960 regarding the nature of the toxin suggested that it might be of fungal origin. In fact, the toxin-producing fungus was identified as Aspergillus flavus and the toxin was given the name Aflatoxin by virtue of its origin.

This discovery has led to a growing awareness of the potential hazards of these substances as contaminants of food and feed causing illness and even death in humans and other mammals. Studies revealed that aflatoxins are produced primarily by some strains of A. flavus and by most, if not all, strains of A. parasiticus.

There are four major aflatoxins : B1, B2, G1, G2. Whereas the B designation of aflatoxins B1 and B2 resulted from the exhibition of blue fluorescence under UV-light, while the G designation refers to the yellow-green fluorescence of the relevant structures under UV-light. In addition, two metabolic products, aflatoxin M1 and M2, that are of significance as direct contaminants of foods and feeds. These were first isolated from milk of lactating animals fed aflatoxin preparations; hence, the M designation.

These toxins have closely similar structures and form a unique group of highly oxygenated, naturally occurring hetero-cyclic compounds.

Agalactia Failure of the mother to secrete enough milk to feed a suckling infant.

Agar Dried extracts from various seaweeds, including Gelidium and Gracilaria spp. It is a partially soluble non-starch polysaccharide, composed of galactose units. It swells with water to form a gel, and is used in soups, jellies, ice-cream, and meat products. It is also used as the basis of bacteriological culture media, as an adhesive, for sizing silk and as a stabilizer for emulsions. Also called agar-agar, Macassar gum, vegetable gelatine.

Agene Nitrogen trichloride, used at one time as a bleaching and improving agent for wheat flour in bread making. It can react with the amino acid methionine in proteins to form the toxic compound methionine sulphoximine, and is no longer used.

Ageusia Loss or impairment of the sense of taste.

Agglomeration The process of producing a free-flowing, dust-free powder from substances such as dried milk powder and wheat flour, by moistening the powder with droplets of water and then redrying in a stream of air. The resulting agglomerates can readily be wetted.

Aiguillette A thin strip or slice of cooked poultry, meat, or fish.

Air classification A way of separating the particles of powdered materials in a current of air, on the basis of their weight and size or density. Particularly applied to the fractionation of the endosperm of milled wheat flour; smaller particles are richer in protein. Various fractions range from 3% to 25% protein.

Air curtain Continuous stream of air directed downwards or sideways across the doorway of a refrigerated space to reduce the ingress of heat when the door is open.

Air speed Rate of movement of air expressed as the distance moved in unit time, as distinct from air velocity.

Air velocity The rate of movement of air, in a given direction, expressed as distance travelled in unit time.

Airlock Small compartment at the entrance to a cold store that prevents direct access and so reduces ingress of heat.

Aitchbone Cut of beef from the upper part of the leg. Sometimes incorrectly called the edgebone.

Alactasia Partial or complete deficiency of the enzyme lactase in the small intestine, resulting in an inability to digest the sugar lactose in milk, and hence intolerance of milk and milk products.

Alanine An amino acid. BCAAs are used as a source of energy for muscle cells. During prolonged exercise, BCAAs are released from skeletal muscles and their carbon backbones are used as fuel, while their nitrogen portion is used to form another amino acid, Alanine. Alanine is then converted to Glucose by the liver. This form of energy production is called the Alanine-Glucose cycle, and it plays a major role in maintaining the body's blood sugar balance.

Albedo The white pith of the inner peel of citrus fruits, accounting for some 20-60% of the whole fruit. It consists of sugars, cellulose, and pectins, and is used as a commercial source of pectin.

Albumin A specific class of relatively small proteins which are soluble in water and readily coagulated by heat. Ovalbumin is the main protein of egg-white, lactalbumin occurs in milk, and plasma or serum albumin is one of the major blood proteins. Serum albumin concentration is sometimes measured as an index of protein-energy malnutrition.

Often used as a non-specific term for proteins.

Albuminolds Fibrous proteins that have a structural or protective rather than enzymic role in the body. Also known as scleroproteins. The main proteins of the connective tissues of the body. There are three main types: 1. collagens in skin, tendons, and bones are resistant to enzymic digestion with trypsin and pepsin, and can be converted to soluble gelatine by boiling with water;

2. Elastins in tendons and arteries, which are not converted to gelatine on boiling;

3. Keratins, the proteins of hair, feathers, scales, horns, and hoofs, which are insoluble in dilute acid or alkali, and are resistant to all animal digestive enzymes.

Alcohol Chemically alcohols are compounds with the general formula $C_nH_{2n+1}OH$. The alcohol in alcoholic beverages is ethyl alcohol; pure ethyl alcohol is also known as absolute alcohol. The energy yield of alcohol is 7 kcal / gram.

Alcoholism Physiological addiction to alcohol, associated with persistent heavy consumption of alcoholic beverages. In addition to the addiction, there may be damage to the liver, stomach, and pancreas, as well as behavioural changes and peripheral nerve damage.

Alderman's walk The name given in London to the longest and finest cut from the haunch of venison or lamb.

Aldosterone A steroid hormone secreted by the adrenal cortex which controls the excretion of salts and water by the kidneys.

Aleatico A grape variety widely used for wine making, although not one of the classic varieties; makes fragrant sweet red wines.

Alecost An aromatic herbaceous plant, Tanacetum balsamita, related to tansy, used in salads and formerly used to flavour ale.

Aleurone layer The single layer of large cells under the bran coat and outside the endosperm of cereal grains. About 3% of the weight of the grain, and rich in protein, as well as containing about 20% of the vitamin B 1, 30% of the vitamin B 2 and 50% of the niacin of the grain. Botanically the aleurone layer is part of the endosperm, but in milling it remains attached to the inner layer of the bran.

Alewives River herrings, Pomolobus pseudoharengus, commonly used for canning after salting.

Alexanders A herb, black lovage with a celery-like flavour.

Algae Simple plants that do not show differentiation into roots, stems, and leaves. They are mostly aquatic — either seaweeds or pond and river-weeds. Some seaweeds, such as dulse and Irish moss, have long been eaten, and a number of unicellular algae, including Chlorella, Scenedesmus, and Spirulina spp. have been grown experimentally as novel sources of food.

Alginates Salts of alginic acid found in many seaweeds as calcium salts or the free acid. Chemically, alginic acid is a non-starch polysaccharide Composed of mannuronic acid units.

Iron, magnesium, and ammonium salts of alginic acid form viscous solutions and hold large amounts of water. They are used as thickeners, stabilizers and gelling, binding, and emulsifying agents in food manufacture, especially in ice-cream and synthetic cream. Trade name Manucol.

Aligote A grape variety widely used for wine making, although not one of the classic varieties. Burgundy's second-ranking white grape; the wines need drinking within three years.

Alimentary canal The digestive tract, comprising the mouth, oesophagus, stomach, duodenum, and small and large intestines.

Alkali A compound that takes up hydrogen ions and so raises the pH of a solution.

Alkaloids Naturally occurring organic bases which have marked pharmacological actions in man and other animals. Many are found in plant foods, including potatoes and tomatoes, or as the products of fungal action, although they also occur in animal foods, and some are formed in the body. A number of alkaloids are used medically, e.g. morphine, colchicine, quinine, and atropine.

Allantoin The oxidation product of uric acid which is the end-product of purine metabolism in most animals apart from man and apes, which excrete uric acid.

All-bran Trade name for a breakfast cereal prepared from wheat bran, and hence a rich source of non-starch polysaccharide. A rich source of vitamins B 1, B 2, B 6, B 12, niacin, folate, iron, and copper.

Allergen A food allergen is the part of a food that stimulates the immune system of food allergic individuals. A single food can contain multiple food allergens. Carbohydrates or fats are not allergens.

Allergy Adverse reaction to foods caused by the production of antibodies.

Alliance Sauce made from reduced white wine with tarragon vinegar and egg yolk.

Allicin A sulphur-containing compound partially responsible for the flavour of garlic.

Allinson bread A wholewheat bread named after Allinson, who advocated its consumption in England at the end of the nineteenth century, as did Graham in the USA. Now trade name for a wholemeal loaf.

Allotriophagy An unnatural desire for abnormal foods; also known as cissa, cittosis, and pica.

Allspice Dried fruits of the evergreen plant Pimenta officinalis, also known as pimento or Jamaican pepper. The name allspice derives from the aromatic oil, which has an aroma similar to a mixture of cloves, cinnamon, and nutmeg. Used to flavour meat and in baking.

Allumettes Potatoes cut into thin 'matchsticks' and fried, also known as straw potatoes. Also used sometimes for narrow fingers of pastry.

Almond A nut, the seeds of Prunus amygdalus var. dulcis. All varieties contain the glycoside amygdalin, which forms hydrogen cyanide when the nuts are crushed. The bitter almond, used for almond oil, may yield dangerous amounts of cyanide.

A rich source of protein, copper, niacin, and vitamins B 2, E; iron and zinc; B 1

Aloe Dried juice from the leaves of Aloe perryi, used in medicines. Contains a glycoside, aloe-emodin or rhabarberone, aloe oil, and aloin or barbaloin.

Alpha lipoic acid A sulfur bearing phytonutrient with antioxidant properties; amplifies effects of other antioxidants. It is an insulin potentiator that may be, in some respects, the very best insulin mimicker. An analogy of what ALA does is that if ALA was an individual, he would be the one which yells at the muscle cells to pick up the key, open the door, and help bring in the creatine.

ALA is a sulfur bearing compound with antioxidant properties. It plays a role in energy metabolism. ALA amplifies the ability of other antioxidants to combat free redicals and enhance recovery. Also may enhance insulin sensitivity, improving the body's ability to add lean mass and reduce fat.

Alpha tocopherol The most common type of vitamin E. It is considered the major lipid soluble chain-breaking antioxidant. It mainly protects cell membranes and lipoproteins from free radical oxidation.

Alpha-laval centrifuge A continuous bowl centrifuge for separating liquids of different densities and for clarifying. Widely used for cream separation.

Alpha-linolenic acid A long-chain polyunsaturated fatty acid with 18 carbons, 3 double bonds, in the omega-3 family. It is considered essential because it cannot be manufactured by the human body.

Aluminium The third most abundant element in the earth's crust but with no known biological function. Present in small amounts in many foods but only a small proportion is absorbed. Aluminium salts are found in the abnormal nerve tangles in the brain in Alzheimer's disease, and it has been suggested that aluminium poisoning may be a factor in the development of the disease, although there is little evidence.

Aluminium is used in cooking vessels and as foil for wrapping food, as well as in cans and tubes; it is a soft flexible metal, resistant to oxidation and deterioration, although it is dissolved by alkalis. The 'silver' beads used to decorate confectionery are coated with either silver foil or an alloy of aluminium and copper.

Baking powders containing sodium aluminium sulphate as the acid agent were used at one time, and aluminium hydroxide and silicates are commonly used in antacid medications.

Alveographe A device for measuring the stretching quality of dough as an index of its protein quality for baking. A standard disc of dough is blown into a bubble and the pressure curve and bursting pressure are measured to give the stability, extensibility, and strength of the dough.

Alzheimer disease Alzheimer disease is a fatal, progressive, degenerative disease of the brain that affects about 11% of all individuals over the age of 60 years .

Amaranthus A genus of plants cultivated for their leaves and sometimes for their seeds also. The leaves are promoted in South East Asia as a good source of carotene.

Amaretto Almond-flavoured liqueur made by infusion of apricot kernels.

Ambergris A waxy concretion obtained from the intestine of the sperm whale, containing cholesterol, ambrein, and benzoic acid. Used in drugs and perfumery.

Amberlite Group of polystyrene resins used to absorb specific ions in water softening and purification in general.

Ambient temperature Temperature of the surroundings, usually of the air outside or in a space under study, or of the air in the space surrounding an object or product, or of the surrounding air outside a refrigerated space or enclosure.

Aminapaptidase An enzyme secreted in the pancreatic juice which removes amino acids sequentially from the free amino terminal of a peptide or protein, until the final product is a dipeptide. Since it works at the end of the peptide chain, it is an exopeptidase.

Amincaciduria Excretion of abnormal amounts of one or more amino acids in the urine, usually as a result of a genetic disease.

Aminctransferase Any enzyme that catalyses the reaction of transamination.

Amines A class of chemicals containing nitrogen of which one nitrogen atom at least in the molecule can take on a positive charge. Ammonia is the simplest example of an amine. Because of the positively charge nitrogen atom, amines are bases and solutions in water have an alkaline reaction. The amines of most relevance in fish technology are the methylamines, trimethylamine and dimethylamine, and biogenic amines.

Amino acids Amino acids function as the building blocks of proteins. Chemically, amino acids are organic compounds containing an amino group and a carboxyl group. Amino acids are classified as essential, nonessential and conditionally essential. If body synthesis is inadequate to meet metabolic need, an amino acid is classified as essential and must be supplied as part of the diet. Essential amino acids include leucine, isoleucine, valine, tryptophan, phenylalanine, methionine, threonine, lysine, histidine and possibly arginine. Nonessential amino acids can be synthesized by the body in adequate amounts, and include alanine, aspartic acid, asparagine, glutamic acid, glutamine, glycine, proline and serine. Conditionally essential amino acids become essential under certain clinical conditions.

Aminogram A diagrammatic representation of the amino acid composition of a peptide or protein. A plasma aminogram is the composition of the free amino acid pool in blood plasma.

Ammonia A chemical with a characteristic pungent, irritating, odour. Has the chemical formula NH. It is a gas at normal ambient temperatures and pressures, but can be liquefied by pressure. The change in state on conversion from liquid to gas with a change in pressure is accompanied by a decrease in temperature and makes ammonia a useful refrigerant. Ammonia is not present in living fish, or only in trace amounts, but is formed in muscle tissue by biochemical processes in the tissues soon after death of the animal, and as result of bacterial degradation of nitrogen-containing compounds during spoilage of fish. This is particularly evident in elasmobranchs because of the degradation of urea prevalent in the muscle tissue of this class of fish.

Amniocentesis A procedure preformed during the 15th week of pregnancy or later to collect material for prenatal testing. A slender needle is inserted into the woman's abdomen and into the uterus and amniotic fluid from around the fetus is withdrawn.

Amoebiasis Infection of the intestinal tract with pathogenic amoeba from contaminated food or water, causing profuse diarrhoea, intestinal bleeding, pain, jaundice, anorexia, and weight loss.

Amomum A group of tropical plants including cardamom and melegueta pepper, which have pungent and aromatic seeds.

Amour Purple-coloured liqueur, flavoured with citrus fruits and violets.

Amphetamine A chemical at one time used as an appetite suppressant; addictive, and a common drug of abuse, its use is strictly controlled by law.

Amydon A traditional starchy material made by steeping wheat flour in water, then drying the starch sediment in the sun, used for thickening broths, etc.

Amygdalin 1. A glycoside in almonds and apricot and cherry stones which is hydrolysed by the enzyme emulsin to yield glucose, hydrocyanic acid, and benzaldehyde. It is therefore highly poisonous, although it has been promoted, with no evidence, as a nutrient, laetrile or so-called vitamin B 17. Unfounded claims have been made for its value in treating cancer.

2. French name for cakes and sweets made with almonds.

Amylodyspepsia An inability to digest starch.

Amylograph A device to measure the viscosity of flour paste as it is heated from 25°C to 90°C, which serves as a measure of the diastatic activity of the flour.

Amyloins Carbohydrates that are complexes of dextrins with varying proportions of maltose.

Amylopectin The branched chain form of starch. About 75-80% of starch is in this form, the remainder is amylose.

Amylopeptic A general description of enzymes that are able to split starch to give soluble products.

Anabolic Metabolic condition in which new molecules are synthesized.

Anaemia A shortage of red blood cells, leading to pallor and shortness of breath, especially on exertion. Most commonly due to a dietary deficiency of iron, or excessive blood losses resulting in iron losses greater than can be met from the diet. Other dietary deficiencies can also result in anaemia, including deficiency of vitamin B 12 or folic acid, vitamin E, and rarely vitamin C or vitamin B 6.

Anaerobic Does not require oxygen.

Anchor worm Parasites of the genus Lernaeocera found attached to the gills of fish. As an example, L. branchialis, which parasitises cod and haddock, is about 25mm long when stretched out, and about 2mm in diameter. Though they have a simple worm-like shape, these parasites are copepod crustacea, but very dege-nerated in form as a consequence of their life style.

Anchovy Small oily fish, Engraulis spp., usually semi-preserved with 10-12% salt and sometimes benzoic acid. A good source of protein and niacin; calcium, iron, and zinc. Anchovy butter is prepared from pounded fillets of anchovy mixed with butter as a savoury spread; anchovy paste from pounded fillets of anchovy mixed with vinegar and spices.

Androstenedione An androgen. Androgens are produced in two sites in the male body - most originate in the testes, but some potent male hormones are produced by the adrenal glands, located just above the kidneys. Androstenedione is a pivotal adrenal steroid that's actually one step closer to being converted to testosterone than DHEA. It raises your testosterone levels and can provide more strength.

Androstenedione really does raise testosterone above normal levels and could be hazardous. Side effects include acne, male baldness, and a decrease in 'good' cholesterol, which may lead to heart disease. This supplement might be reclassified as a steroid by the FDA, and it is banned in the NFL, Olympics, and other major sports organizations.

Anemia Anemia is a condition in which a deficiency in the size or number of erythrocytes or the amount of hemoglobin they contain limits the exchange of oxygen and carbon dioxide between the blood and the tissue cells. Most anemias are caused by a lack of nutrients required for normal erythrocyte synthesis, principally iron, vitamin B-12, and folic acid. Others result from a variety of conditions, such as hemorrhage, genetic abnormalities, chronic disease states or drug toxicity.

Angel fillet See Filleting.

Angelica Crystallized young stalks of the umbelliferous herb Angelica archangelica, which grows in southern Europe. They are bright green in colour, and are used to decorate and flavour confectionery goods. The roots are used together with juniper berries to flavour gin, and the seeds are used in vermouth and Chartreuse. Essential oils are distilled from the roots, stem, and leaves.

Angostura The best known of the bitters, widely used in cocktails; a secret blend of herbs and spices, including the bitter aromatic bark of either of two trees of the orange family. Invented in 1824 by Dr Siegert in the town of Angostura in Bolivia, originally as a medicine, and now made in Trinidad. A few drops of Angostura in gin makes a 'pink gin'.

Angular stomatitis A characteristic cracking and fissuring of the skin at the angles of the mouth, a symptom of vitamin B 2 deficiency, but also seen in other conditions.

Anisakis, anisakiasis A genus of nematode parasites found in the viscera and muscle tissue of many species of fish, but especially in herring. They are capable of infecting humans and causing gastric illness if raw or lightly processed fish, for example cold smoked fish or matje herring, are consumed. The disease caused is anisakiasis. Freezing to below -18°C followed by frozen storage for 24 hours kills the parasites and renders lightly processed products made from the treated raw material safe for onsumption.

Anise pepper A pungent spice from the Sichuan region of China; the flavour develops gradually after biting into the pepper.

Aniseed The dried fruit of Pimpinella anisum, a member of the parsley family, which is used to flavour baked goods, meat dishes, and drinks, including anise, anisette, and ouzo.

Anisette A sweet liqueur flavoured with aniseed.

Annatto See Dye.

Anorexia nervosa An eating disorder characterized by refusal to maintain a minimally normal weight for height and age. The condition includes weight loss leading to maintenance of body weight 15 percent below normal; an intense fear of weight gain or becoming fat, despite the individual's underweight status; a disturbance in the self-awareness of one's own body weight or shape; and in females, the absence of at least three consecutive menstrual cycles that would otherwise be expected to occur.

Anosmia Lack or impairment of the sense of smell.

Anserine A dipeptide found in muscle, of unknown function. It consists of ß-alanine and methylhistidine.

Antabuse Trade name for the drug disulfiram, used in the treatment of alcoholism. It inhibits the further metabolism of acetaldehyde arising from the metabolism of alcohol, and so causes headache, nausea, vomiting, and palpitations if alcohol is consumed.

Antacids Bases or buffers that neutralize acids, used generally to counteract excessive gastric acidity and to treat indigestion. Antacid preparations generally contain such compounds as sodium bicarbonate, aluminium hydroxide, mag-nesium carbonate, or magnesium hydro-xide.

Anthocyanins Violet, red, and blue water-soluble colours extracted from flowers, fruits, and leaves. They can react with iron or tin, and so may cause discoloration in canned foods.

Anthoxanthins Alternative name for flavonoids.

Anthropometry Body measurements used as an index of physiological development and nutritional status; a non-invasive way of assessing body composition. Weight for age provides information about the overall nutritional status of children; weight for height is used to detect acute malnutrition; height for age to detect chronic malnutrition. Mid-upper arm circumference provides an index of muscle wastage in undernutrition. Skinfold thickness is related to the amount of subcutaneous fat as an index of over- or under-nutrition.

Antibiotic ice Ice containing a small amount of antibacterial substance, e.g. a tetracycline, added to extend the shelf life of chilled fish by decreasing the rate of growth of spoilage bacteria. Not generally recommended as an effective treatment or for its potential contribution to the general development of antibiotic resistance, and its use is not permitted in many countries.

Antibiotics Antibiotics are used to prevent and treat bacterial diseases in humans and animals. In animal agriculture antibiotics are also used to improve the rate of growth and the feed efficiency of animals so they produce more meat or milk on less feed.

Antibody Protein produced in body tissues in response to the presence of a specific antigen.

Anti-caking agents Compounds added in small amounts to powdered foodstuffs to prevent clumping or caking - e.g. anhydrous disodium hydrogen phosphate is added to salt and icing sugar, aluminium calcium silicate or calcium or magnesium carbonate to table salt, calcium silicate to baking powder.

Anticoagulants Compounds that prevent or slow the process of blood clotting or coagulation, either in samples of blood taken for analysis or in the body. One of the most commonly used such substances is heparin, which is formed in the body.

People at risk of thrombosis are often treated with Warfarin and similar compounds as an anticoagulant, to reduce the risk of intravenous blood clot formation. These act by antagonizing the action of vitamin K in the synthesis of blood clotting proteins, and people taking anticoagulants should be careful not to take supplements containing large amounts of vitamin K. It is most unlikely that the vitamin K in foods would be enough to have any such adverse effect.

Antiemetic Drug used to prevent or alleviate nausea and vomiting. antigalactic Drug that reduces or prevents the secretion of milk in women after parturition.

Antigen A foreign substance that, when introduced into the body, stimulates an immune response.

Antimetabolite Compound that inhibits a normal metabolic process, acting as an analogue of a normal metabolite. Some are useful in chemotherapy of cancer, others are naturally occurring toxins in foods, frequently causing vitamin deficiency diseases by inhibiting the normal metabolism of the vitamin.

Antimony Toxic metal of no known metabolic function, and therefore not a dietary essential. Antimony compounds are used in treatment of some parasitic diseases.

Antimycotics Substances that inhibit the growth of moulds and fungi, such as sodium and calcium propionates, methyl hydroxybenzoate, quaternary ammonium chloride, sodium benzoate, sorbic acid.

Antioxidant An antioxidant is a substance that counters the negative effect of oxygen. In foods anti-oxidants prevent the oxidation of different food components. Oxidation of foods results for example in an undesirable brown colour and/or off-odours and off-flavours known as rancidity. Antioxidants are substances that react with oxygen and thus counteract the negative effects. Examples are vitamin C and vitamin E.

In the body antioxidants protect key cell components by neutralizing the damaging effects of "free radicals," natural byproducts of cell metabolism. Free radicals form when oxygen is metabolized, or burned by the body. They travel through cells, disrupting the structure of other molecules, causing cellular damage. Such cell damage is believed to contribute to aging and various health problems. Example of anti-oxidants which are active in the body are the vitamins A, C and E, and polyphenols.

Antisense A piece of DNA that produces the mirror image, or antisense messenger RNA, that is exactly opposite in sequence to one that directs the cells to produce a specific protein. Since the antisense RNA binds tightly to its image, it prevents the protein from being made.

Anti-spattering agents Compounds such as lecithin, sucrose esters of fatty acids, and sodium sulpho-acetate derivatives of mono- and diglycerides, which are added to oils and fats used for frying to prevent potentially dangerous spattering. They function by preventing the coalescence of water droplets.

Anti-staling agents Substances that retard the staling of baked products, and soften the crumb, e.g. sucrose stearate, polyoxyethylene monostearate, glyceryl monostearate, stearoyl tartrate.

Aortic stenosis The aortic valve is stiffened and has a narrowed opening. It does not open and close properly, which increases strain on the heart because the left ventricle has to pump harder to send blood out to the body.

Apgar Activity, pulse, grimace, appearance, respiration.

Aphagia Inability to swallow. Difficulty in swallowing is dysphagia.

Aphagosis Inability to eat.

Apoenzyme The protein part of an enzyme which requires a coenzyme for activity, and is therefore inactive if the coenzyme is absent.

Apolipoprotein The protein part of lipoproteins without the associated lipid.

Apollinaris water An alkaline, highly aerated, mineral water containing sodium chloride and calcium, sodium, and magnesium carbonates, from a spring in the valley of Ahr.

Apoptosis Programmed cell death in response to determined signaling pathways with energy consumption.

Apositia Aversion to food.

Appearance The attributes of a product assessed by the sight. It encompasses properties such as shape; colour; gloss/ dullness; translucency/opacity; surface texture.

Apple Fruit of the tree Malus sylvestris and its many cultivars and hybrids; there are more than 2000 varieties in the British National Fruit Collection. Crab apples are grown mainly for decoration, and for pollination of fruit-bearing trees, although the sour fruit can be used for making jelly. Cooking apples are generally sourer varieties than dessert apples, and normally have flesh which crumbles on cooking; cider apples are sour varieties especially suited to the making of cider. One apple provides 2.2 g of dietary fibre and supplies 40 kcal.

Apple sauce Pulped stewed apple; in the UK it is made from sour apples, as an accompaniment to pork and goose; in the USA also used for stewed apple as a dessert.

Apraxia Disorder of motor planning caused by damage to the motor control areas of the brain; inability to execute volitional movements. Limb apraxia and apraxia of speech are characterized by difficulty sequencing and coordinating movements in the absence of paralysis or weakness of muscles, usually resulting in highly inconsistent performance.

Apricot Fruit of the tree Prunus armeniaca. Apricot kernels are used to prepare almond oil.A rich source of vitamin A; copper; niacin and iron.

Arachidonic acid A long-chanin polyunsaturated fatty acid from the omega-6 family. AA contains 20 carbons and 4 double bonds. It is the brain's principal omega-6 fatty acid. Arachidonic acid is found primarily in animal fats and is often too high in modern diets. This fatty acid can be converted into the powerful locally acting hormones, prostaglandins and thromboxane via the cycloxygenase pathway, and leukotirenes via de lipoxy-genase pathway. One example is PGE2, which has inflammatory activity. PG and related substrances are involved in many other biological processes from gene expression to blood clotting. Alpha-linolenic acid, EPA, and DHA can counter the effects of arachidonic acid.

Arachis oil Oil extracted from the groundnut or peanut, Arachis hypogea, 20% saturated, 50% mono-unsaturated, 30% polyunsaturated, less than 1% linolenic acid.

Arbute Fruit of the southern European strawberry tree; resemble strawberries in appearance but have a grainy texture and little taste.

Argininaemia A genetic disease affecting the metabolism of the amino acid arginine, and hence the normal formation of urea and elimination of end-products of protein metabolism. Depending on the severity of the condition, affected infants may become comatose and die after a moderately high intake of protein. Treatment is by severe restriction of protein intake. Sodium benzoate may be given to increase the excretion of nitrogenous waste as hippuric acid.

Arginine A conditionally essential amino acid with anabolic and immune system supportive effects. It is required for growth, immune function, wound healing and many aspects of protein metabolism. Arginine is necessary for the production of growth hormone. It is also a precursor for nitric oxide, a critical substance that helps regulate the function of cardiovascular, nervous and immune systems and which is essential for muscle growth.Arginine is essential for sperm formation, so it plays a vital role in male fertility.

Argol Crust of crude cream of tartar which forms on the sides of wine vats, also called wine stone. It consists of 50-85% potassium hydrogen tartrate and 6-12% calcium tartrate, and will be coloured by the grapes, so white argol comes from white grapes and red argol from red grapes. Used in vinegar fermentation, in the manufacture of tartaric acid and as a mordant in dyeing.

Ariboflavinosis Deficiency of riboflavin characterized by swollen, cracked, bright red lips, an enlarged, tender, magenta-red tongue, cracking at the corners of the mouth, congestion of the blood vessels of the conjunctiva and a characteristic dermatitis with filiform excrescences.

Armagnac Brandy made from white wine from one of three defined areas of France: Bas-Armagnac, Haut-Armagnac, or Tenareze.

Arogel Trade name for a potato starch preparation which is stable to heating and is used as a thickener in gravies, sauces, and canned foods.

Arrowroot Tuber of the Caribbean plant Maranta arundinacea, mainly used to prepare arrowroot starch, a particularly pure form of starch, which contains only a trace of protein and is free from vitamins. It is used to thicken sauces and in bland, low-salt, and proteinrestricted diets, and as an infant food in some Caribbean islands.

Arsenic A toxic metal, with no known metabolic function. Organic arsenic derivatives have been used as pesticides and in treatment of diseases such as syphilis, leprosy, and yaws. Arsenic can accumulate in crops treated with arsenical pesticides, and in fish and shellfish living in arsenic-polluted water.

Arteriosclerosis Thickening and calcification of the arterial walls, leading to loss of elasticity, occurring with ageing and especially in hypertension.

Artificial cream A non-dairy substitute for cream, made from vegetable oils, with stabilizers, emulsifiers, and sometimes flavourings and colours added.

Artificial nose See Electronic nose.

Asafoetida A resin extracted from the oriental umbelliferous plant Narthex asafoetida, with a bitter flavour and strong garlic-like odour, used widely in oriental and Middle-Eastern cooking, and in small amounts in sauces and pickles.

Ascites Abnormal accumulation of fluid in the peritoneal cavity, occurring as a complication of cirrhosis of the liver, congestive heart failure, cancer, and infectious diseases. Depending on the underlying cause, treatment may sometimes consist of a high-energy, high-protein, low-sodium diet, together with diuretic drugs and fluid restriction.

Ascorbic acid Also known as Vitamin C. A water soluble vitamin, and an antioxidant. Your body cannot store Vitamin C, so you must supplement it regularly. It is not resistant to heat, so cooking will destroy it. Vitamin C functions primarily in the formation of collagen, the chief protein substance of your body's framework. It also helps in the production of vital body chemicals. Vitamin C also is a detoxifier.

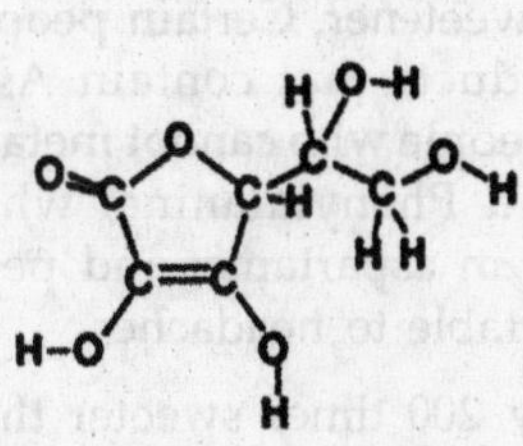

Fig. Ascorbic acid (Vitamin C)

Ascorbyl palmitate An ester of ascorbic acid and palmitic acid which is used as an anti-staling compound in bakery goods.

Aseptic Without contamination by micro-organisms, i.e. sterile.

Aseptic canning A heat sterilisation process in which the ingredients or potential contents are sterilised outside the container and the container is then filled aseptically and seamed. Used for liquid foods and small solids in suspension, but not widely for fish products, e.g. tuna salad, the advantage being shorter, less severe process conditions for more heat sensitive food constituents such as the mayonnaise.

Aseptic filling The filling of cans or other containers with food that has already been sterilized, the process thus having to be carried out under aseptic conditions. Continuous sterilization as the food passes along narrow pipes allows more rapid heating, with less effect on the quality of the food, than sterilization by heating after canning.

Aseptic packaging Is a system wherein the food product and the container are sterilised separately, and the containers are filled and sealed in a sterile environment.

Ash The residue remaining after all the organic matter in a sample has been burned off. Often included in the proximate composition of a product.

Asparagine A non-essential amino acid, chemically the ß-amide of aspartic acid.

Aspartame An artificial sweetener. Known by the trade name NutraSweet, it is an artificial sweetener. Certain people should avoid products that contain Aspartame. They are people who cannot metabolize the amino acid Phenylalanine, which is an ingredient in aspartame; and people who are suspectable to headaches.

It is nearly 200 times sweeter than sugar. Aspartame is produced from the amino acids L-asartic acid and L=phenylalannine. It has been approved for use in certain goods since 1981. Heat causes aspartame to lose its sweetness, therefore, you can't bake with it. Unlike some sweetners, it has no aftertaste.

Aspergilius A family of moulds, important in the spoiling of stored nuts and grains because A. flavus produces aflatoxins. A. oryzae is grown commercially as a source of takadiastase.

Aspic jelly A clear jelly made from fish, chicken, or meat stock, sometimes with added gelatine, flavoured with lemon, tarragon, vinegar, sherry, peppercorns, and vegetables, used to glaze foods such as meat, fish, and game. The name may be derived from the herb espic or spikenard.

Aspirin Can reduce soreness after a hard workout, or lower your risk of heart disease. Can also reduce your risk for throat or stomach cancer.

Assay Technique for measuring a biological response; a test.

Associated foods Undercooked or raw hamburger has been implicated in many of the documented outbreaks, however E. coli O157:H7 outbreaks have implicated alfalfa sprouts, non-pateurised fruit juices, dry-cured salami, lettuce, game meat, and cheese curds. Raw milk was the vehicle in a school outbreak in Canada.

Astaxanthin The red pigment present in the shells of crustacea and the flesh and eggs of salmonids. The astaxanthin in salmonid tissues is obtained by absorption of the pigment from crustacea in the diet of the fish. Astaxanthin, and/or the chemically-related pigment canthaxanthin are added to the feeds of farmed salmonids to give the desired red colour. The pigment is also produced by microorganisms such as microalgae that can be cultivated as sources of pigment for fish feeds.

Asthma The difficulty experienced in breathing due to excessive contraction of the involuntary muscle in the walls of bronchial tubes leading into the lungs, with consequent narrowing of the tubes. The muscle reacts excessively to a wide range of stimuli such as infections, exertion, and, most importantly, allergens, the substances that cause allergies. If the attack is prolonged it is complicated by plugging of the small airways by abnormal secretion, and it is this that can make asthma a threat to life. Asthma commonly starts in childhood, and about half those affected improve or recover around puberty. The condition can be alleviated by drug treatment.

Astringency The action of unripe fruits and cider apples, among other foods, to cause a contraction of the epithelial tissues of the tongue. It is believed to result from a destruction of the lubricant properties of saliva by precipitation by tannins.

Atheroma The fatty deposit composed of various lipids, complex carbohydrates, and fibrous tissue which forms on the inner wall of blood vessels in atherosclerosis.

Atherosclerosis A condition that exists when too much cholesterol builds up in the blood and accumulates in the walls of the blood vessels.

ATP The coenzyme that acts as an intermediate between energy-yielding metabolism and energy expenditure as physical work and in synthetic reactions. ADP is phosphorylated to ATP linked to oxidations; in energy expenditure ATP is hydrolysed to ADP and phosphate ions.

Aubergine The fruit of Solanum melongena, a native of South East Asia, widely cultivated and eaten as a vegetable, also known as egg-plant and field egg. Half a medium-sized aubergine is a source of folate and vitamin C; provides 2.8 g of dietary fibre; supplies 20 kcal.

Auditory discrimination The ability to recognize and identify sounds, to hear likenesses and differences.

Auditory figure-ground skills The ability to hear a message even though there is background noise, the ability to separate the important sounds from the background sounds.

Auditory memory The ability to retain and recall information.

Auditory perception The ability to receive and comprehend words and concepts through hearing.

Auditory processing disorder It interferes with an individual's ability to analyze or make sense of information taken in through the ears. Difficulties with auditory processing do not affect what is heard by the ear, but do affect how this information is interpreted, or processed by the brain.

Augmentative communication Systems of adapted technology that encourage and enhance verbal and nonverbal communication.

Aurantiamarin A glucoside present in the albedo of the bitter orange which is partly responsible for its flavour.

Autism Pervasive developmental disorder marked by severe interruption of social interaction. Individuals with autism are often described as being severely withdrawn, being rigidly dependent on routine, avoiding social contact, and showing repetitive, stereotypic behaviors.

Autoclave A vessel in which high temperatures can be achieved by using high pressure; the domestic pressure cooker is an example. At atmospheric pressure water boils at 100°C; at 5 lb above atmospheric pressure the boiling point is 109°C; at 10 lb, 115°C; at 15 lb, 121°C, and at 20 lb, 126°C.

Autoclaves have two major uses. In cooking, the higher temperature reduces the time needed. At these higher temperatures, and under moist conditions, bacteria are destroyed more rapidly, so permitting sterilization of foods, surgical dressing and instruments, etc.

Autolysis Breakdown of proteins, lipids and other compounds in fish by the action of enzymes present in the fish especially causing deteriorative changes. The process begins immediately after death and the rate depends on temperature. Also loosely called self digestion.

Automated A system where machines handle and control the processing from raw materials to the finished product.

Automatic smoke producer See Smoking kiln.

Autosome A chromosome that is not involved in sex determination. The autosomes of males are similar to those of females.

Autotrophes Organisms that can synthesize all the compounds required for growth from simple inorganic salts, as distinct from heterotrophes, which must be supplied with complex organic compounds. Plants are autotrophes, whereas animals are heterotrophes. Bacteria may be of either type; heterotrophic bacteria are responsible for food spoilage and disease.

Avicel Trade name for microcrystalline a-cellulose. It is natural cellulose which has been partially hydrolysed with acid, and reduced to a fine powder. It disperses in water and has the properties of a gum. it is used in oily foods such as cheese and peanut butter, as well as in syrups and honey, sauces and dressings.

Avidin A protein in egg white which binds the vitamin biotin, so rendering it unavailable to the body. Cooking denatures proteins, and denatured avidin in cooked eggs does not bind biotin.

Avitaminosis The absence of a vitamin; may be used specifically, as, for example, avitaminosis A, or generally, to mean a vitamin deficiency disease.

Avocado Fruit of the tree Persica americana, also known as the avocado pear or alligator pear, because of its rough skin and pear shape, although it is not related to the pear. It is unusual among fruits for its high fat content, of which 7-14% is linoleic acid, and also for the fact that it does not ripen until after it has been removed from the tree.

Half an avocado is a rich source of vitamin C and copper; a good source of vitamin B 6 ; a source of protein and iron; contains 26 g of fat of which 20% is saturated; provides 2.6 g of dietary fibre; supplies 265 kcal.

Azo dyes In chemistry, azo compounds generally have a molecular formula of the form R-N=N-R', in which R and R' can be either aromatic or aliphatic. The N=N group is called an azo or diimide functional group. This helps to stabilise the N=N group by making it part of an extended delocalised system. This also has the effect of making many azo compounds coloured, as delocalised or *conjugated* systems often absorb visible frequencies of light.

The name azo comes from *azote,* an old name of nitrogen that originates in French and is derived from the Greek *a* (not) + *zoe* (to live). Aromatic azo compounds are usually stable and have vivid colors such as red, orange, and yellow. Therefore, they are used as dyes, which are called azo dyes. Some azo compounds, eg. Methyl orange, can also be used as acid-base indicators, due to their ability to function as weak acids, and the different colours of the acid and salt forms. Azobenzene is another typical aromatic azo compound. Their colour originates from absorbance in the visible region of the spectrum due to the delocalization of electrons in the benzene and azo groups forming a conjugated system, whereby the N=N group is the chromophore.

Aliphatic azo compounds are rather unstable. At an elevated temperature or by irradiation, two carbon-nitrogen bonds are cleaved simultaneously with the loss of nitrogen gas to generate carbon-centered radicals. Owing to this process, some aliphatic azo compounds are utilized as radical initiators. Azobisisobutylonitrile is a typical one and is widely used in industrial processes and in laboratory experiments. Aromatic azo compounds can be synthesized by using an azocoupling reaction, that is, an electrophilic substitution reaction on aromatic rings with diazonium salts. Diazonium salts decompose at temperatures warmer than about 5 degrees Celsius, so the reaction must take place in solution under freezing conditions:

B

B 1 A vitamin which maintains energy levels, supports brain function. Aids in digestion. Necessary for metabolism of sugar and starch to provide energy. Maintains a healthy nervous system. Alcohol can cause deficiencies of this vitamin and all the B-complex vitamins.

B 12 Necessary for carbohydrate, protein and fat metabolism. Important to amino acid and fatty acid synthesis; essential for hemoglobin and nerve cell growth and maintenance. The anti-stress vitamin, sometimes prescribed for stress reduction.

B 2 A vitamin which helps with energy production and amino acid production. Helps body obtain energy from protein, carbohydrates, and fats. Helps maintain good vision and healthy skin.

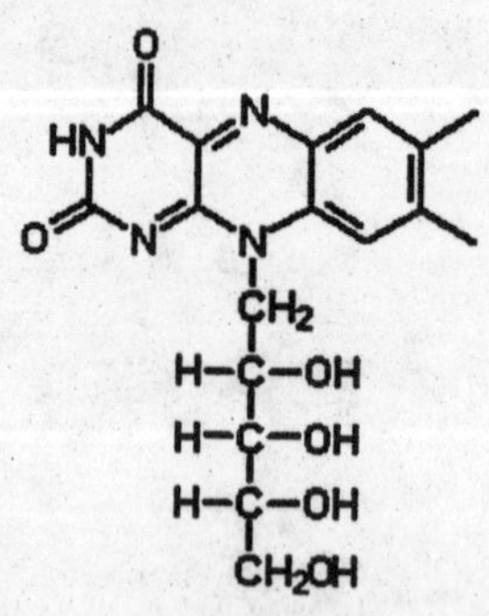

Fig. B 2 (Riboflavin)

B 3 A vitamin Important in carbohydrate metabolism, formation of testosterone and other hormones, formation of red blood cells and maintaining the integrity of all cells. Helps body utilize protein, fats, and carbohydrates. Necessary for a healthy nervous system and digestive system. It also lowers elevated blood cholesterol levels when taken in large amounts of more than 1,000 milligrams a day.

B 5 A vitamin which supports carbohydrate, protein and fat metabolism; hemoglobin synthesis. Helps release energy from protein, carbohydrates, and fat. Needed to support a variety of body functions, including the maintenance of a healthy digestive system.

B 6 A vitamin which supports glycogen and nitrogen metabolism; production and transport of amino acids; production and maintenance of red blood cells Essential for the body's utilization of protein. Needed for the production of red blood cells, nerve tissues, and antibodies. Women taking oral contraceptives have lower levels of B-6.

B Complex Vitamins A group of eleven known vitamins that work together in your body. All play vital roles in the conversion of food into energy. Essential for the normal functioning of the nervous system, and the maintenance of good digestion. Helps promote healthy skin, hair, and eyes. These are water soluble vitamins, which means they cannot be stored by your body and must be replaced every day.

Babaco The seedless fruit of the tree Carica pentagona, related to the pawpaw, discovered in Ecuador in the 1920s, introduced into New Zealand in 1973, and more recently into the Channel Islands. A rich source of vitamin C.

Babassu oil Edible oil from the Brazilian palm nut, similar in fatty acid composition to coconut oil, and used for food and in soaps and cosmetics.

Babcock test A test for the fat content of milk.

Baby foods General term to include infant formula milk and weaning foods.

Bacillus Singular for a rod-shaped bacterium. Also used as the name of a genus of bacteria, including the species Bacillus thuringiensis.

Bacillus stearothermophilus See Thermophilic and Canning.

Bacillus thuringiensis Bacterium that produces a protein called Bt toxin, a biological insecticide. Bt toxin is used to control insect pests by dusting the crop with Bt bacteria. When ingested, Bt toxin kills certain insect larvae, but is regarded as harmless to humans, pets and most beneficial insects such as bees. Inserting a copy of the Bt gene into plants enables them to produce Bt toxin protein. Such plants can resist some insect pests.

Bacon Cured meat from the back, sides, and belly of a pig; variety of cuts with differing fat contents.

A rich source of protein, niacin, vitamin B1, B2.

Gammon is bacon made from the top of the hind legs; green bacon has been cured but not smoked.

Bacoreta A variety of tuna fish with dark flesh.

Bacteria Unicellular micro-organisms, ranging between 0.5 and 5 [mu]m in size. They may be classified on the basis of their shape: spherical, rodlike, spiral, comma-shaped, corkscrew-shaped, or filamentous. Other classifications are based on whether or not they are stained by Gram's stain, aerobic or anaerobic, and autotrophic or heterotrophic. Some bacteria form spores which are relatively resistant to heat and sterilizing agents.

Bacteria are responsible for much food spoilage, and for disease, but they are also made use of, for example in the pickling process and fermentation of milk, as well as in the manufacture of vitamins and amino acids and a variety of enzymes and hormones.

Between 45 and 85% of the dry matter of bacteria is protein, and some can be grown on petroleum residues, methane, or methanol for use in animal feed.

Bacterial counts The numbers of bacteria in a sample. The term needs to be qualified by the identifier for the group of bacteria or the species of bacterium being counted. For example, Total viable count is the number of bacteria present in a sample seen as colonies on a non-specific solid bacteriological growth medium under the conditions of test. Selective culturing conditions and culture media are used for individual species or genera of bacteria, for example Salmonella, or for closely related organisms of sanitary significance such as coliforms. Bacterial counts are typically specified as number of organisms/g or colony forming units/g.

Bacterial filter A filter fine enough to prevent the passage of bacteria, which permits removal of bacteria from solutions. Viruses are considerably smaller, and will pass through a bacterial filter.

Bacteriological standard A standard that specifies maximum bacterial counts in a product.

Bacterium Class of single-cell organisms. One member, E. coli, is commonly used in recombinant DNA technology for producing proteins and other chemicals.

Bagasse The residues from sugar-cane milling, consisting of the crushed stalks from which the juice has been expressed; it consists of 50% cellulose, 25% hemicelluloses, and 25% lignin. It is used as a fuel, for cattle feed, and in the manufacture of paper and fibre board. The name is sometimes also applied to the residues of other plants, such as beet, which is sometimes incorporated into foods as a source of dietary fibre.

Bagel A circular bread roll with a hole in the middle, made from fermented wheat flour dough, which is boiled before being baked. A Jewish speciality; variants include onion bagels, which are topped with crisply fried onion before baking and poppy-seed bagels, which are topped with poppy seeds.

Baked alaska Dessert of ice cream on a sponge base and covered with meringue, baked in a hot oven for a very short time, so that the meringue is cooked but the ice cream remains frozen.

Baking Cooking in an oven by dry heat.

Baking additives Materials added to flour products for a variety of purposes, including bleaching the flour, ageing, slowing the rate of staling, and improving the texture of the finished product.

Baking powder A mixture that liberates carbon dioxide when moistened and heated. The source of carbon dioxide is sodium bicarbonate, and an acid is required. This may be cream of tartar or calcium acid phosphate, sodium pyrophosphate, or sodium aluminium sulphate. Legally, baking powder must contain not less than 8% available, and not more than 1.5% residual, carbon dioxide. Golden raising powder is similar, but is coloured yellow, and must contain not less than 6% available, and not more than 1.5% residual, carbon dioxide.

Baking soda Or bicarbonate of soda, chemically sodium bicarbonate, the source of carbon dioxide in baking powder.

Ball mill A vessel in which material is ground by rolling with heavy balls, used especially for hard materials.

Ballottine A kind of galantine of meat, poultry, game, or fish, boned, stuffed, and rolled into a bundle; also small balls of meat or poultry.

Balm A herb with hairy leaves and a lemon scent, therefore often known as lemon balm. Used for its flavour in fruit salads, sweet or savoury sauces, etc., as well as for preparation of herb teas. Claimed to have calming medicinal properties, and promoted at one time as an elixir of life and a cure for impotence; it is rich in tannins.

Bambarra groundnut Also known as the Madagascar peanut or earth pea, Voandseia subterranea. It resembles the true groundnut, but the seeds are low in oil. They are hard and require soaking or pounding before cooking.

Bamboo shoots Thick, pointed young shoots of Bambusa vulgaris and Phyllostachys pubescens eaten in Eastern Asia. A source of vitamin B 1.

Banana Fruit of the genus Musa; cultivated kinds are sterile hybrids, and so cannot be given species names. Dessert bananas have a high sugar content and are eaten raw; plantains have a higher starch and lower sugar content and are picked when too hard to be eaten raw.

A good source of vitamin A,B6,C, copper.

Banding Light stripes on smoked fish where the tenterstick or banjo around which the fish is hung, or the mesh on which it has been laid, has prevented contact with smoke during the smoking process.

Bangers Colloquial English term for sausages. When served with mashed potatoes, the dish is known as bangers and mash.

Bannock A flat round cake made from oat, rye, or barley meal and baked on a hearth or griddle. Pitcaithly bannock is a type of almond shortbread containing caraway seeds and chopped peel.

Bap Traditionally a soft, white, flat, flour-coated Scottish breakfast roll. Now also used for any relatively large soft-crusted roll, made from white, brown, or wholemeal flour.

Barbecue Originally Caribbean name for a wooden frame used to smoke and dry meat over a slow, smoky fire; the whole animal was placed on a spit over burning coals. Now outdoor cooking of meat, sausages, etc., on a charcoal or gas fire; also the fire on which they are cooked.

Barbecue sauce Chopped onions fried in butter, made into a sauce with tomato paste and seasoned with sugar, vinegar, mustard, and Worcestershire sauce, served with barbecued meat and sausages.

Barbera A grape variety widely used for wine making, although not one of the classic varieties; makes the dark fruity and often sharp red wines of northern Italy.

Barcelona nut Spanish variety of hazel nut. A rich source of copper.

Barding To cover the breast of a bird with slices of fat before roasting, to prevent the flesh from drying.

Barley Grain of Hordeum vulgare, one of the hardiest of the cereals; mainly used for brewing beer. The whole grain with only the outer husk removed requires several hours' cooking; the commercial product is usually pearl barley where most of the husk and germ is removed. Barley flour is ground pearl barley; barley flakes are the flattened grain. A source of niacin, vitamin B6, folate, zinc, copper, and iron.

Barley sugar Sugar confectionery made by melting and cooling sugar, originally made by boiling with a decoction of barley.

Barmene Trade name for yeast extract, prepared from autolysed brewer's yeast, plus vegetable juices, used for flavouring.

Barrel A standard barrel contains 36 gallons.

Basal metabolic rate The energy cost of maintaining the metabolic integrity of the body, nerve and muscle tone, respiration and circulation. It depends on the amount of metabolically active body tissue, and hence can be calculated from body weight, using different factors for males and females, and at different ages. For children the. BMR also includes the energy cost of growth. Experimentally, BMR is measured as the heat output from the body, or the rate of oxygen consumption, under strictly standardized conditions, 12-14 hours after the last meal, completely at rest and at an environmental temperature of 26-30°C, to ensure thermal neutrality. Measurement of metabolic rate under less rigorously controlled conditions gives the resting metabolic rate.

For people with a sedentary lifestyle and relatively low physical activity, BMR accounts for about 70% of total energy expenditure. The energy costs of different activities are generally expressed as the physical activity ratio, the ratio of energy expenditure in the activity to BMR.

Base On the DNA molecule, one of the four chemical units that are linked in a series to make a strand of DNA. The four DNA bases are: adenine (A), cytosine (C), guanine (G), and thymine (T). In RNA, uracil (U) substitutes for thymine. See DNA finger printing, nucleotide.

Basil An aromatic herb Ocimum basilicum and 0. minimum; other members of the genus Ocimum are also used as seasoning.

Basmati Long-grain Indian variety of rice; much prized for its delicate flavour.

Bath bun A small English cake made from milk-based yeast dough, with dried fruit and a topping of sugar crystals, attributed to Dr W. Oliver of Bath.

Bath cheese A small English cheese, made from cow's milk with the subsequent addition of cream.

Batter Thick liquid mixture of flour, milk, and eggs, used to coat fish before frying, fried alone to make pancakes or griddle cakes, or baked to make Yorkshire pudding.

BCAA's Leucine, Valine, and Isoleucine are called "branch chain" aminos due to their molecular structure, and are important essential amino acids well known for their anticatabolic benefits. They are called BCAA's because they structurally branch off another chain of atoms instead of forming a line. Studies have shown that BCAA's postively affect skeletal muscle growth, enhance fat loss, help to stimulate protein synthesis and inhibit its breakdown, so BCAA's have powerful anabolic and anticatabolic effects on the body. They may also potentiate the release of some anabolic hormones, such as growth hormone. Regular ingestion of BCAA's help to keep the body in a state of postive nitrogen balance. In this state, your body much more readily builds muscle and burns fat. Studies have shown that athletes taking extra BCAA's have shown a loss of more bodyfat than those not taking BCAA's.

BCAAs are used as a source of energy for muscle cells. During prolonged exercise, BCAAs are released from skeletal muscles and their carbon backbones are used as fuel, while their nitrogen portion is used to form another amino acid, Alanine. Alanine is then converted to Glucose by the liver. This form of energy production is called the Alanine-Glucose cycle, and it plays a major role in maintaining the body's blood sugar balance.

Beans Seeds of the family leguminosae, eaten as food.

Beef Flesh of the ox; flesh from young calves is veal. A rich source of protein, niacin, iron, copper, and vitamin B 12 ; a good source of vitamin B 2 and copper; a source of vitamins B 1, B 6, and selenium.

Beer Alcoholic beverage made by the fermentation of cereals; traditionally barley, but also maize, rice, and sorghum. The first step is the malting of barley: it is allowed to sprout, when the enzyme amylase hydrolyses some of the starch to dextrins and maltose. The sprouted barley is dried, then extracted with hot water to produce wort. After the addition of hops for flavour, the wort is allowed to ferment. Two types of yeast are used in brewing: top fermenting yeasts which float on the surface of the wort and bottom or deep fermenters.

Beestings The first milk given by the cow after calving, the colostrum, rich in immunoglobulins.

Beetroot The root of Beta vulgaris, eaten cooked or pickled. Known simply as beet in North America. The violet-red pigment, betanin, is used as a food colour. One small beetroot is a good source of folate; provides 1.6 g of dietary fibre; supplies 18 kcal.

Belly burst A defect in fish in which the belly walls are perforated by the action of gut enzymes dissolving the belly walls. It occurs in ungutted fish, particularly when the fish have been feeding heavily and digestive enzymes are very active. Prior to perforation of the belly wall, the belly wall will be seen on opening the belly to be stained and softening.

Beluga Russian name for the white sturgeon, whose roe forms the most prized caviar.

Bemax Trade name for a wheat germ preparation. A rich source of vitamins B 1 and E, folate, copper, and zinc; a good source of iron; a source of protein, vitamins B 2, B 6, and niacin.

Benedictine A French liqueur invented in about 1510 by the monks of the Benedictine Abbey of Fecamp in France. The Abbey was closed, and the recipe lost after the French Revolution, then rediscovered about 1863. It is based on double-distilled brandy, flavoured with some 75 herbs and spices; it contains 40% alcohol and 30% sugar; 300 kcal/100 mL.

Benzo[a]pyrene See Polycyclic aromatic hydrocarbons.

Benzoates Salts of benzoic acid.

Benzoic acid An organic acid used as the acid or its sodium or potassium salts as a food additive to preserve the product by inhibiting growth of microorganisms. Generally, it is permitted as a food additive in only a small range of fishery products such as dried fish.

Bercy Sauce made from white wine, light stock, shallots, and herbs.

Beriberi The result of severe and prolonged deficiency of vitamin B 1, still a problem in parts of South East Asia where the diet is high in carbohydrate and poor in vitamin B 1. In developed countries vitamin B1 deficiency is associated with alcohol abuse; while it may result in beriberi, more commonly the result is central nervous system damage, the Wernicke-Korsakoff syndrome. In beriberi there is degeneration of peripheral nerves, starting in the hands and feet and ascending the arms and legs, with a loss of sensation and deep muscle pain. There is also enlargement of the heart, which may lead to oedema, and death results from heart failure. Fatal heart failure may develop without the nerve damage being apparent. The name is derived from the BahasaMalay word for sheep, to describe the curious sheep-like gait adopted by sufferers.

Beta amyloid protein A protein constituted by more than 40 aminoacids that is the main compound of senile plaques, the typical histology lesion of Alzheimer disease and Down syndrome. It appears to be a source of reactive oxygen species.

Beta carotene A phytonutrient carotenoid with antioxidant and provitamin A activity. In addition to providing the body with a safe source of Vitamin A, beta carotene works with other natural protectors to defend your cells from harmful free radical damage. This is an important micrinutrient in helping the body with metabolic functions, such as recovery from exercise. Beta-Carotene is the plant derived form of Vitamin A, which is important for several bodily functiosn, including eyesight, immune function and even bone growth. Beta-Carotene is nontoxic. Good sources of Beta Carotene include Carrots, Sweet Potatoes, Pumpkins, Cantaloupes, Acricots, Spinach and Broccoli.

Beta Hydroxy Beta Methylbutyrate It is a compound mande in the body and a metabolite of the essential amino acid Leucine. Studies have found that HMB has a decrease in stress-induced muscle protein breakdown. Studies also found that HMB may enhance increases in both muscle size and strength when combined with resistance training.

There are a number of theories why you may need HMB. The first, is that under stressful conditions, the body may not make enough HMB to satisfy the increased needs of tissues. It could also be that stress may alter enzymes or concentration of certain biochemicals that decrease normal HMB production. Another theory is that HMB may regulated enzymes sresponsible for muscle tissue breakdown.

Betanin The violet-red pigment of beetroot.

Betel Leaf of the creeper Piper betel, which is chewed in some parts of the world for its stimulating effect, due to the presence of the alkaloids arecoline and guvacoline. The leaves are chewed with the nuts of the areca palm, Arecha catechu, which is therefore often called the betel palm, and the nut is called betel nut. The Indian delicacy pan is based on betel leaf and areca nut, together with aromatic spices and herbs.

Bhujia Vegetable dishes, highly spiced. In the final stage of cooking they are fried in ghee which has been heated with onions, chillies, garlic, and ginger.

Biersuppe Soup made from light beer, thickened with potato flour and flavoured with cinnamon and lemon peel.

Biffins Apples that have been peeled, partly baked, then pressed and dried.

Bifidus factor A carbohydrate in human milk which contains nitrogen and stimulates the growth of Lactobacillus bifidus in the intestine. In turn, this organism lowers the pH of the intestinal contents and suppresses the growth of E. coli and other pathogenic bacteria.

Bigos Casserole of sauerkraut, cooked meat, game, sausages, ham, etc., with vodka and wine.

Bilberry The berry of wild shrubs of the genus Vaccinium, not generally cultivated. Variously known as whortleberry, blaeberry, whinberry, huckleberry. A rich source of vitamin C

Bile Fluid produced by the liver and stored in the gall bladder before secretion into the small intestine via the bile duct. It contains the bile salts, which function in the emulsification and hence digestion of fats, bile pigments and cholesterol. It is alkaline, and hence neutralizes the acid from the stomach as the food reaches the small intestine. Relatively large amounts of vitamin B 12 and folic acid are secreted in the bile and then reabsorbed from the small intestine. Most of the cholesterol and bile salts are also reabsorbed from the small intestine.

Bile salts Salts of cholic and deoxycholic acid and their glycine and taurine conjugates, secreted in the bile; they assist the digestion of fats by emulsifying them.

Bilgy fish Fish with a foul smell as a consequence of rapid growth of anaerobic bacteria in the flesh in unfrozen fish. The condition occurs, for example, when fish are pressed hard against the side or floor of a container so that air is excluded, and the fish are also warm by being against the sides or floor of an unchilled surface. Also called stinkers.

Biltong Strips of dried meat, salted, spiced, and dried in air for 10-14 days.

Binary ice See Icing.

Bind To add liquid, fat, or egg to a mixture to hold it together.

Binge-purge syndrome A feature of the eating disorder bulimia nervosa, characterized by the ingestion of excessive amounts of food and the excessive use of laxatives.

Bioassay Biological assay; measurement of biologically active compounds by their ability to support growth of micro-organisms or animals.

Biocides Chemicals used to kill unwanted organisms: herbicides, insecticides, fungicides.

Biocytin The main form of the vitamin biotin in most foods, bound to the amino acid lysine.

Bio-electrical impedance A method of measuring the proportion of fat in the body by the difference in the resistance to passage of an electric current between fat and lean tissue.

Biogenic amines A type of amine formed from amino acids in foods by the action of bacteria. Many have physiological effects in humans. They occur in very spoiled fish, especially when spoilage has occurred at high temperatures, above 10ºC. Biogenic amines found in spoiled fish include histamine, cadaverine, agmatine, putrescine, spermine, spermidine and tyramine. Histamine is of concern because it is associated with scombrotoxicosis.

Biological control Managing pest populations by purposefully manipulating beneficial natural enemies - predatory or parasitic insects that kill pest insects, or microbes that cause insect diseases.

Biological oxygen demand A way of assessing bacterial contamination of water, milk, etc. by micro-organisms which take up oxygen for their metabolism.

Biological value An attempt to measure how efficiently protein us used in the body. Biological Value is derived from providing a measure intake of protein, then noting the nitrogen uptake versus nitrogen excretion. The actually process is much more complicated though. In theory, a BV value of 100 is maximal. Some studies claim they have a higher BV than 100, but they refer to a chemical score, not the biological value of whey.

Biopesticide A biopesticide is any material of natural origin used in pest control derived from living organisms, such as bacteria, plant cells or animal cells.

Biopterin The coenzyme for a number of enzymes, including phenylalanine, tyrosine, and tryptophan hydroxylases. Not a dietary requirement, since it can readily be synthesized in the body. Rare patients with a variant form of phenylketonuria cannot synthesize biopterin, and have to receive supplements.

Biopulping Experimental way of using a fungus to pretreat wood chips before making paper pulp. Biopulping reduces both energy use and water-polluting by-products.

Bioremediation Using organisms to remove toxins from the environment. Examples: a fungus to detoxify a wood preservative that contaminates soil at sawmills; a bacterium to help clean up oil spills.

Bios A name given to a factor in cell-free extract of yeast which was essential for the growth of yeast, by Wildiers in 1901. Three components were subsequently identified: inositol, ß-alanine and biotin. Of these, only biotin is a vitamin and essential for human beings.

Biotechnology The simplest definition of biotechnology is "applied biology." The application of biological knowledge and techniques to develop products. It may be further defined as the use of living organisms to make a product or run a process. By this definition, the classic techniques used for plant and animal breeding, fermentation and enzyme purification would be considered biotechnology. Some people use the term only to refer to newer tools of genetic science. In this context, biotechnology may be defined as the use of biotechnical methods to modify the genetic materials of living cells so they will produce new substances or perform new functions. Examples include recombinant DNA technology, in which a copy of a piece of DNA containing one or a few genes is transferred between organisms or "recombined" within an organism.

Biotin A vitamin that helps with energy metabolism, fatty acid and nucleic acid synthesis.

Birch beer A non-alcoholic carbonated beverage flavoured with oil of wintergreen or oils of sweet birch and sassafras.

Bird's nest soup Made from the dried gelatinous coating of the nest of various species of southern Asian swifts with chicken broth, ham, egg white, and spring onion.

Biriani Highly spiced rice dish flavoured with saffron and layered with meat.

Birnbrot Bread with filling of dried pears and other dried fruits.

Biscuit A baked flour confectionery dried down to low moisture content. The name is derived from the Latin bis coctus, meaning cooked twice. Known as cookie in the USA, where 'biscuit' means a small cake-like bun.

Biscuit check The development of splitting and cracking in biscuits immediately after baking.

Biscuit cure See Salting.

Bishop A medieval beverage of hot spiced and sweetened wine.

Biskoids Trade name for saccharin.

Bisque Thick rich soup, generally made from fish or shellfish stock.

Bitki Meatballs made from raw minced beef with fried onions, grilled or fried.

Bitot's spots Irregularly shaped foam-like plaques on the conjunctiva of the eye, characteristically seen in vitamin A deficiency, but not considered to be a diagnostic sign without other evidence of deficiency.

Bitter, bitterness, bitter taste bitter Describes one of the basic tastes characterised by solutions of quinine and caffeine. Spoiled fish can have bitter taste, perhaps due to the production of bitter peptides by degradation of proteins by spoilage bacteria. Urea, present in elasmobranchs has a bitter taste. Bitterness is the attribute of a substance that has a bitter taste.

Bitterballen Dutch; fried meatballs flavoured with Worcestershire sauce and nutmeg.

Bitters Extracts of herbs, spices, roots, and bark, steeped in, or distilled with, spirits. Originally prepared for medicinal use; now used mainly to flavour spirits and cocktails, or as aperitifs.

Biuret test A chemical test for proteins which depends on the formation of a violet colour when copper sulphate reacts with a peptide bond in alkaline solution.

Bixin See Dye.

Bixin A carotenoid pigment found in the seeds of the tropical plant Bixa orellana; the crude extract is the colouring agent annatto.

Black cumin Seeds of Nigella sativa, used as a spice; unrelated to cumin.

Black PN A black food colour, also known as brilliant black BN.

Black pudding Traditional European dish made with sheep or pig blood and suet, originally together with oatmeal, liver and herbs, stuffed into membrane casings shaped like a horseshoe. Although it is already cooked, it is usually sliced and fried. In Germany and France it is made without cereal. An exceptionally rich source of iron; a rich source of protein; a good source of niacin and copper; a source of vitamin B 1.

Black spot Blackening of the shell of crustacea, especially shrimp, prawns and lobsters, due to the enzymatic formation of the black pigment melanin. Typically starts at the base of the legs, spreads over the carapace then onto the tail. It is associated with spoilage though discoloration begins before spoilage odours or flavours are noticeable in the tail meat. It can be inhibited by treating the crustacea with sodium sulphite, a generally permitted food additive for crustacea. The process is also called melanosis.

Black tongue disease A sign of niacin deficiency in dogs, the canine equivalent of pellagra, historically important in the isolation of the vitamin.

Black velvet Mixture of equal parts of sparkling wine and stout.

Blackberry Berry of the bramble, Rubus fruticosus. A good source of vitamin C; a source of folate and copper.

Blackberry odour An unpleasant, sulphide odour found occasionally in some species of fish. The odour is due to the presence in the flesh of the fish of dimethylsulphide formed from dimethyl-propiothetin in the feed of the fish. The odour usually appears in the spring, and in particular localities, when the fish are feeding on species of pteropods, pelagic gastropod molluscs. Sometimes, termed weedy. Cod, especially from Spitzbergen and Labrador, tend to develop this condition when feeding on the pteropod Spiratella helicina, and mackerel are known to develop a similar condition when feeding on the pteropod Spiratella retroversa.

Blackcurrant Fruit of the bush Ribes nigra, of special interest because of its high vitamin C content. The British National Fruit Collection has 120 varieties. A rich source of vitamin C; a source of iron and copper.

Blackening 1. A black discoloration found in canned fish and shellfish. Caused by defects in the plating or lacquering of the insides of steel cans such that sulphides present in the product can interact with the exposed steel to form black iron sulphide. The blackening might be seen only on the walls of the can, but can extend into the contents. Not the same as black spot.

2. Intense black, localised, discoloration arising from the copepod parasite Sarcotaces articus, the ink bag parasite, found in blue ling, Molva dypterigia. It is a black liquid, and if the sac is cut during filleting the surrounding flesh is stained black. Otherwise the parasite can be carefully excised without causing staining.

Blanc de blancs White wine made from white grapes. By contrast, blanc de noirs is white wine made from black grapes; but the skin is removed before fermentation begins.

Blanching Process of immersing in hot water or heating in steam at 95°C for 1-5 minutes to reduce enzyme activity.

Blancmange powder Usually a cornflour base with added flavour and colour, mixed with hot milk to make a dessert.

Blanco White wines.

Blanco Spanish; white wines.

Bland diet A diet that is non-irritating, does not over-stimulate the digestive tract and is soothing to the intestines; generally avoiding alcohol, strong tea or coffee, pickles, spices, and high intake of dietary fibre.

Bland Lacking a strong taste or character.

Blank See Fish finger.

Blanquette White stew of chicken, veal, lamb, or sweetbreads, with a rich cream sauce.

Blast freezer and blast freezing See Freezer.

Blawn fish Scottish; fresh fish, rubbed with salt and hung in a windy passage for a day, then grilled.

Bleach figure A measure of the extent of bleaching of the flour from the relative paleness of the extracted pigments.

Bleaching The removal or destruction of colour. In the context of food it usually refers to the bleaching of flour. It also refers to the bleaching of oils, a stage in the purification by which dispersed impurities and natural colouring materials are removed by activated or fuller's earth.

Bleaching, bleached The loss of colour from the skin of fish that is then referred to as bleached. It occurs in fish stored in water or seawater, in water thawing, or when the fish have been stored in melting ice for a time.

Bleeding Removing the blood from fish soon after capture on the fishing vessel or after harvesting at the fish farm. Recommended for white fleshed fish to avoid blood spots and blood discoloration in the fillet, which becomes a brown discoloration after freezing and thawing. Also recommended for farmed fish such as salmon, trout and other species, particularly to avoid undesirable dark discoloration after smoking. White fleshed fish for chilling should be gutted quickly, cleanly and neatly soon after catching, the gut cavity washed, and the fish stowed in ice belly down to allow the gut cavity to drain. White fleshed fish frozen at sea should be kept cool before being gutted, or headed and gutted, left to bleed chilled for up to 30 minutes, and washed thoroughly before being frozen. Fish left ungutted for more than a few hours cannot be bled properly because the blood clots in the blood vessels. In handling farmed salmon, it is common practice to enhance bleeding by cutting the gill arches during killing, and the fish are kept chilled prior to gutting. It is impractical to bleed small pelagic fish caught in large numbers at a time, e.g. herring, mackerel, pilchard and anchovy.

Blend To evenly combine ingredients using a spoon or spatula.

Bleu, au Fish such as trout, cooked immediately after they are caught, by simmering in white wine with herbs, or in water containing salt and vinegar.

Blewits Edible wild fungus, Tricholoma saevum, also known as bluetail.

Blind experiment In a single blind experiment, the subjects do not know whether they are receiving an experimental treatment or a placebo. In a double blind experiment, neither the researchers nor the participants are aware of which subjects receive the treatment - until after the study is completed.

Blind staggers Acute vitamin B 1 deficiency in horses and other animals, caused by eating bracken, which contains thiaminase, which destroys the vitamin.

Blinis Small yeast pancakes made from buckwheat flour, served with salt herring, smoked salmon, or caviar, and sour cream.

Blintzes Pancakes, stuffed with either curd cheese or minced meat.

Bloaters Salted, cold-smoked herrings.

Block fillet See Filleting.

Block ice See Icing.

Block press See Laminated block.

Blood curd See Curd.

Blood pickle See Salting.

Blood plasma The liquid component of blood, accounting for about half the total volume of the blood, Plasma is a solution of nutrients and various proteins, mainly albumin and various globulins, including the immunoglobulins which are responsible for much of the body's defence against infection, as well as some adverse reactions to foods. When blood has clotted, the resultant fluid is known as serum.

Blood spots Superficial red blood marks on fish, especially noticeable on white fish and salmon fillets, which should be readily removed by washing.

Blown May refer to a filled can after heat processing that has swollen ends and is indicative of spoilage of the contents, or it may refer to a fish oil that has been slightly oxidised by blowing air through it.

Blue spot A blue discoloration of abalone meat. The colour is probably due to blood, which is blue in colour in abalone, lodging in the meat after death of the animal. Also referred to as blueing.

Blueberry Fruit of Vaccinium corymbosum or V. augustifolium grown mainly in north America.

Blueing A blue discoloration, or blueing, sometimes occurs during the canning of crab, and the processing of abalone.

Blush wine Californian term for rose wines.

Boar, wild Meat of Sus scrofa. Hunted in parts of Europe, farmed on small scale in UK.

Bobby veal Veal from calves younger than 3 months.

Bocadillo Spanish; a sandwich made by slicing a long crusty loaf or roll lengthways.

Bocal Wide-mouthed glass jar used for bottling or pickling fruit and vegetables.

Body image The mental picture a person has of his/her body as well as the individual's associated thoughts, feelings, judgments, sensations, awareness, and behavior. Body image is not a static concept. It is developed through interactions with people and the social world, changing across life span in response to changing feedback from the environment.

Body mass index An indicator of obesity primarily for adults. Calculated by dividing the weight of a person in kilograms by the square of the person's height in meters. BMI of 25 to 29.9 is considered "overweight," and 30 or higher is considered "obese."

Body surface area Heat loss from the body is related to surface area and basal metabolic rate and energy expenditure are sometimes expressed per unit body surface area. It is commonly calculated according to the formulae of: Du Bois: area (cm2) = 71.84 x weight0.425(kg) x height0.725(cm) Meeh: area (cm2) = 11.9 x weight?kg).

The surface area of adults is about 18,000 cm2 (men) or 16,000 cm2.

Body type The proportion and arrangement of body mass as determined by genetics. Variation in bone structure, muscle, and distribution of body fat are a few examples. Each person has different inherited characteristics that cannot be changed by behavior.

Body-building food A term used indiscriminately, but generally referring to proteins. The Code of Practice in advertising suggests that no claim should be made for the body-building properties of a food unless a reasonable amount of protein is present in a normal portion.

Bog butter Norsemen, Finns, Scots and Irish used to bury firkins of butter in bogs to ripen and develop a strong flavour.

Bog myrtle A wild plant with a strong resinous flavour. The leaves and seeds are used to flavour soups and stews.

Boiled sweets Sugar and water boiled at such a high temperature that practically no water remains and a vitreous mass is formed on cooling.

Boiler Danish, Norwegian; dumplings.

Boiling point The temperature at which the saturated vapour pressure of a liquid equals the external atmospheric pressure. As a consequence, bubbles form in the liquid and the temperature remains constant until all the liquid has evaporated. As the boiling point of a liquid depends on the external atmospheric pressure, boiling points are usually quoted for standard atmospheric pressure.

Boletus Edible wild mushroom, Boletus edulis or B. granulatus, also known as the yellow mushroom or cep.

Bologna Italian smoked pork and veal sausage, also known as polony.

Bombay duck Fish found in Indian waters, Harpodon nehereus or Saurus ophiodon, eaten either fresh or after salting and curing. The name is a corruption of the local name of the fish, bombil.

Bombay halva Indian dessert made from semolina, almonds and pistachio nuts.

Bombe A mould with a tight-fitting lid, used to shape ice cream and sometimes fruit. Bombe glacee is a hemispherical frozen dessert made from two separate layers of different ice creams.

Bommerlunder A German schnapps.

Bonbon General term for sugar confectionery.

Bondiola Cured shoulder of pork.

Bondon Soft cheese shaped in the form of a bun.

Bone Bones consist of an organic matrix composed of collagen and other proteins and crystalline mineral, mainly hydroxyapatite (calcium phosphate and calcium hydroxide), together with mag-nesium phosphate, fluorides, and sulphates.

Bone broth Prepared by prolonged boiling of bones to break down the collagen and extract it as gelatine. Of little nutritive value, since it consists of 2-4% gelatine, with little calcium.

Bone charcoal Charcoal produced by heating pieces of bone sufficiently to burn off the organic matter, leaving the carbon deposited on a framework of calcium carbonate. It is used to purify solutions because it will absorb colouring matter and other impurities. Also known as animal charcoal.

Bone separator See Minced fish.

Boneless fish meat See Minced fish.

Boneless kipper See Kipper and Kippering.

Boneless Strictly means, without bones, but it may mean, nominally without bones, having a specified very low bone content, as in the case of a laminated block or mince block, or it may refer to some smoked fish products, such as the so-called boneless kipper, that has most of the large bones removed and is also called a boned kipper.

Bone-meal Prepared from degreased bones and used as a supplement in both animal and human foods as a source of calcium and phosphate. Also used as a plant fertilizer as a slowly released source of phosphate.

Boniatillo Caribbean (Cuba); creamed dessert made from sweet potato and eggs.

Bonnag Manx; soda bread.

Bonne femme Cooked in a simple or 'housewifely' style, with a garnish of fresh vegetables or herbs, usually including mushrooms. Applied especially to fish dishes and cream soups.

Bonne-bouche Small savoury appetizers.

Bontrae Trade name for textured vegetable protein prepared by spinning or extrusion.

Boonchi Caribbean name for yard-long beans or asparagus beans, Vigna sesquipedalis.

Boquerones Spanish; fresh anchovies, floured and joined together by making a small incision at the tail of one and slipping the tails of 3 or 4 others through. Fried in the shape of a fan.

Borage A herb, Borago officinalis. The flowers and leaves have a cucumberlike flavour and are used to flavour drinks.

Borax The sodium salt of boric acid.

Borborygmos Audible abdominal sound produced by excessive intestinal motility.

Bordeaux Red and white wines produced in the Bordeaux region of France; red Bordeaux wines are called claret in the UK.

Boric acid Chemically H_3BO_4, derived from the element boron, boric acid has been used in the past as a preservative in bacon and margarine, but boron accumulates in the body. It was formerly used as an anti-infective agent and eye-wash but there was a high incidence of toxic reactions.

Boron It is a trace mineral. Studies show that Boron helps the body retain minerals, such as Calcium and Magnesium. Large amounts of Boron, over 10 milligrams a day, can be toxic, particularly to the organs that manufacture testosterone. You can find traces of Boron in all the food groups, even in wine, with the greatest concentration in prunes, raisins, parsley flakes, and almonds. A 1987 study showed that Boron could dramatically increase testosterone levels, however, the study was for postmenopausal women who had testosterone deficiencies. Once their boron-rich diets brought their testosterone levels back up to normal, those levels stabilized, and they didn't get any higher no matter how many more prunes or parsley flakes that they ate. Thus, it is somewhat unproven that boron can help build muscle mass by increasing your testosterone levels. However, a lack of boron in your diet may have a 'negative' impact on energy utilization.

Borscht Beetroot soup, served hot or cold.

Boston brown bread An American spiced pudding, steamed in the can.

Botargo A relish or dip prepared from fish roe.

Boti kabab Small pieces of meat, marinated and cooked rapidly under intense heat, basted with butter or ghee.

Bottle house An old English term for a manufacturer of glass containers as distinct from tableware.

Bottle The traditional wine bottle holds 700, 720, or 750 mL of wine, depending on the variety; within the EU wine bottles are being standardized at 700 mL.

A two-bottle size is a magnum; four-bottle is a Jeroboam or double magnum, six is a Methuselah, twelve a Salmanzar, and twenty is a Nebuchadnezzar.

Botulinum cook See Canning and Commercial sterility.

Botulism, botulinum organism, botulinum toxin A type of food poisoning, often fatal, in the form of a paralysing disease of the central nervous system caused by consumption of a powerful neurotoxin produced by the organism Clostridium botulinum, the botulinum organism, some types of which are occasionally carried by fish. Death can follow within hours of consumption of the botulinum toxin. Outbreaks have arisen from the consumption of uncooked, partially cooked, fermented, canned and smoked fish products. Fish products that are intended to be eaten without further cooking such as some raw products, pickled, fermented, canned, some types of smoked product, and products packed anaerobically are those for which most care needs to be taken to avoid the hazard of botulism. In practice, if fish are adequately iced immediately after catching and not kept above 3°C at all stages of processing until eaten, or subjected to an effective botulinum cook during heat sterilisation in containers where the integrity of the seals and seams remain remain intact, or cooked to destroy the toxin, they will remain safe. Botulism is also known as botulinal or botulinum poisoning. See Clostridium botulinum, Canning and Smoking.

Boucanning A Caribbean process by which meat was preserved by sun-drying and smoking while resting on a wooden grid known as a boucan.

Bouillabaisse Fish stew or soup flavoured with saffron, spices, and herbs; speciality of the Mediterranean region. So named because it is repeatedly boiled.

Bouillon A plain, unclarified beef or veal broth; also used synonymously with stock.

Boula boula American; turtle soup mixed with pureed fresh green peas.

Bound water The bound water content of a food is the water that is never available for freezing and cannot be frozen out of the food. It is water chemically bound to specific sites such as the carbonyl and amino groups of proteins.

Bouquet garni A small bundle of parsley, thyme, marjoram, and bay leaves, tied together with cotton and added to the dish being cooked. Now also the same mixture of herbs in a porous paper sack. Also known as a faggot.

Bourbon American whiskey made by distilling fermented maize mash. Sour mash bourbon is made from mash that has yeast left in it from a previous fermentation.

Bournvita Trade name for a preparation of malt, milk, sugar, cocoa, eggs, and flavouring, to make a beverage when mixed with milk.

Bourride Fish soup similar to bouillabaisse but without saffron, flavoured with aïoli.

Bovine Formal word for any animal in the cattle family.

Bovine cartilage A source of mucopolysaccharides which have anti-inflamatory and joint protective properties.

Bovine colostrum Usually from cows, a dairy product that has similar properties to human colostrum. Normally, adults cannot absorb colotrum's antobodies and growth factors the way a newborn can. But it still has superior nutritional values which may make it a useful supplement.

Bovine growth hormone/bovine somatotropin Two names for one protein hormone produced in the pituitary gland of cattle. BGH/BST can also be produced by inserting a copy of the gene for BGH/BST into laboratory bacteria. Such recombinant BGH/BST is also referred to as rBGH/rBST. Purified from pituitary glands or from bacterial cultures, BGH/BST injected into dairy cows can increase milk production up to 20 percent. Compare to porcine somatotropin (PST).

Bovine spongiform encephalopathy (BSE) Bovine spongiform encephalopathy, or BSE, is also known as "mad cow disease." It is a rare, chronic degenerative disease affecting the brain and central nervous system of cattle. Cattle with BSE lose their coordination, develop abnormal posture and experience changes in behavior. Clinical symptoms take 4-5 years to develop, followed by death in a period of several weeks to months unless the affected animal is destroyed sooner.

Bovril Trade name for a preparation of meat extract, hydrolysed beef, beef powder and yeast extract, used as a beverage, a flavouring agent, and for spreading on bread.

Boxing Normally refers to the practice of packing whole fish in boxes in ice in the fishroom of a fishing vessel. Used more broadly it could include the packing of chilled fish or fish products such as fillets in boxes, frequently of the expanded polystyrene type, for overland distribution or for air freight.

Bracken Young unopened leaves (fronds) of bracken, eaten as a vegetable and regarded as a delicacy in the Far East. Known as fiddleheads in Canada. They contain an antagonist of vitamin B 1 ; cattle and horses eating large amounts suffer from blind staggers due to acute vitamin B 1 deficiency; also contain a number of known or suspected carcinogens.

Bradycardia An unusually slow heartbeat, less than 60 beats/min. Such a low rate may be normal in trained athletes.

Bradyphagia Eating very slowly.

Brain Sugar Obsolete name for galactose.

Brain Traditionally the brains of sheep and calves are stewed and eaten; probably not advisable because of the risk of transmitting the agents responsible for various degenerative brain diseases, including scrapie and bovine spongiform encephalopathy (BSE).

Braise A method of cooking in a closed container, with very little liquid, normally in an oven.

Bran plus Trade name for untreated wheat bran with germ.

Bran The outer layers of cereal grain, which are largely removed when the grain is milled. The germ is discarded at the same time, and there is a considerable loss of iron and other minerals, and particularly of the B vitamins, as well as of dietary fibre.

Brandy A spirit distilled from wine, and containing 37-44% alcohol by volume. The name is derived from the German brandtwein, meaning burnt wine, corrupted to brandy wine. Most wine-producing countries also make brandy.

The age of brandy is generally designated as 3-star; VSOP; Napoleon; XO, Extraordinary Old. Cognac and Armagnac are brandies made in defined regions of France.

Fruit brandies are either distilled from fruit wines or are prepared by soaking fruit in brandy.

Brandy butter Hard sauce made from butter, caster sugar, and brandy, traditionally served with Christmas pudding and mince pies.

Brandy sauce English; sauce made from egg yolk, cream, sugar, and brandy, traditionally served with Christmas pudding.

Brandy snaps Crisp toffee-like biscuits made from flour, butter, syrup, and powdered ginger, baked, then rolled into cylinders while still warm and pliable.

Brassica Genus of vegetables that includes broccoli, Brussels sprouts, cabbage, cauliflower, kale, kohl rabi, mustard, and swedes.

Bratwurst Pork sausage with many regional specialist varieties; may be served boiled, grilled, or fried.

Brawn Made from pig meat, particularly the head, boiled with peppercorns and herbs, minced and pressed into a mould. Mock brawn differs in that other meat by-products are used.

Brazil nuts From wild trees of Bertholletia excelsa. An exceptionally rich source of selenium; a rich source of vitamins B 1 and E; a good source of protein, niacin, and calcium; a source of iron.

Bread A staple food made basically by baking a mixture of flour and water. Ordinary leavened bread is made by mixing a dough of flour, water, yeast, sugar, salt, and sometimes other ingredients. The dough is kneaded and left to rise twice, a process that can take several hours, before being baked. White loaves, rolls, French sticks, and brown bread are made from different types of wheat flour. Rye bread is another variety, which has a stronger more bitter flavour and can be black, brown, or white.Bread has been baked since the earliest times, evidence of barley cakes having been found in Neolithic dwellings.

Bread sauce Thick white sauce made from bread and milk in which an onion has been boiled; a traditional accompaniment to poultry.

Bread, starch-reduced Bread is normally 9-10% protein and about 50% starch; if the starch is reduced, either by washing some of it out of the dough or by adding extra protein, the bread is referred to as starchreduced, and is often claimed to be of value in slimming and diabetic diets. Legally, the name 'starch-reduced bread' may be applied only to bread containing less than 50% carbohydrate, and any claims for its value as a slimming aid are strictly controlled.

Breadfruit Large spherical, starchy fruit of the tree Artocarpus communis or A. incisa. Staple though seasonal food in the Caribbean, eaten roasted whole when ripe, or boiled in pieces when green.

Breading, breaded products See Coating.

Breadspreads General term for fats used to spread on bread, including butter, margarine, and low-fat spreads that may not legally be called margarine.

Breakfast cereal Legally defined as any food obtained by the swelling, roasting, grinding, rolling, or flaking of any cereal. Products are described under their individual names.

Breathing Of red wine, opening the bottle some time before serving to allow oxidation and development of the full mature flavour.

Bredsoy Trade name for an unheated full-fat soya flour.

Bresaola Thinly sliced dried salted beef

Bretonne sauce Used with eggs and fish, made with onions or leeks and white wine.

Brewers' grains Cereal residue from brewing, containing about 25% protein; used as animal feed.

Brewing The process by which beers, ales, and lagers are made. In the West the basic ingredient of beer is barley, while in Africa millet or maize may be used, and rice beer is made in Japan . In beer-making the barley or other grain is germinated, and the young seedlings are then dried to produce malt. The malt is ground, and placed in a mash tub with water and cereal, where enzymes in the malt convert the starch into fermentable sugars. The resulting liquid, called wort, is transferred to a brewing kettle, where flavourings, particularly hops, are added, and the mixture is boiled. The mixture is then filtered, and fermentation is stimulated by the introduction of yeast. Traditionally, the liquid at this stage is filtered again and placed in wooden barrels, where fermentation continues. Ales and stouts are usually fermented at 15-20°C, while lagers are fermented at 6-8°C for longer periods. In keg beers and lagers, fermentation is stopped after only a short period by placing the liquid in sealed metal barrels and introducing carbon dioxide.

Brie Soft white cheese originally from the Brie region of France, made from cows' milk and moulded into a flat disc; it has a white rind, is ripened for 3-4 weeks, and deteriorates rapidly.

Brillat-savarin A French gourmet whose name is given to a consomme, baba, and several other dishes.

Brine bath (brine tank, brine tub, brine vat) See Brining.

Brine freezing See Freezer.

Brine immersion freezer See Freezer.

Brine Solution of a salt in water sometimes used as a secondary refrigerant for chilling, freezing or cold storage; sodium chloride or calcium chloride is used, but direct contact with food is restricted to the former. A solution of sodium chloride in water used for brining prior to smoking, drying or some other form of processing. The amount of sodium chloride in solution is termed the brine strength, as measured by the density of the solution in brineometer degrees, where 100° is a saturated solution containing 264 g salt per litre of brine at 16°C. Temperature is specified because the amount of salt that can be dissolved depends on temperature. An 80° brine is a common working strength for batch brining of fish for smoking.

Brining Immersion of fish in brine in a brine bath for salting before smoking, drying, canning or some other form of processing. Dye may be added to the brine for smoking. An 80° brine is a common working strength for batch brining of fish for smoking. In a continuous briner the fish are conveyed at a speed that can be adjusted to give the required residence time in the brine bath to achieve the desired salt concentration and salt content. The brine is maintained almost at saturation point by passing through a bed of salt and continuously recirculated through a sieve to remove debris. The uptake of salt is a complex process depending on many factors, such as time of immersion in brine, strength and temperature of brine, size, shape, thickness and fat content of the prepared fish, whether the fish is frozen and thawed or unfrozen, presence or not of skin or degree of scoring of the skin, ratio of brine to fish, and homogeneity or mixing of the brine.

For a particular product, in brine of constant strength and temperature, the uptake of salt is roughly proportional to the square root of the brining time. Fish increase in weight due to retention of brine, the more so in brines weaker than 50°; use of saturated brine may result in the formation of salt crystals on the drier areas of the smoked product; 70 to 90° brine is recommended for most products. Brines must be kept to strength, changed at least once a day, and scales and other debris removed frequently. Fish in contact with stale brine will be contaminated with microorganisms. See Brine.

Brining time See Brining.

Brisket The meat covering the breast bone of the animal.

Brislings Young sprats, Clupea sprattus. An extremely rich source of vitamin D and a rich source of vitamin A.

Brittle Easily broken into small pieces.

Brix A table of specific gravity based on the Balling tables, calculated in grams of cane sugar in 100 g of solution at 20°C; degree Brix = % sugar. It is used to refer to the concentration of sugar syrups used in canned fruits.

Broasting A cooking method in which the food is deep fried under pressure, which is quicker than without pressure, and the food absorbs less fat.

Broche, en Also en brochette; roasted or grilled on a spit or skewer.

Broiling Cooking by direct heat over a flame, as in a barbecue; American term for grilling. Pan broiling is cooking through hot dry metal over direct heat.

Bromatology The science of foods, from the Greek broma, food.

Bromelains Enzymes in the pineapple and related plants of the family Bromelidaceae, which hydrolyse proteins. They are available as by-products from commercial pineapple production, usually from the stems, and are used to tenderize meat, to treat sausage casings, and to .chill-proof beer. Similar enzymes are found in figs.

Brominated oils Oils from a variety of sources, including peach and apricot kernels, olive and soya oils which have been reacted with bromine. They are used to help to stabilize emulsions of flavouring substances in soft drinks. Also known as weighting oils.

Bromine An element, chemically related to iodine, chlorine, and fluorine, not known to have any function in the body, and not a dietary essential.

Brooklime A wild plant that grows in very wet marshy conditions. The large, round, fleshy leaves can be added to salads.

Broonie Orkney; gingerbread made with oatmeal.

Brose A Scottish dish made by pouring boiling water onto oatmeal or barley meal; fish, meat, and vegetables may be added.

Broth Thick soup.

Broussiko Greek sweet fortified wine from the Cyclades.

Brown Adipose Tissue A type of specialized fat that is known to convert fat calories into heat, thereby disspating excess calories.

Brown betty American; pudding made from apple and breadcrumbs; similar to apple charlotte.

Brown colours Three brown colours are used in foods: Brown FK (E-154), synthetic, which is used to colour kippers; Chocolate-brown HT (E-155), synthetic, and caramel (E-150).

Brown sugar Brown sugar has a molasses film on the sugar crystals, which imparts the brown colour and characteristics flavour of this sugar. It contains approximately 2% moisture and requires storage protection against moisture loss.

Essentials of food science- Vickie A. Vaclavik

Brownie American cake made with chocolate.

Browning reactions Chemical reactions in foods which result in the formation of a brown colour.

Browning Refers to a number of discoloration conditions found in fish. Brown discoloration, especially of dried or canned fish products, caused by the Maillard reac-tion or non-enzymic browning reaction.

It is a deteriorative reaction between amino groups of proteins or amino acids and carbonyl groups of sugars during storage, particularly of dried and canned foods, that results in flavour changes and loss of some nutritive value as well as discoloration. Pre-cooking and decanting of the cook-out can help to avoid the problem in canned fish.

Brugnon Hybrid fruit, a cross between plum and peach. Resembles nectarine, and name sometimes used in France for nectarines.

Brunoise Mixture of diced or shredded vegetables used as base for a soup, sauce, or garnish.

Bruschetta Toasted bread moistened with olive oil.

Brussels sprouts Leaf buds of Brassica oleracea gemmiferra. Nine sprouts (90 g) are a rich source of folate and vitamin C.

Bruxism Teeth grinding, the habitual involuntary grinding or clenching of the teeth.

BT Bt is the abbreviation for Bacillus thuringiensis , a gram-positive bacterium that is common in soil. These bacteria show toxic effects to some classes of insects. In contrast to that Bt-toxin shows no toxic effects on mammals and birds. The toxic effect of these bacteria is used to create plants that possess resistance against a broad range of insects. The genes that produce the toxin are introduced into the crop through vectors, organisms that are able to introduce their genes into plants. In that way the plant later shows the same ability for toxin-production as Bacillus thuringiensis . In that way the plants produce their own insecticide. However, not all harmful insects are affected by that toxin; for example, it has no toxic effect on aphids, but just a reduced amount of chemical insecticides is still needed. Conversely, it has some effects on non-target insects, which is one point of criticism against the application of these transgenic crops. On the other hand, chemical insecticides are also not selective and kill non-target organisms as well.

Bubble and squeak English; originally cold boiled beef fried with cooked potatoes and cabbage. More commonly a fried mixture of left-over cabbage and potatoes. Colcannon is a similar Irish dish.

Buckling A hot-smoked herring.

Buck's fizz Sparkling wine mixed with orange juice; known in the USA as a mimosa.

Buckwheat A plan native to Asia that has fragrant white flowers and small triangular seeds. The edible seeds are often ground into flour. According to animal studies, buckwheat is better than casein for promoting muscle growth and body growth and decreasing blood lipds. For persons allergic to wheat gluten, it provides a gluten-free food with uses similar to gains. Roasted buckwheat is known as Kasha. Buckwheat is usually available as flour.

Budino Italian name for puddings, often containing cheese, fruit, or nuts.

Buffalo currant Two varieties of N. American currant: Ribes odoratum, which has a distinctive smell, and R. aureum, the golden or Missouri currant.

Buffer Substances that prevent a change in the pH when acid or alkali is added. Salts of weak acids and bases are buffers and are commonly used to control the acidity of foods. Amino acids and proteins also act as buffers. The pH of blood is maintained by physiological buffers including phosphates, bicarbonate, and proteins.

Bulgur The oldest processed food known. Prepared precooked wheat, originally from the Middle East. Wheat is soaked, cooked, and dried, then lightly milled to remove the outer bran and cracked. It is eaten in soups and cooked with meat . Also called ala, burghul, cracked wheat, and American rice.

Bulimia nervosa An eating disorder characterized by rapid consumption of a large amount of food in a short period of time, with a sense of lack of control during the episode and self-evaluation unduly influenced by body weight and shape. There are two forms of the condition, purging and non-purging. The first type regularly engages in purging through self-induced vomiting or the excessive use of laxatives or diuretics. Alternatively, the non-purging type controls weight through strict dieting, fasting or excessive exercise.

Buljol Caribbean; salad of salt cod, chilli, tomato, and avocado.

Bulking Storing loose whole fish, intermixed with ice, in layers, typically 40 to 50 cm deep, especially in the fishroom of a fishing vessel. See Chilling, Icing, Boxing, Shelfing.

Bullace Fruit of the wild damson, Prunus insititia; similar to sloe, and very acidic.

Bull's eye Cushion-shaped, mint-flavoured boiled sweet, with dark and light stripes.

Bully beef The name given by troops during the First World War to corned beef.

Bulrush A wild plant common in ponds and marshes. The young sprouts and shoots can be eaten in salads, the pollen is used as a flavouring, and the roots and unripe flower heads may be boiled as a vegetable.

Bun Sweetened bread roll; correctly made with yeast dough, although sometimes applied to small cakes made with baking powder, or to cream buns, which are made with choux pastry. Also applied to the rolls used for hamburgers.

Buni Coffee beans left in the field to dry; generally hard and of poor quality.

Burgundy Red and white wines produced in the Burgundy region of France.

Burnet Salad burnet, a wild plant growing in grassland on chalky soil. The leaves have the flavour of cucumber, and can be used to flavour fruit wines, vinegar, and butter, and are used in salads. Also called pimpernel.

Burning foot syndrome Nutritional melalgia; severe aching, throbbing, and burning pain in the feet, associated with nerve damage, observed in severely under-nourished prisoners of war in the Far East. It results from long periods on a diet poor in protein and B vitamins, and may be due specifically to a deficiency of pantothenic acid.

Bushel A traditional dry measure of capacity, equivalent to 80 lb of distilled water at 17°C with a barometer reading of 30 inches, i.e. 8 imperial gallons (36.4 L); used as a measure of corn, potatoes, etc. The American bushel is 3% larger.

Butifarra Spanish; spiced pork sausage containing pine nuts, almonds, cumin seed, and cinnamon. Butifarrones are smaller.

Butt A cask for beer or wine, containing 108 imperial gallons (491 L).

Butter Made from separated cream by churning; legally not less than 80% fat of which around 60% is saturated, a small proportion (3%) polyunsaturated, the rest being mono-unsaturated. Lactic butter, which is preferred in some countries, is made by first ripening the cream with a bacterial culture to produce lactic acid and increase the flavour. This is normally unsalted or up to 0.5% salt added. Sweet cream butter may be salted up to 2%. Butter supplies 72 kcal (300 kJ) per g; a rich source of vitamin A.

Buttermilk The residue left after churning butter, 0.1-2% fat, with the other constituents of milk proportionally increased. It is slightly acidic, with a distinctive flavour due to the presence of diacetyl and other substances. Usually made by adding lactic bacteria to skim milk; 90-92% water, 4% lactose with acidic flavour from lactic acid, it is similar to skim milk in composition. Dried buttermilk is used in bakery products and ice cream.

Butyric acid A short-chain saturated fatty acid containing four carbon atoms. It occurs as the triglyceride in 5-6% of butter fat, and in small amounts in other fats and oils.

C

CA Chronological age.

Cabbage Leaves of Brassica oleracea capitata. A rich source of vitamin C; a good source of folate; a source of vitamin B 1.

Cabernet franc A grape variety widely used for wine making, although not one of the classic varieties. Also known as bouchet.

Cabernet sauvignon One of the 9 'classic' grape varieties used for wine making, used for some of the great red wines of Bordeaux, and widely grown throughout the world.

Cabinet pudding Moulded pudding made from bread and butter or sponge cake, with custard; sometimes glace cherries and egg are added.

Cacciatora, alla In the hunter's style, generally game or poultry with onions, herbs, and tomatoes in a wine sauce.

Cacciatore Small Italian salami.

Cacen-gri Welsh; soda scones made with currants and buttermilk.

Cachelos Spanish dish of potatoes, cabbage, ham, and chorizo.

Cachexia The condition of extreme emaciation and wasting seen in patients with advanced diseases such as cancer and AIDS. Due to both an inadequate intake of food and the effects of the disease in increasing metabolic rate and the breakdown of tissue protein.

Cachou Small scented tablets for sweetening the breath.

Cadmium A mineral of no known function in the body, and therefore not a dietary essential. It accumulates in the body throughout life, reaching a total body content of 20-30 mg. It is toxic, and cadmium poisoning is a recognized industrial disease. In Japan cadmium poisoning has been implicated in itai-itai disease, a severe and sometimes fatal loss of calcium from the bones, as the disease occurred in an area where rice was grown on land irrigated with contaminated waste water. Accidental contamination of drinking water with cadmium salts also leads to kidney damage, and enough cadmium can leach out from cooking vessels with cadmium glaze to pose a hazard.

Caecum The first part of the large intestine, separated from the small intestine by the ileo-colic sphincter. It is small in carnivorous animals and very large in herbivores, since it is involved in the digestion of cellulose. In omnivorous animals, including man, it is of intermediate size.

Caffeine Caffeine is a naturally-occurring substance found in the leaves, seeds or fruits of over 63 plant species worldwide and is part of a group of compounds known as methylxanthines. The most commonly known sources of caffeine are coffee and cocoa beans, cola nuts and tea leaves. Caffeine is a pharmacologically active substance and, depending on the dose, can be a mild central nervous system stimulant. Caffeine does not accumulate in the body over the course of time and is normally excreted within several hours of consumption.

Caffeol A volatile oil in coffee beans, giving the characteristic flavour and aroma.

Cafiroleta Caribbean; dessert made from boniatillo and coquimol.

Cake Baked from flour with added fat, sugar, and eggs. Plain cakes are made by rubbing the fat and sugar into the flour, with no egg; sponge cakes by whipping with or without fat; rich cakes contain dried fruit.

Calabasa West Indian or green pumpkin, with yellow flesh.

Calabrese An annual plant, a variety of broccoli which yields a crop in the same year as it is sown. Also called American, Italian, or green sprouting broccoli.

Calamary Mediterranean squid.

Calamondin A citrus fruit resembling a small tangerine, with a delicate pulp and a lime-like flavour.

Calas 1. American rice fritters.

2. Akkra.

Calcidiol The 25-hydroxy-derivative of vitamin D, also known as 25-hydroxy-cholecalciferol, the main storage and circulating form of the vitamin in the body.

Calciferol Used at one time as a name for ercalciol, made by the ultra-violet irradiation of ergosterol. Also used as a general term to include both vitamers of vitamin D.

Calciol The official name for cholecalciferol, the naturally occurring form of vitamin D.

Calcionetta Dessert made from fried ravioli filled with spiced sweetened chestnut puree.

Calcitriol The 1,25-dihydroxy-derivative of vitamin D, also known as 1,25-dihydroxycholecalciferol; the active metabolite of the vitamin in the body.

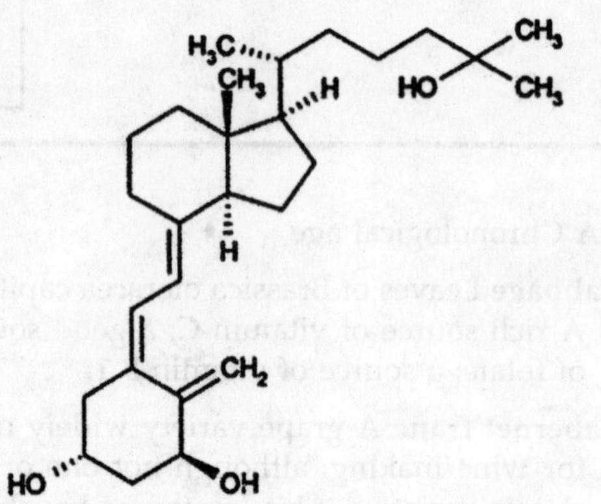

Fig. Calcitriol

Calcium acid phosphate Also known as monocalcium phosphate and acid calcium phosphate or ACP, $Ca(H_2PO_4)_2$. Used as the acid ingredient of baking powder and self-raising flour, since it reacts with bicarbonate to liberate carbon dioxide. Calcium phosphates are permitted food additives (E-341).

Calcium Most abundant mineral in the body; essential for the formation and repair of bone and teeth, but also essential to nerve transmission, muscle contraction, blood clotting and other metabolic activities as well. Long term calcium deficiency is linked to degenerative bone diseases.

Calculi Stones formed in tissues such as the gall bladder, kidney, or ureters. Renal calculi may consist of uric acid and its salts or of oxalic acid salts. Oxalate calculi may be of metabolic or dietary origin and people at metabolic risk of forming oxalate renal calculi are advised to avoid dietary sources of oxalic acid and its precursors. Rarely, renal calculi may consist of the amino acid cystine.

Caldereta Spanish; stewed fish or meat, named from the caldera or cauldron in which it is cooked.

Caldo Portuguese, Spanish; broth.

Calfos Trade name for a prepared 'bone meal used as a source of calcium and phosphate in foods.

calf's foot jelly Stock made by boiling calves' feet in water; it sets to a stiff jelly on cooling; largely water, so of little nutritional value.

Calipash The gelatinous green fat attached to the upper shell of the turtle; the yellow fat attached to the lower shell is calipee.

Calisay Spanish; liqueur made from chinchona bark, aged in oak casks.

Callaloo Caribbean name for leaves of both taro and Chinese spinach, and for the soup made from them.

Calorie The unit of measure used to describe the amount of energy provided by food. Scientifically, the amount of heat needed to raise the temperature of one gram of water one degree Celsius. The amount of calories one must consume in order for adequate energy to be available to maintain normal body functions is known as Caloric need.

Calorimeter An instrument for measuring the amount of oxidizable energy in a substance, by burning it in oxygen and measuring the heat produced. The energy yield of a foodstuff in the body is equal to that obtained in a bomb calorimeter only when the metabolic end-products are the same as those obtained by combustion. Thus, proteins liberate 5.65 kcal.

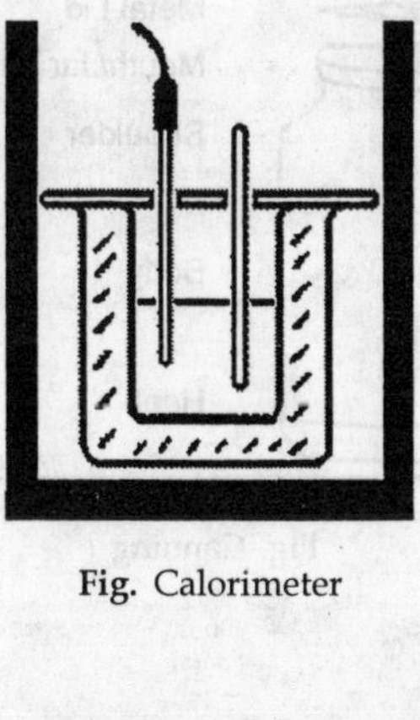

Fig. Calorimeter

Calorimetry The measurement of energy expenditure by the body. Direct calorimetry is the direct measurement of heat output from the body, as an index of energy expenditure, and hence energy requirements. The subject is placed inside a small, thermally insulated room, and the heat produced is measured. Few such difficult studies have been performed, and only a limited range of activities can be studied under these confined conditions.

Indirect calorimetry is a means of estimating energy expenditure indirectly, rather than by direct measurement of heat production.

Calphos Trade name for a prepared bone-meal used as a source of calcium and phosphate in foods.

Calvados Apple brandy made by distillation of cider. Appellation controllee calvados is from the Calvados region in Normandy, and is double distilled.

Calzone Crescent-shaped turnover of leavened dough, filled with ham and cheese; may be fried or baked. Sometimes called pizza calzone.

Cambridge sauce English; substitute for mayonnaise, made with oil, vinegar, and pounded yolk of hard-boiled eggs, flavoured with capers, anchovies, and herbs.

Camembert A major soft, French cheese made from cows' milk, originating from Auge, Normandy, France. Covered with a white mould (Penicillium candidum or P. camembertii) which participates in the ripening process; deteriorates after a few days.

Camomile Either of two herbs, Anthemis nobilis or Matricaria recutica. The essential oil is used to flavour liqueurs; camomile tea is a tisane prepared by infusion of the dried flower heads, and the whole herb can be used to make a herb beer.

Campari Bitter aperitif wine.

Campden process The preservation of food by the addition of sodium bisulphite (E-222), which liberates sulphur dioxide. Also known as cold preservation, since it replaces heat sterilization.

Campden tablets Tablets of sodium bisulphite (E-222), used for sterilization of bottles and other containers and in the preservation of foods.

Campylobacter A genus of pathogenic organisms which are the most commonly reported cause of gastro-enteritis in the UK, although it is not known what proportion of cases are foodborne. Campylobacteriosis has been associated with the consumption of undercooked meats, milk that has been inadequately pasteurized or contaminated by birds, and contaminated water. Helicobacter pyloris was formerly classified as a campylobacter.

Can See Canning and Heat-treated.

Canapes Small rectangular slices of bread, toasted or fried, spread with meat or fish pate; originally served as an accompaniment to winged game, and spread with forcemeat made from the entrails of the birds. The name derives from the French canape 'sofa', since these were traditionally eaten in the drawing room, before dinner was served in the dining room.

Canbra oil Oil extracted from selected strains of rapeseed containing not more than 2% erucic acid.

Cancer A wide variety of diseases characterized by uncontrolled growth of tissue. Dietary factors may be involved in the initiation of some forms of cancer, and a high-fat diet has been especially implicated. There is some evidence that antioxidant nutrients such as carotene, vitamins C and E, and the mineral selenium may be protective.

Canderel Trade name for tablets of the sweetener aspartame.

Candied peel Preserved fruit peel, commonly of citrus fruits, used in confectionery and cake making. It is prepared by softening the peel, then boiling with sugar syrup.

Candling The visual examination of fish fillets for nematodes, bones, and other defects by laying the fillet on a translucent plate over a strong light. The translucent plate and light source are usually combined in a light box.

Candy Crystallized sugar made by repeated boiling and slow evaporation.

A general term for sugar confectionery.

Cane sugar Sucrose extracted from the sugar cane Saccharum officinarum; identical with sucrose prepared from any other source, such as sugar beet.

Canihua Seeds of Chenopodium pallidicaule, grown in the Peruvian Andes; nutritionally similar to cereals.

Canning A means of obtaining long-term microbiological stability for non-dried foods without the use of refrigeration, by prolonged heating in hermetically sealed

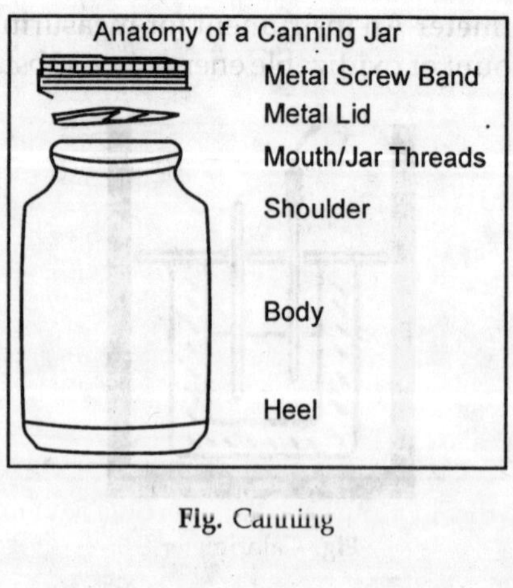

Fig. Canning

containers, such as cans or retortable pouches, to render the contents of the container sterile. This entry does not deal with the detailed stages of the entire canning process, e.g. the pre-processing steps such as preparation of the fish, brining, salting, smoking, marinating, or the process steps of pre-cooking, exhausting, sealing, seal integrity, retorting, and post-process cooling. The process depends on a thermal process lethality, in terms of exposure time to a specific high temperature, sufficient to effectively eliminate the most dangerous and heat resistant pathogens, particularly Clostridium botulinum. Thermal processes are calibrated in terms of the equivalent time the thermal centre of the product, i.e. the point of the product in the container most distant from the heat source or cold spot, spends at 121.1°C, and this thermal process lethality time is termed the F0value. Low acid foods having a pH greater than 4.5, which includes most canned fish products, must receive a minimum thermal process of F03 (121.1°C for 3 minutes), known as a botulinum cook, or an equivalent heat sterilisation process. The heat process F03 is designed to achieve a 12 log cycle reduction in spores of Clostridium botulinum, which means a less than one in 1012chance of survival. This is regarded as an acceptably low level possibility of survival of this dangerous pathogen and is the basis of so-called commercial sterility. Canners often use F values ranging from 6 to 14 for low acid products to provide an additional safety margin to compensate for inaccuracies in temperature measurement. Canned foods destined for hot countries, or manufactured in hot countries, may require a much higher thermal process, such as F greater than 10. In order to maintain the sterility of the contents of the sealed container, it is essential for product safety that the integrity of the container and its seals are maintained to prevent ingress of microbial contamination after heat treatment. Scrupulous post-process hygiene is paramount particularly during the cooling phase. The main purpose of microbiological testing of heat processed foods is to assess post-processing contamination.

Canned fish products fall into two basic categories with respect to selection of the appropriate thermal process high acid, less than pH 4.5; low acid, greater than 4.5. Fish marinades and pickles containing acetic, citric, tartaric or lactic acids, or vinegar such that the pH of the product is below 4.5 will not support the growth of spore forming, human pathogens. The organisms that can grow in such acid conditions are destroyed by relatively mild heat treatments, e.g. raising the temperature of the product at its cold spot to 90°C and immediate cooling. The low acid category includes most canned fish products, including those canned in tomato sauce, and all require a full heat sterilisation process.Typically, the fatty fish species such as herring, mackerel, sprat, pilchard, anchovy, salmon and tuna, crustacean shellfish such as shrimp, prawn, krill and crab, lobster and crayfish, molluscan shellfish such as abalone, cockle, mussel, oyster, whelk and winkle, and some other species, are canned. White fish do not usually can well. Usually the heat process results in significant changes in colour, texture and flavour of the product compared to the chilled or frozen item. For some products such as crustaceans and tuna, and some mackerel products, the fish or shellfish is pre-cooked before canning. Some problems may occur with the canning of fish products, such as excessive formation of curd when frozen salmon are canned, or excessive cook out or water loss from flesh during cooking. Discoloration may occur such as browning reactions and greening in tuna. Blackening may occur, particularly with eels, abalone and albacore tuna, especially when material is held in frozen storage prior to canning, and it is due to the precipitation of iron sulphide during heat processing. Undesirable colour changes in shellfish during canning often involve metal ions. Crustaceans must be canned very soon after death because spoilage is rapid and volatile sulphur compounds, such as hydrogen sulphide produced during spoilage and during heating can react to form metal sulphides. In the presence of iron, tin and copper, dark staining or blackening may occur in the can. A blue discoloration, or blueing, sometimes occurs, especially in crab, and a blue/green discoloration can occur in the frill of abalone during the canning, part of a problem with abalone collectively referred to a blue spot. Pink patches may appear after brining and canning stale abalone. Struvite crystals are occasionally found in cans of salmon, tuna, mackerel and crustacea and mistaken for glass by the consumer. A wide range of containers is available other than the widely used conventional, three-piece, three-seam, open-top, tin-plate can. Both tin-plate and aluminium cans can be drawn from a single plug of metal to have a single seam at the processors-end that may have a ring-pull end. The internal surfaces of metal cans are lacquered to prevent internal corrosion and to prevent the fish adhering to the surface of the can. The lacquers may be modified, e.g. to confer acid resistance. The external surfaces may also be lacquered where there is risk of external corrosion, for example, at sea or in coastal areas. Glass jars are often used for fish pastes and spreads. The flexible pouch or retortable pouch enables the degree of heat processing required to make the pack sterile to be reduced because, compared with a conventional can, it has a higher surface area and is thinner. But, it requires an over-pressured retorting regime to prevent rupture of the seals and bursting of the pouch. Typically, the pouch is made from a three layered, bonded laminate of heavy gauge outer plastic polymer, bonded to an intermediate layer of aluminium foil, bonded in turn to an inner plastic polymer layer that can be heat-sealed. The pouch may be protected with an outer cardboard carton. The so-called plastic can is thermoformed from mutilayered coextruded plastics that can withstand the temperatures of normal canning processes, but it also requires an over-pressured retorting regime to prevent rupture of the seals. These containers can be lidded with a ring pull metal end seamed onto the plastic body or with a foil laminate heat sealed to the rim. The thermal conductivity is lower than a metal can, requiring a longer heat process time to achieve the same thermal process lethality than metal cans of the same dimensions. See Heat-treated and Aseptic canning.

Canola A variety of rape which is low in glucosinolates.

Canthaxanthin An orange red pigment chemically related to astaxanthin that can be incorporated in feeds for salmonids to impart a red colour in the flesh. The colour imparted to the flesh is not identical to that produced by astaxanthin, but canthaxanthin can be synthesised more easily than can astaxanthin.

Canthaxanthin is not a permitted food additive in some countries.

Cape gooseberry Fruit of the Chinese lantern Physalis peruviana, P. pubescens, or P. edulis; herbaceous perennial resembling small cherry, surrounded by dry, bladder-like calyx, also known as golden berry, Peruvian cherry, and ground tomato.

Caper Unopened flower buds of the subtropical shrub Capparis spinosa or C. inermis with a peppery flavour; commonly used in pickles and sauces. Unripe seeds of the nasturtium can be pickled and used as a substitute.

Capercaillie A large game bird, also known as wood grouse or cock of the wood.

Capillary fragility A measure of the resistance to rupture of the small blood vessels, Which would lead to leakage of red blood cells into tissue spaces. Deficiency of vitamin C can lead to increased capillary fragility. There is some evidence that the flavonoids reduce capillary fragility, at least when given intravenously, and they may have a clinical use. Although flavonoids have been called vitamin P, there is no evidence-that they are a dietary essential, nor indeed that they can be absorbed from the intestinal tract.

Capon A castrated cockerel, which has a faster rate of growth, and more tender flesh, than the cockerel. Surgery has generally been replaced by chemical caponization, the implantation of pellets of female sex hormone.

Caponata Sicilian; fried aubergine, in a tomato sauce containing capers, olives, celery, and anchovy, garnished with slices of tuna, crawfish, etc.

Cappelletti Small hat-shaped pasta envelopes stuffed with minced meat, etc.; a variant of ravioli.

Cappucino Coffee with a head of frothy milk or whipped cream.

Capric acid One of the medium-chain fatty acids, containing ten carbon atoms, found in the triglycerides of coconut oil and in goat and cow butter.

Caproic acid One of the short-chain fatty acids, containing six carbon atoms, found in the triglycerides of coconut oil and goat and cow butter.

Caprylic acid One of the medium-chain fatty acids, containing eight carbon atoms, found in the triglycerides of goat and cow butter, coconut oil, and human fat.

Carambola Or star fruit, long ribbed fruit of Averrhoa carambola and A. bilimbi.

Caramel Brown material formed by heating carbohydrates in the presence of acid or alkali; also known as burnt sugar. It can be manufactured from various sugars, starches, and starch hydrolysates and is used as a flavour and colour in a wide variety of foods: soft drinks, alcoholic beverages, baked goods, sauces, canned meats, and stews.

Caramel cream A dessert made either by gently heating eggs in milk and allowing to cool or by using cornflour and milk, topped with a caramel sauce. Also known as crème caramel and as flan.

Caraway Dried ripe fruit of Carum carvi, an aromatic spice, used to flavour the liqueur kümmel, some types of aquavit, on bread and rolls, and in seed cake.

Carbohydrate Carbohydrates are organic compounds that consist of carbon, hydrogen and oxygen. They vary from simple sugars containing from three to seven carbon atoms to very complex polymers. Only the hexoses and pentoses and their polymers play important roles in nutrition. Carbohydrates in food provide 4 calories(17 kJ) per gram.

Plants manufacture and store carbohydrates as their chief source of energy. The glucose synthesized in the leaves of plants is used as the basis for more complex forms of carbohydrates. Classification of carbohydrates relates to their structural core of simple sugars, saccharides. Principal monosaccharides that occur in food are glucose and fructose. Three common disaccharides are sucrose, maltose and lactose. Polysaccharides of interest in nutrition include starch, dextrin, glycogen and cellulose.

Carbohydrate loading Practice of some endurance athletes in training for a major event; it consists of exercising to exhaustion, so depleting muscle glycogen, then eating a large carbohydrate-rich meal so as to replenish glycogen reserves with a higher than normal proportion of straight-chain glycogen.

Carbon dioxide Colourless gas, heavier than air, having a slight smell and used in liquid or in solid form as a refrigerant. Dry ice sublimes at a temperature of -78°C and may be used to maintain a low ambient air temperature in a vehicle or in a container during the transport of chilled or frozen fish.

Carbonnade Stewed or braised meat, often with beer in the sauce; originally the French term for meat grilled over hot coals.

Carboxyglutamate A derivative of the amino acid glutamate which is found in prothrombin and other enzymes involved in blood clotting. Its formation requires vitamin K. It also occurs in the protein osteocalcin in bone, where it has a function in ensuring the correct crystallization of bone mineral. Again its formation is dependent on vitamin K.

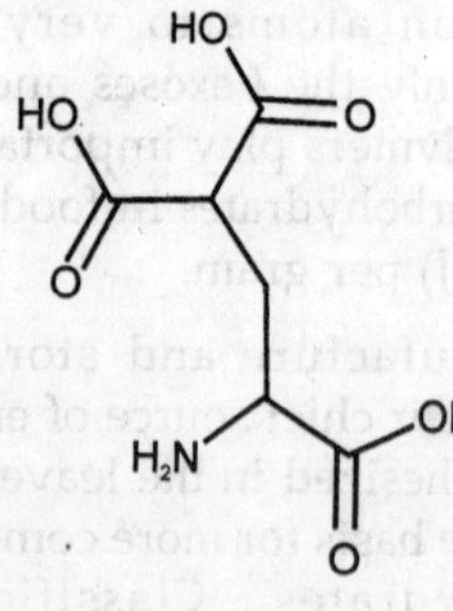

Fig. Carboxyglutamate

Carboxypeptidase An enzyme secreted in the pancreatic juice which is involved in the digestion of proteins.

Carcinogen An agent that causes cancer. Carcinogens can be chemicals, radiation, and some viruses. Chemical carcinogens include tar, aniline, and azo dyes. Large doses of radiation and several viruses can also cause cancer.

Carciofi Tender globe artichokes, eaten whole.

Cardamom The seeds and dried, nearly ripe, fruit of Elettaria cardamomum, a member of the ginger family. An aromatic spice used as a flavouring in sausages, bakery goods, sugar confectionery, and whole in mixed pickling spice. It is widely used in Indian cooking, and as one of the ingredients of curry powder. Arabic coffee is flavoured with ground cardamom seeds.

Cardoon Leafy vegetable; both the fleshy root and the ribs and stems of the inner leaves are eaten. Sometimes called chard, although distinct from true chard or spinach beet.

Caries Dental decay caused by attack on the tooth enamel by acids produced by bacteria that are normally present in the mouth. Sugars in the mouth promote bacterial growth and acid production; sucrose specifically promotes plaque-forming bacteria, which cause the most damage. A moderately high intake of fluoride increases the resistance of tooth enamel to acid attack.

Carignan A grape variety widely used for wine making, although not one of the classic varieties. Probably the commonest grape of France, a prolific cropper producing uninspiring wines.

Cariogenic Causing tooth decay by stimulating the growth of acid-forming bacteria on the teeth. The term is applied to sucrose and other fermentable carbohdrates.

Carmine Brilliant red colour derived from cochi-neal.

Carmoisine A red colour, also known as azorubine.

Carnatz Rumanian; croquettes of seasoned minced beef and pork, served with a tomato sauce.

Carnitine Non-structural amino acid that transports fatty acids into muscle cells for use as energy fuel. Carnitine is water-soluble and can be made in the body from the amino acids lysine and methionine with the assistance of vitamins C, B6 and Niacin, which act as coenzymes in the process.

Carnitine as discovered in meat in 1905, and was once called Vitamin T, because they thought it was a vitamin at first. 98% of te body's carntine exists in the heart and skelatal muscles. Carnitine is synthesized in the liver from Lysiine and Methionine, but half of the body's daily requirement for Carnitine comes from foods sources, including meat, poultry, fish and some diary products. Without supplemental carnitine, some people cannot use fat as energy.

Carnosine A dipeptide, ß-alanyl-histidine, found in the muscle of most animals. Its function is not known.

Carob gum The gum extracted from the carob, used as an emulsifier and stabilizer (E-410) as well as in cosmetics and as a size for textiles. Also known as locust bean gum.

Carob Seeds and pod of the tree Ceratonia siliqua, also known as locust bean and St John's bread: It contains a sweet pulp which is rich in sugar and gums, as well as containing 21% protein and 1.5% fat. It is used as animal feed, to make confectionery, and is used for the preparation of carob gum.

Caroline Small choux pastry filled with puree of meat or pate.

Carophyll Trade name for apo-8-carotenal, a carotene derivative.

Carotene The red and orange pigments of many plants, obvious in carrots, red palm oil, and yellow maize, but masked by chlorophyll in leaves. Three main types of carotene in foods are important as precursors of vitamin A: [alpha]-, [beta]- and [gamma]-carotene, which are also used as food colours (E-160a). Plant foods contain a considerable number of other carotenes, most of which are not precursors of vitamin A.

Carotene is mostly converted into vitamin A (retinol) in the wall of the intestine, but some is absorbed unchanged. 6 [mu]g of [beta]-carotene, and 12 [mu]g of other provitamin A carotenoids, are nutritionally equivalent to 1 [mu]g of preformed vitamin A. About 30% of the vitamin A in Western diets, and considerably more in diets in less-developed countries, comes from carotene.

Carotenoids Carotenoids are familiar to all of us through the orange-red colours of popular foods like oranges, tomatoes and carrots, and the yellow colours of many flowers.

I

II

III

IV

V

VI

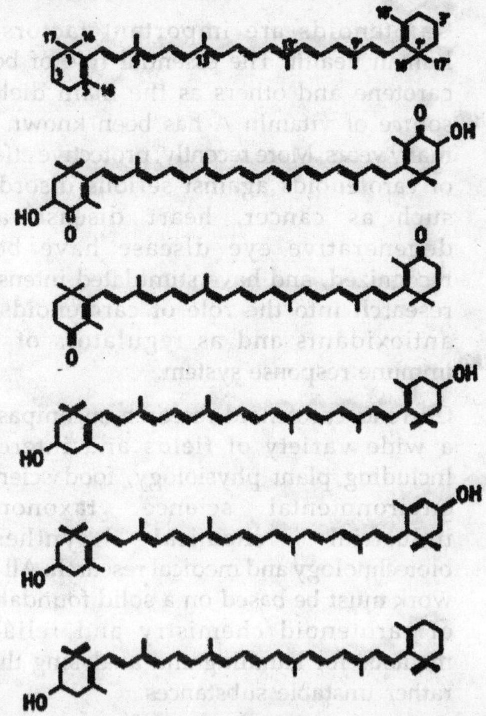

Fig. Carotenoids

They are also added as colorants to many manufactured foods, drinks and animal feeds, either in the form of natural extracts or as pure compounds manufactured by chemical synthesis. The production of carotenoids by biotechnology is of increasing interest. Carotenoids are essential to plants for photosynthesis, acting in light-harvesting and, especially, in protection against destructive photo oxidation. Without carotenoids photosyn-thesis in an oxygenic atmosphere would be impossible.

But carotenoids are not simply pigments of terrestrial plants. They occur widely in bacteria, fungi and algae, where they can be useful taxonomic markers. The production of carotenoids in seaweed runs to hundreds of million tons per year.

Some animals use carotenoids for coloration, especially birds, fish and a wide variety of invertebrate animals, where complexation with protein may modify their colour to blue, green or purple.

Carotenoids are important factors in human health. The essential role of beta-carotene and others as the main dietary source of vitamin A has been known for many years. More recently, protective effects of carotenoids against serious disorders such as cancer, heart disease and degenerative eye disease have been recognized, and have stimulated intensive research into the role of carotenoids as antioxidants and as regulators of the immune response system.

Currents carotenoid research encompasses a wide variety of fields and interests including plant physiology, food science, environmental science, taxonomy, industrial chemical synthesis, biotechnology and medical research. All the work must be based on a solid foundation of carotenoid chemistry and reliable methods for handling and analysing these rather unstable substances.

Carp Freshwater fish, Cyprinus carpio.

Carpaccio Thinly sliced raw beef fillet, served as an antipasto or hors d'oeuvre.

Carpal tunnel syndrome Painful disorder of wrist and hand due to compression of the median nerve in the carpal tunnel. Claimed to be relieved by high intakes of vitamin B6, some 50-100 times the reference intake, but there is little evidence.

Carrageen Edible seaweeds, Chondrus crispus, also known as Iberian moss or Irish sea moss, and Gigartina stellata; stewed in milk to make a jelly or blancmange. A source of carrageenan.

Carrageenan A polysaccharide extracted from red algae, especially Chondrus crispus and Gigartina stellata. One of the plant gums, it binds water to form a gel, increases viscosity, and reacts with proteins to form emulsions. It is used as an emulsifier and stabilizer in milk drinks, processed cheese, low-energy foods, etc. (E-407).

Carrot The root of Daucus carota, commonly used as a vegetable.

Cartilage The hard connective tissue of the body, composed mainly of collagen, together with chondromucoid (a protein combined with chondroitin sulphate) and chondroalbuminoid. New bone growth consists of cartilage on which calcium salts are deposited as it develops.

Cartose Trade name for a steam hydrolysate of maize starch, used as a carbohydrate modifier in milk preparations for infant feeding. It consists of a mixture of dextrin, maltose, and glucose.

Case hardening Undesirable surface hardening of fish due to drying too quickly during the initial stages that prevents adequate drying of the inner layers of fish. If too high a temperature and/or airspeed is used during drying a relatively impermeable outer layer is formed on the surface of the fish during the initial period of drying which effectively halts the drying process, and the moist interior of the fish will spoil rapidly. The same undesirable condition can occur during the initial period of smoking if the temperature and/or airspeed are too high.

Casein Primary protein found in milk, along with whey protein. Casein is the insoluble protein fraction of milk. It is absorbed more slowly than whey and provides the body with amino acids over an extended period in comparison to whey protein, the soluble protein fraction in milk.

Caseinogen An obsolete name for the form in which casein is present in solution in milk; when it was precipitated it was then called casein.

Cashew nut Fruit of the tropical tree Anacardium occidentale , generally eaten roasted and salted. The nut hangs from the true fruit, a large fleshy but sour apple-like fruit, which is very rich in vitamin C.

Casilan Trade name for a casein preparation used as a protein concentrate.

Caspase A family of proteins with protease activity that are responsible of the execution of apoptotic mechanisms.

Cassareep Caribbean; boiled-down juice squeezed from grated cassava root, flavoured with cinnamon, cloves, and brown sugar; used as a base for sauces. It can also be fermented with molasses.

Cassata Ice cream made in a mould with layers of diced fruit, nuts, and macaroons. Cassata Siciliana is sponge cake containing layers of ricotta cheese, grated chocolate, and crystallized fruit.

Cassava The tuber of the tropical plant Manihot utilissima. It is the dietary staple in many tropical countries, although it is an extremely poor source of protein; the plant grows well even in poor soil, and is extremely hardy, withstanding considerable drought. It is one of the most prolific crops, yielding, e.g. in Nigeria, 13 million kcal/acre, compared with yam, 9 million, and sorghum or maize, 1 million. Cassava root contains cyanide, and before it can be eaten it must be grated and left in the open to allow the cyanide to evaporate. The leaves can be eaten as a vegetable, and the tuber is the source of tapioca.

Casserole Lidded container designed for slow cooking of meat or fish and vegetables in the oven; also the food so cooked.

Cassia The inner bark of a tree grown in the Far East, used as a flavouring, similar to cinnamon.

Cassina A tea-like beverage made from cured leaves of a holly bush, Ilex cassine, containing 1-1.6% caffeine and 8% tannin.

Cassolette Individual dish containing a single portion of a savoury mixture; sometimes lined with potato or puff pastry.

Cassoulet Stew of haricot beans with assorted meats. The name is derived from the earthenware vessel in which it is cooked, a cassole d'Issel.

Castagnacci Corsican; fritters made from chestnut flour.

Castor oil Oil from the seeds of the castor oil plant, Ricinus spp. The oil itself is not irritating, but in the small intestine it is hydrolysed by lipase to release ricinoleic acid, which is irritant to the intestinal mucosa and therefore acts as a purgative. The seeds also contain the toxic lettin, ricin.

Cat's claw An herb used in South American folk medicine for its anti-inflamatory and immune system protective properties.

Catabolic Metabolic condition in which muscle is broken down and energy is released.

Catabolism Protein breakdown in muscles.

Catadromous fish Fish that live in fresh water and go to sea to spawn, such as eels.

Catalane, A la Dish with a garnish of aubergine and rice.

Catalase An enzyme that catalizes de conversion of hydrogen peroxide to water and oxygen.

Catalyst An agent that participates in a chemical reaction, speeding the rate, but itself remains unchanged. Catalysts are used, for example, in the hydrogenation of vegetable oils. Enzymes and coenzymes are biological catalysts.

Cathepsins A group of intracellular enzymes in animal tissues which hydrolyse proteins. They are involved in the normal turnover of tissue protein, and the softening of meat when game is hung.

Cation Chemical term for a positively charged ion.

Catmint The wild catmint is distinct from the cultivated catmint of gardens; both have a minty smell which is liked by cats. The leaves are used to prepare herb teas and may be added to stews; young shoots can be eaten raw in salads.

Caul Membrane enclosing the foetus; that from sheep or pig used to cover meat while roasting.

Cauliflower The edible flower of Brassica olearacea bortytis, normally creamy-white in colour, although some cultivars have green or purple flowers. Horticulturally, varieties that mature in summer and autumn are called cauliflower, and those that mature in winter broccoli, but commonly both are called cauliflower.

Caviar(e) The salted hard roe of the sturgeon, Acipenser spp.; three main types, sevruga, asetra, and beluga, the prime variety. A rich source of protein and iron; a source of calcium.

Celacol Trade name for methyl, hydroxyethyl, and other cellulose derivatives. celeriac A variety of celery with a thick root which is eaten grated in salads or cooked as a vegetable, Apium graveolens var. rapaceum, also known as turnip-rooted or knob celery.

Celery Edible stems of Apium graveolens var. dulce.

Celery pepper Milled black pepper mixed with ground celery seed, used as a flavouring in salads and cooking.

Celery salt A mixture of ground celery seed and salt, used as a condiment.

Celiac Disease A digestive disease that damages the small intestine and interferes with absorption of nutrients from food. Common gluten sensitivity symptoms include diarrhea, increased appetite, bloating, weight loss, irritability and fatigue. Gluten is found in wheat, rye, barley and sometimes oats.

Cell Smallest unit of living matter able to grow and reproduce independently. Cells contain DNA for storing information, ribosomes for making proteins, and mechanisms for converting energy.

Cellabiose A disaccharide composed of 2 glucose units linked by a 1,4-ß linkage, which is not hydrolysed by mammalian digestive enzymes. It is a product of the hydrolysis of cellulose.

Cellofas Trade name for derivatives of cellulose: Cellofas A is methylethylcellulose, Cellofas B is sodium carboxyme-thylcellulose.

Cellophane Trade name for the first of the transparent, non-porous films, made from wood pulp. Still widely used for wrapping foods and other commodities.

Cellulase An enzyme that hydroiyses cellulose to its constituent monosaccharide and disaccharide units. It is present in the digestive juices of some wood-boring insects and in various micro-organisms, but not in mammals.

Cellulose A polysaccharide of glucose units which is not hydrolysed by mammalian digestive enzymes. It is the main component of plant cell walls, but does not occur in animal tissues. It is digested by the bacterial enzyme cellulase, and hence only ruminants and animals that have a large intestine have an adequate population of intestinal bacteria to permit them to digest cellulose to any significant extent. There is little digestion of cellulose in the human large intestine; nevertheless, it serves a valuable purpose in providing bulk to the intestinal contents, and is one of the major components of dietary fibre or non-starch polysaccharides.

Celtuce Stem lettuce, Lactuca sativa; enlarged stem eaten raw or cooked, with a flavour between celery and lettuce; the leaves are not palatable.

Centrifugation A process by which liquid samples are spun around at high speed to cause the accelerated settling of particles in suspension.

Cep Cèpe; edible wild fungus, Boletus edulis, also known as boletus.

Cephalopods Cuttlefish, squid, octopus.

Ceplapro A protein-rich baby food in a granular form, made from degerminated maize flour, wheat, defatted soya flour, and skim milk powder, with added vitamins and minerals (US).

Cereal Any grain or edible seed of the grass family which may be used as food; e.g. wheat, rice, oats, barley, rye, maize, and millet. Collectively known as corn in the UK, although in the USA corn is specifically maize. Cereals provide the largest single foodstuff in almost all diets; in some less-developed countries up to 90% of the total diet may be cereal, and in the UK bread and flour provide 25-30% of the total energy and protein of the average diet.

Cereal coffee Beverages prepared from roasted cereal grains.

Cerebrosides Complex lipids, containing the carbohydrate galactose, which is important in nerve membranes and the myelin sheath of nerves.

Cerelose A commercial preparation of glucose containing about 9% water.

Ceruloplasmin A copper-containing protein in blood plasma, the main circulating form of copper in the body.

Cestodes Tapeworms. Members of the zoological class Cestoda of the phylum Platyhelminthes. See Diphyllobothriasis.

Chafing dish Metal dish on a trivet over a spirit lamp, used for heating or cooking at the table.

Chalasia Abnormal relaxation of the cardiac sphincter muscle of the stomach so that gastric contents reflux into the oesophagus, leading to regurgitation.

Chalky fish Fish exhibiting an abnormal chalky white appearance and a watery texture. It has been described particularly in halibut in North America, and is associated with fish in which there has been a rapid drop in pH after capture.

Champagne Sparkling wine from the Champagne region of northeastern France, made by a second fermentation in the bottle. Sparkling wines from other regions, even when made in the same way, cannot legally be called champagne, but are known as being made by the methode champenoise.

Champignon Button mushrooms, champignons de Paris.

Chanakhi Mutton casserole with potatoes, onions, green beans, tomatoes, aubergines, and peppers.

Chanfaina Stew made from the offal of goat or kid, cooked together with the head and feet, with artichokes, chard, lettuce, and peas.

Chanterelle Edible wild fungus, Cantharellus cibarius.

Chao shao roast pork Chinese; strips of pork marinated then roasted.

Chapatti unleavened whole-grain wheat or millet bread, baked on an ungreased griddle. Phulka are small chapattis; roti are chapattis prepared with maize flour. Two chapattis (60 g) are a source of vitamin B 1 and copper; provide 4.2 g of dietary fibre; if made with added fat contain 8 g of fat; supply 210 kcal (880 kJ); if made without added fat contain 0.6 g of fat; supply 130 kcal (545 kJ).

Chapon Crust of bread rubbed with garlic and added to salad for flavour, then removed.

Chaptalization Addition of sugar to grape must during fermentation to increase the alcohol content of the final wine.

Characterize Describe the distinguishing traits.

Charcoal Finely divided carbon, obtained by heating bones or wood in a closed retort to carbonize the organic matter. Used to purify solutions because it will absorb colouring matter and other impurities; wood charcoal is commonly used as a fuel for barbecues.

Charcuterie Pork butchery; shop where various cold meats are sold, also the meats.

Chardonnay One of the nine 'classic' grape varieties used for wine making, widely grown throughout the world. Chardonnay wines are among the white wines best adapted to maturing in oak barrels.

Charlotte Dessert made from stewed fruit encased in, or layered alternately with, bread or cake crumbs, e.g. apple charlotte. In charlotte russe there is a cream mixture in the centre. surrounded by cake.

Charollias Breed of cattle noted for the quality and leanness of the meat.

Charqui South American (especially Brazilian); dried meat, normally prepared from beef, but may also be made from sheep, llama, and alpaca in Peru. Strips of meat cut lengthways and pressed after salting, then air-dried. The final form is flat, thin, flaky sheets, so differing from the long strips of biltong. Also called jerky.

Chartreuse 1. A liqueur invented in 1605 and still made by the Carthusian monks, named for the great charterhouse which is the mother house of the order, near Grenoble in southern France. It is reputed to contain more than 200 ingredients. There are three varieties: green Chartreuse is 55%, yellow 43% and white 30% alcohol.

2. A dish turned out of a mould; more usually fruit enclosed in jelly.

Chasnidarh Sweet-sour foods.

Chasseur, a la In the hunter's style; dish with a garnish of mushrooms cooked with shallots and white wine. Chasseur sauce is a highly seasoned sauce based on white wine, with mushrooms and shallots, served with meat or game.

Chastek paralysis Acute deficiency of vitamin B 1 in foxes and mink fed on diets high in raw fish, which contains thiaminase.

Chateau salad French-Canadian; cooked frogs' legs with shredded lettuce and watercress, garnished with hard-boiled egg and lemon.

Chateaubriand Thick steak cut from beef fillet. Named originally in 1822 in honour of the Comte de Chateaubriand.

Chaudfroid Sauce used to coat meat, fish, or poultry to be served cold.

Chausson Turnover; pastry folded over jam or fruit before baking.

Chayote Tropical squash, Sechium edule; prickly skin and a single edible seed.

Cheddar Hard cheese dating from sixteenth century prepared by a particular method; originally from the Cheddar area of Somerset, England; matured for several months or even years. Red Cheddar is coloured with annatto; fat-reduced versions are now made.

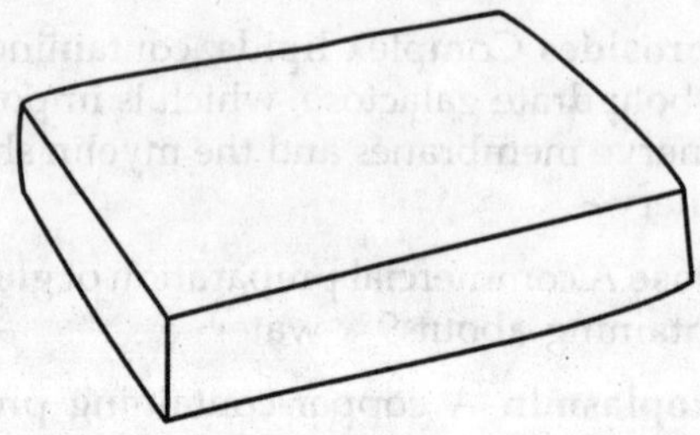

Fig. Cheddar

Cheddaring In the manufacture of cheese, after coagulation of the milk, heating of the curd, and draining, the curds are piled along the floor of the vat, where they consolidate to a rubbery sheet of curd. This is the cheddaring process; for cheeses with a more crumbly texture the curd is not allowed to settle so densely.

Cheese analogues Cheese-like products made from in or soya and vegetable fat.

Cheese Prepared from the curd precipitated from milk by rennet, purified chymosin, or lactic acid. Cheeses other than cottage and cream cheeses are cured by being left to mature with salt, under various conditions that produce the characteristic flavour of each type of cheese. Although most cheeses are made from cow's milk, goat's milk and sometimes ewe's milk can be used to make speciality cheeses. These are generally soft :heeses.

There are numerous variants including more than 100 from England and Wales alone. Some varieties are regional specialities, and legally may only be made in a defined geographical area; others are defined by the process rather than the region of production. The strength of flavour of cheese increases as it ages; mild or mellow cheeses are younger, and less strongly flavoured, than mature or extra-mature cheeses.

Cheeses differ in their water and fat content and hence their nutrient and energy content, ranging from 50-80% water in soft cheeses to less than 20% in hard cheese with semihard cheeses around 40% water. They retain much of the calcium of the milk and many contain a relatively large amount of sodium from the added salt. Blue-veined cheeses derive the colour from the growth of the mould Penicillium roquefortii, during ripening. Holes arise during ripening from gases produced by bacteria.

Traditionally, hard cheeses must contain not less than 40% fat on a dry weight basis, and the fat must be milk fat. However, a number of low-fat variants of traditional hard cheeses, and vegetarian cheeses, are now made. A rich source of vitamin B 12, and calcium; a good source of protein.and vitamin A. Cottage cheese is soft, uncured, white cheese made from pasteurized skim milk by lactic acid starter, heated, washed, and drained. Contains more than 80% water. Also known as pot cheese, Schmierkäse and, in the USA, as Dutch cheese.

Baker's or hoop cheese is made in the same way as cottage cheese, but the curd is not washed, and it is drained in bags, giving it a finer grain. It contains more water and acid than cottage cheese.

Cream cheese is unripened soft cheese made from cream with varying fat content; at 50% fat.

Processed cheese is made by milling various hard cheeses with emulsifying salts, whey, and water, and pasteurizing to extend the shelf life of the cheese. Typically 40% water. A soft version with 50% water is used. as a spread.

Feta is Greek and general Balkan; white, soft, crumbly, salted cheese made from goat's or ewe's milk.

Swiss cheese is an American name for any hard cheese that contains relatively large bubbles of air, such as the Swiss Emmentaler and Gruyère cheeses.

Cheese, austrian Classified according to butterfat content as percentage of dry weight as: doppelfett 65%, üverfett 55%, vollfett 45%, dreiviertelfett 35%, halbfett 25%, viertelfett 15%, and mager under 15% fat.

Cheese, vegetarian Cheese in which animal rennet has not been used to precipitate the curd. Precipitation is achieved using lactic acid alone, or a plant enzyme or biosynthetic chymosin. Truly vegetarian cheese is made from vegetable protein rather than milk.

Cheese, whey Made from whey by heat coagulation of the proteins.

Cheesecake A flan or tart filled with curd or cream cheese. Traditional Middle-European cheesecake is baked; most cheesecakes now sold are uncooked, set with gelatine, and topped with fruit.

Cheilosis Cracking of the edges of the lips, one of the clinical signs of vitamin B 2 deficiency.

Chelating agent Chemicals capable of combining with metal ions to render them ineffective. They are also called sequestrants. Examples of chelating agents used in the food industry are ethylene-diaminetetraacetic acid (EDTA), usually as the mixed calcium and sodium salt, and sodium polyphosphate. They are used as food additives to complex metal ions, particularly calcium and magnesium that could give rise to defects. For example they are added to canned fish to prevent development of struvite crystals, and to canned crabmeat to inhibit blueing of the meat.

Chelsea bun A yeast bun that originated in Chelsea, London, in the eighteenth century.

Chemical ice Ice containing a preservative, e.g. a solution of antibiotics or other chemicals; used to preserve fish.

Chemical inertness Substances or elements which do not react easily with any other substances or elements. Chemically inactive materials.

Chemical score A measure of protein quality based on chemical analysis.

Chenin blanc One of the nine 'classic' grape varieties used for wine making; the great white grape of the middle Loire valley.

Chenopods Seeds of two species of Chenopodium eaten in the Peruvian Andes: C. quinoa and C. pallidicaule. Other species of Chenopodium have been considered for poultry feed, including Russian thistle, summer cypress, and garden orache.

Cherry Fruits of Prunus spp.; a source of vitamin C. .

Cherry, ground Fruit of Physalis pruinosa, similar to Cape gooseberry. Grows wild, and can be eaten raw but is more usually boiled or made into a preserve; also called strawberry tomato or dwarf Cape gooseberry.

Cherry, maraschino Cherries preserved in maraschino.

Cherry, morello A sour cherry that cannot be eaten uncooked; used for cooking and making liqueurs and jams.

Cherry, West Indian The fruit of a small bush native to the tropical and sub-tropical regions of America, Malpighia punicifolia. It is the richest known source of vitamin C; the edible portion of the ripe fruit contains 1000 mg, and the green fruit 3000 rag/100 g. Also known as Barbados or Antilles cherry, acerola.

Chervil A herb, Anthriscus cerefolium, with parsley-like leaves, used in the fresh green state as a garnish and fresh or dried to flavour salads and soups. The turnip-rooted chervil, Chaerophyllum bulbosum, a hardy biennial vegetable cultivated for its roots.

Cheshire cheese Oldest English cheese dating from Roman Britain; crumbly, may be pale yellow, blue-veined, or coloured orange with annatto; matured 2-6 weeks; approx. 30% water, 24% protein, 30% fat. Cheshire cat is an old English measure of cheese.

Chest sweetbread The thymus of an animal, as distinct from the gut sweetbread, which is the pancreas.

Chestnut 1 Spanish or sweet chestnut from trees of Castanea spp. Unlike other common nuts it contains very little fat, being largely starch and water. Seven nuts (75 g) provide 5.3 g of dietary fibre and are a good source of copper; a source of vitamins B 1 and B6; contain 2 g of fat of which 18% is saturated; supply 135 kcal (570 kJ). 2Water chestnut, seeds of Trapa natans, also called caltrops or sinharanut; eaten raw or roasted.

3 Chinese water chestnut, also called matai or waternut; tuber of the sedge, Eleocharis

tuberosa or dulcis; white flesh in a black, horned shell.

Cheyda A dry and highly spiced mixture of deep-fried rice, dahl, chickpeas, and small pieces of chickpea batter, with peanuts and raisins, seasoned with sugar and salt, a common North Indian snack food, also known as Bombay mix.

Chicharrones Caribbean; fried pork crackling.

Chick embroyo extract This type of extract contains short amino acid chains called oligopeptides, plus additional essential amino acids and trace elements. The peptides in the extract possess both cell stimulating and protective properties. The peptides indicated that they were activating growth factor receptors in them. The extract works as a general tonic and stimulates the adrenal gland, normalizing its function. It also acts as a mild stimulant and an antidepressant; and it improves libido, erectile function, sparmatogenesis and other aspects of sexual function in men and women. It also improves sleep and promotes weight loss.

Chicken Domestic fowl, Gallus domesticus. A rich source of protein and niacin; a good source of copper and selenium; a source of iron and vitamins B1, B2, and B6. There are differences between the white (breast) and dark (leg) meat, the former being lower in fat but also lower in iron and vitamin B2.

Chicken kiev Boned breast of chicken stuffed with garlic butter, coated in batter and breadcrumbs, and baked.

Chicken maryland Chicken fried in breadcrumbs or batter, served with fruit fritters.

Chickpea Also known as garbanzo; seeds of Cicer arietinum, widely used in Mediterranean and Middle-Eastern stews and casseroles. Pureed chickpea is the basis of humus and deep-fried balls of chickpea batter are felafel. A rich source of copper; a good source of folate; a source of protein, vitamin A, and iron.

Chicle The partially evaporated milky latex of the evergreen sapodilla tree; it contains gutta and resin, together with carbohydrates, waxes, and tannins. The same tree also produces the sapodilla plum.

Chicory Witloof or Belgian chicory, Cichorum intybus; looks like a dandelion; the root is harvested and grown in the dark to produce bullet-shaped heads of young white leaves (chicons). Also called succory; red variety is radicchio. The leaves are eaten as a salad or braised as a vegetable and the bitter root, dried and partly caramelized, is often added to coffee as a diluent to cheapen the product.

Chiffonade Shredded green vegetables.

Chilled foods Perishable products stored at temperatures between 0 and 7°C.

Chilled sea water Sea water in fixed tanks, especially at sea installed in the hold, chilled by adding ice, for quickly and uniformly chilling, especially pelagic fish and salmon, by immersion. Portable tank systems have also been developed. In a full tank the proportions of sea water, ice and fish should be about 1:1:4 and the water may be circulated by pump or agitated by compressed air to maintain a uniform temperature and prevent temperature stratification. The tank is usually filled one third to one quarter full with equal volumes of sea water and ice before the fish are loaded. The sea water can be lowered to about -1°C without freezing the fish to obtain the maximum benefit , and the catch can be cooled more rapidly than by stowage in ice. Using the champagne system, it is possible to cool herring from 15°C to 0°C within 2 to 3 hours. Some species take up unacceptable amounts of water and salt when kept in sea water, the eyes become cloudy and the gills are bleached as the blood is leached out. Consequently, stowage in CSW is usually limited to a maximum of 3 to 4 days.

Chilled water stowage The storage of fish in chilled fresh or salt water, the chilling being effected by either mechanical refrigeration or the addition of ice.

Chilli sauce Piquant sauce made from tomatoes, with spices, onions, garlic, sugar, vinegar, and salt. Similar to tomato ketchup but containing more cayenne, onion, and garlic.

Chilli con carne Minced beef stewed with onion, chilli, and red kidney beans; originally Mexican and south-western USA.

Chilling Fish and shellfish are among the most perishable of foods. Spoilage, which begins immediately after death, can be retarded by reducing the temperature of the fish by either chilling or freezing as soon as possible after capture. In chilling, the aim is to cool the fish quickly to a temperature close to 0°C without freezing it. A simple and effective method is to use ice. With good boxing and icing practice, using a fish to flake ice ratio no greater than 3:1 and stowage in a fishroom operating at 0 to 2°C, fish can be chilled effectively in temperate climates from an initial temperature of 10°C to 0°C within about 8 hours. More rapid cooling can be achieved by using slush ice or binary ice initially. For example, when mixed with binary ice made from 3% brine with an ice fraction of 35% and allowed to drain so that the fish to ice ratio is about 1:1, fish will cool from 10°C to 0°C in about 1.5 hours. Alternatively, chilled water stowage either in tanks of chilled sea water (CSW) using ice for chilling or in mechanically refrigerated sea water (RSW) can be used particularly for marine pelagic species, or in chilled or refrigerated freshwater for freshwater species. The temperature of sea water can be lowered to about -1°C without freezing the fish and the catch can be cooled more rapidly in chilled or refrigerated sea water than stowage in ice. Using CSW and agitation with compressed air it is possible to cool herring from 15°C to 0°C within 2 to 3 hours. Under optimal conditions, the refrigeration capacity of an RSW system should be sufficient to cool the fish to 0°C within a few hours. At temperatures less than -1°C, the fish begin to freeze, and reducing the temperature to the range -1.5°C to -3°C is known as superchilling or partial freezing. Cooling fish or fish products to a temperature just above and as close to 0°C as possible by forced circulation of cold air over the fish is known as air blast chilling. Typically, this is done in the duct or tunnel of an air blast chiller in which the stream of cool air is guided over trolleys in a batch chiller, or over a conveyor carrying the product in a continuous chiller.

Chillproofing A treatment to prevent the development of haziness or cloudiness due to precipitation of proteins when beer is chilled. Treatments include the addition of tannins to precipitate proteins, materials such as bentonite to adsorb them and proteolytic enzymes to hydrolyse them.

China Bitter aperitif wines containing quinine.

Chincona Spanish; bitter aperitif wines containing quinine.

Chine A joint of meat containing the whole or part of the backbone of the animal.

Chining To sever the rib bones from the backbone by sawing through the ribs close to the spine.

Chino Spanish; small conical strainer used to sieve chopped vegetables to make gazpacho.

Chipolata Small sausage.

Chips Chipped potatoes; pieces of potato deep fried in fat or oil. Known in French as pommes frites or just frites; in the USApotato crisps are known as chips, and chips are called French fries or just fries. A a rich source of vitamins C and B 1 ; a source of protein, niacin, and iron.

Chiretto Rose wines.

Chirga Skinned chicken rubbed with cayenne, paprika, and lime juice, coated with a sauce of ginger, onion, and pimiento in yoghurt, then roasted, basting with ghee. It has a red colour.

Chitin The organic matrix of the hard parts of the exoskeleton of insects and crustaceans, and present in small amounts in mushrooms. It is an insoluble and indigestible non-starch polysaccharide, similar to cellulose, but composed of N-acetylglucosamine units rather than glucose. Partial deacetylation results in the formation of chitosans, which are used as protein-flocculating agents.

Chitosan Chitosan is a natural product extracted from Chitin. Chitosan and chitin are waste products of the crab and shrimp industry. It can be used to inhibit fat digestion and as a drug delivery transport agent. It also has been used as a cholesterol lowering substance. Chitosan is marketed as a 'fat blocker'. It appears that it can impede fat absorption by 'gelling' with fat in the small intestine. Side effect of Chitosan is that since it is made from sea food, some people have allergic responses to it. Also, you need a high concentration of Chitosan for it to 'gel' with fat.

Chitterlings The small intestine of ox, calf, or pig.

Chive Small member of the onion family; the leaves are used as a garnish or dried as a herb; mild onion flavour.

Chlorine An element found in biological tissues as the chloride ion; the body contains about 100 g (3 mol) of chloride and the average diet contains 6-7 g, mainly as sodium chloride. Free chlorine is used as a sterilizing agent, e.g. for drinking water.

Chlorophyll The green pigment of plant materials which is responsible for the trapping of light energy for photosynthesis, the formation of carbohydrates from carbon dioxide and water. Both [alpha]- and [beta]-chlorophylls occur in leaves, together with the cariotenoids xanthophyll and carotene. Chlorophyll has no nutritional value, although it does contain magnesium as part of its molecule, and although it is used in breath-fresheners and toothpaste, there is no evidence that it has any useful action.

Chlorophyllide The green colour found in the water after cooking some vegetables; it is a water-soluble derivative of chlorophyll formed by either enzymic action or alkaline hydrolysis.

Chlortetracycline An antibiotic.

Chocolate chocolate Made from cocoa nibs (husked, fermented, and roasted cocoa beans) by refining and the addition of sugar, cocoa butter, flavouring, lecithin, and, for milk chocolate, milk solids. It may also contain vegetable oils other than cocoa butter. A rich source of copper; a good source of calcium; a source of protein, vitamin B 2, iron, and selenium.

Chocolate, drinking Partially solubilized cocoa powder for preparation of a chocolate-flavoured milk drink, containing about 75% sucrose.

Cholagogue A substance that stimulates the secretion of bile from the gall bladder into the duodenum.

Cholent Traditional Jewish and Middle-European casserole of beans and beef, cooked extremely slowly.

Cholesterol Cholesterol is not a fat, but rather a fat-like substance classified as a lipid. Cholesterol is vital to life and is found in all cell membranes. It is necessary for the production of bile acids, vitamin D and steroid hormones. Dietary cholesterol is found only in animal foods. Abundant in organ meats and egg yolks, cholesterol is also contained in meats and poultry. Vegetable oils and shortenings are cholesterol-free.

Cholesterol Blood cholesterol is divided into three separate classes of lipoproteins: very-low density lipoprotein (VLDL); low-density lipoprotein (LDL), which contains most of the cholesterol found in the blood; and high-density lipoprotein (HDL).

LDL seems to be the culprit in coronary heart disease and is popularly known as the "bad cholesterol." By contrast, HDL is increasingly considered desirable and known as the "good cholesterol."

Cholesterol High blood cholesterol is a risk factor in the development of coronary heart disease. Most of the cholesterol that is found in the blood is manufactured by the body, in the liver, at a rate of about 800 to 1,500 milligrams a day.

Cholestyramine Drug used to treat hyperlipidaemia by complexing bile salts in the intestinal lumen and increasing their excretion, so increasing the metabolic clearance of cholesterol.

Choline A derivative of the amino acid serine, formed in the body; an important component of cell membranes. Phosphatidylcholine is also known as lecithine, and preparations of mixed phospholipids rich in phosphatidylcholine are generally called lecithin, although they also contain other phospholipids; lecithin from peanuts and soya beans is widely used as an emulsifying agent (E-322). Choline released from membrane phospholipids is important for the formation of the neurotransmitter acetylcholine, and choline is also important in the metabolism of methyl groups.

Cholla A loaf of white bread made in a twisted form by plaiting together a large and small piece of dough. The dough is made from white flour, enriched with eggs and a pinch of saffron, and the loaf is decorated with poppy seed. It is mentioned in the Bible, translated as 'loaves', and is traditionally used for benediction of the Jewish sabbath and festivals.

Choline A derivative of the amino acid serine, formed in the body; an important component of,cell membranes. Phosphatidylcholine is also known as lecithin, and preparations of mixed phospholipids rich in phosphatidylcholine are generally called lecithin, although they also contain other phospholipids; lecithin from peanuts and soya beans is widely used as an cmulsifying agent (E-322). Choline released from membrane phospholipids is important for the formation of the neurotransmitter acetylcholine, and choline is also important in the metabolism of methyl groups.

Chondroitin Chondroitin is extracted from bovine tracheas or shark cartilage. It is a major component of connective tissue, especially cartilage. Chondroitin can stimulate repair of cartilage cells and also black enzymes that damage joints.

Chondrometer An instrument used to determine the specific weight of wheat. A wet English wheat may weigh 68 kg, and a dry American wheat 84 kg/hectolitre.

Chondrus crispus A seaweed, the source of carrageenan.

Chop A slice of mutton, lamb, or pork containing a part of the bone; commonly the rib, but also cut from the chump or tail end of the loin or neck.

Chopping board Provides a solid base for cutting food or resting hot pans. May be made of plastic, wood or marble.

Chopsuey Chinese dishes based on bean sprouts and shredded vegetables, cooked with shredded quick-fried meat, capped with a thin omelette. Not authentically Chinese, but an invention of Chinese restaurateurs in Western countries. Unlike true Chinese food, the flavours of a chopsuey are all mixed together; one translation is 'savoury mess'.

Chorionic villus sampling CVS A procedure performed between 10 and 12 weeks after a woman's last menstrual period to collect material for prenatal testing. A needle is used or a catheter is inserted into the substance of the placenta but staying outside the amniotic sac.

Chorizo Spanish; coarse textured, red, spiced sausage.

Chorleywood bread process A method of preparing dough for bread making by submitting it to intense mechanical working, so that, together with the aid of oxidizing agents, the need for bulk fermentation of the dough is eliminated. This is a so-called 'no-time' process and saves 1½-2 hours in the process; permits use of an increased proportion of weaker flour, and produces a softer, finer loaf, which stales more slowly. Named after the British Baking Industries Research Association at Chorleywood.

Choux pastry Light, airy pastry, invented by the French chef Carême, used in eclairs and profiterols. The batter is pre-cooked in a saucepan, then baked. The name comes from the French for cabbage, chou, because of the characteristic shape of the cream-filled puffs.

Chow-chow Preserve of ginger, orange peel and fruit in syrup, or a mixed vegetable pickle containing mustard and spices.

Chowder Thick soup made from shellfish or other fish, with pork or bacon. Originally French, now mainly New England and Newfoundland.

Chrane Jewish and Middle-European condiment made from beetroot and horseradish.

Chromium / chrimium picolinate Chromium increases the efficiency of the hormone insulin, which the pancreas releases after you eat carbohydrates or protein. Chromium acts to make the receptor of muscle cells more sensitive to insulin. Insulin also helps muscles use amino acids for building protein rather than breaking them down. Chromium can promote modest muscular gains and decreases in bodyfat. Exercise increases the excretion rate of chromium.

Chromoproteins Proteins conjugated with a metal-containing group, such as the haem group of haemoglobin, which contains iron.

Chromosome An organized structure located in the cell nucleus, consisting of a single molecule of DNA and many associated proteins, which contains the code for specific genetic information. Chromosomes are located within every cell, and are responsible for directing the development and functioning of all the cells in the body. The normal number in humans is 46.

Chrysin An anti-aromatase, which means it stops a lot of excess testosterone from converting to estrogen. Also, a compound with significant antiviral activity - especially in relation to the HIV. A an effect on the benzodiazapene receptors. It is possible that Chrysin taken in the correct dosage could reduce cortisol secretion due to the body's sensing less stress. Chrysin may act as an antiestrogen by inhibiting aromatase activity, limiting the conversion of testosterone to estrogen.

Chuño Traditional dried potato prepared in the highlands of Peru and Bolivia. The tubers are crushed, pressed, frozen during the night, then dried in the sunshine during the day, a process of freeze-drying.

Churros Spanish; sausage-shaped, deep-fried doughnuts, dusted with sugar.

Chutney Pickled fruits or vegetables, normally spiced, used as a relish; of Indian origin.

Chylomicrons The droplets of unhydrolysed fat absorbed from the small intestine into the lymphatic system, circulating in the lymph and bloodstream.

Chymase Alternative name for chymosin.

Chyme The partly digested mass of"food as it exists in the stomach.

Chymosin In the abomasums of calves and the stomach of human infants which clots milk by precipitation of the casein. There is no evidence that it plays any part in digestion in the adult. Also known as rennin; however, to avoid confusion with the kidney enzyme renin the name rennin should be replaced by chymosin. Biosynthetic chymosin is used in cheese making.

Chymotrypsin An enzyme involved in the digestion of proteins; secreted as the inactive precursor chymotrypsinogen in the pancreatic juice. It is activated by trypsin, and is an endopeptidase, with a different specificity from trypsin and pepsin.

Cibophobia Dislike of food.

Cider, Cycler Ail alcoholic beverage; fermented apple juice. Dry cider has 2.6% sugars, 3.8% alcohol, and 110 kcal (460 kJ) per 300 mL. Sweet cider has 4.3% sugars and supplies 125 kcal (525 kJ) per 300 mL. Vintage cider has 7.3% sugars, 10.5% alcohol, and supplies 300 kcal (1260 kJ) per 300 mL (half pint).

In the USA cider or fresh cider is the term used for unfermented apple juice; the fermented product is termed hard or fermented cider.

Ciguatera Food poisoning resulting from the consumption of tropical reef-dwelling fish that have become toxic by feeding directly or indirectly on species of toxic dinoflagellates, microscopic marine organisms. The dinoflagellates live on coral reefs attached to algae. Herbivorous fish eat the algae, and the dinoflagellates, and take up the toxin in their bodies. The fish become toxic, but further, are consumed by other species and the toxin is transmitted along the aquatic food chain. The principal, but not only, dinoflagellate incriminated in ciguatera poisoning is Gambierdiscus toxicus. Worldwide, it is one of the most frequent sources of food poisoning from marine fish, in the order of some tens of thousand reported cases a year. The two toxins associated with ciguatera are ciguatoxin and maitotoxin. They are very toxic. The level of concern for human consumption is 35 nanog/100g of fish flesh. The toxins are neurotoxins, the symptoms including parathesia, and weakness may persist for months. Death rates - from respiratory failure - are very variable in incidents from this type of fish poisoning, but are in the range 0.1 to 20%.

Ciguatoxins The toxins responsible for ciguatera.

Cinnamon The aromatic bark of various species of the genus Cinnamomum; it is split from the shoots, cured, and dried. During drying the bark shrinks and curls into a cylinder or 'quill'. It is used as a flavour in meat products, bakery goods, and confectionery, and may be available either as the whole quill or powdered ready for use.

Ceylon or true cinnamon differs from other types and the oil contains mostly cinnamic aldehyde, together with some eugenol. Saigon cinnamon also contains cineol; Chinese cinnamon has no eugenol.

Cinsaut A grape variety widely used for wine making, although not one of the classic varieties. The common bulk wine producing grape of southern France. In South Africa it has been crossed with the pinot noir to produce the pinotage grape.

Cissa An unnatural desire for foods; alternative words, cittosis, allotriophagy, and pica.

Citral An important constituent of many essential oils, especially lemon. Used as the starting material for the synthesis of ionone, which is an intermediate in the chemical synthesis of retinol.

Citrange An American citrus fruit resulting from a cross between the ordinary orange and the trifoliate orange, Poncirus trifoliata. citric acid An organic acid which is widely distributed in plant and animal tissues; it is an important metabolic intermediate, and yields 2.47 kcal (10.9 kJ)/g. It is used as a flavouring and acidifying agent (E-330), and its salts (citrates) are used as acidity regulators. Commercially it is either prepared by the fermentation of sugars by the mould Aspergillus niger or extracted from citrus fruits.

Citranin A flavonone glycoside from the peel of immature Ponderosa lemons.

Citrimax See Hydroxycitrate

Citrin A mixture of two flavonones found in citrus pith, hesperidin and eriodictin. A constituent of what is sometimes called vitamin P.

Citron The first of the citrus fruits to become known to Europeans; Citrus medica; can be grown only in warm regions. The fruit has a very thick peel and solid, sweet, and acid-free pith with practically no juice. It is used for preparing candied peel.

Citronella Lemon-scented tropical grass Cymbopogon nardus used in salads and dressings; the essential oil is also used as an insect repellent.

Citrovorum Factor The name given to a growth factor for the microorganism Leuconostoc citrovorum, now known to be one of the main forms of the vitamin folic acid.

Citroxanthin A yellow carotenoid pigment in orange peel which has vitamin A activity. Also known as mutachrome.

Citrullinaemia A genetic disease affecting the formation of urea, and hence the elimination of end-products of protein metabolism. The defect may be mild, or so severe that affected infants become comatose and may die after a moderately high intake of protein. Treatment is usually by restriction of protein intake and feeding supplements of the. amino acid arginine. Sodium benzoate may be given to increase the excretion of nitrogenous waste as hippuric acid.

Citrulline An amino acid formed as a metabolic intermediate, but not involved in proteins, and of no nutritional importance.

Citrus Genus of trees with fleshy, juicy fruits; there is considerable confusion over the names because of hybridization and mutations. Sweet orange Citrus sinensis; various cultivars including Valencia, Washington navel, Jaffa, Shamouti. Sour, bitter, or Seville orange, C. aurantium, is too bitter to eat and is used for marmalade. Lemon, C. limon. Lime, C. aurantifolia. Citron, C. medica, has thick, white inner skin and is used mainly to make candied peel. Pomelo, C. grandis, is the parent of the grapefruit. Grapefruit, C. paradisi, is a hybrid of pomelo and sweet orange. Tangerine, satsuma, mandarin, calamondin, naartje, are small citrus fruits with loose skins. Clementine, a hybrid of tangerine and bitter orange, is sometimes regarded as a variety of tangerine. Mineola is a hybrid of grapefruit and tangerine. Ortanique is a hybrid of orange and tangerine, unique to Jamaica. Citrange is a hybrid of citron and orange. Tangors are hybrids of tangerine and sweet orange. Ugli fruit is a hybrid of grapefruit and tangerine. Tangelo is a hybrid of tangerine and pomelo. All are a rich source of vitamin C and contain up to 10% sugars.

Cittosis An unnatural desire for foods; alternative words are, cissa, allotriophagy, and pica.

Clabbered milk Unpasteurized milk that has soured naturally, becoming thick and curdy. Clabber cheese is curd or cottage cheese.

Clafoutis Batter pudding with black cherries or other fruit. claims, misleading descriptions There are no restrictions on the sales of foods so long as they are not harmful and do not transgress the regulations. What is controlled by law is any claim. It is illegal to claim that a food is capable of preventing, treating, or curing a disease unless it has a specific Product Licence under the Medicines Act. Misleading claims are illegal even if, strictly speaking, they are true.

Clam bake American; beach picnic.

Clams Several types of marine bivalve molluscs, including scallop, large tropical clam, Tridacna gigas, quahog, Mya arenaria, and Venus mercenaria. An exceptionally rich source of vitamin B 12 ; a good source of protein and niacin.

Clapshot Scottish; mashed potato and turnip, seasoned with chives and bacon fat.

Claret Name given in the UK to red wines from the Bordeaux region of France.

Clarete Portuguese, Spanish; light red wines.

Clarification The process of clearing a liquid of suspended particles. It may be carried out by filtration, centrifugation, the addition of enzymes to hydrolyse and solubilize particulate matter, or the addition of flocculating agents.

Clarifixation A method of homogenizing milk in which the cream is separated, homogenized, and remixed with the milk in one machine.

Clarifying Of fats; freeing the fat of water so that it can be used for frying, pastry-making, etc. Clarified fats are less susceptible to rancidity on storage; ghee is clarified butterfat. Also the process of filtering juices before making jellies, etc.

Clarke degrees Units of measurement of water hardness.

Classification Refers to the grouping of fish, shellfish, algae, and indeed all other species, scientifically into a formal, hierarchical series of groups. The smallest group regularly used is species, which is given a specific Latin description, e.g. the cod, Gadus morhua; the herring, Clupea harengus.

Clay A dried mineral clay under the names of sikor, mithi, pakhuri, and khatta, is sometimes used by Asians as a treatment for indigestion and as a vitamin supplement, but can be toxic since it contains varying amounts of arsenic and lead.

Cleaning The process of removing dirt. See Hygiene.

Clementine A citrus fruit, Citrus nobilis var. deliciosa; regarded by some as a variety of tangerine and by others as a cross between the tangerine and a wild North African orange.

Clenbuterol This drug is classified as a Beta-2 agonist, and is used for treating asthma. It is not an anabolic steroid, or a growth hórmone. Currently, it is not approved for sale in the United States, but it is sold in Europe, under various names, such as Spiropent, Monores, and Ventipulmin. Researchers in 1987 noted an unusual effect in animals when using this drug. Their muscles grew, while they lost body fat. The main side effects associated with this drug include tremors, heart rhythm disturbances, headaches, nervousness, excessive sweating and insomnia. Bodybuilders use it because it mimics the actions of Epinephrine. It simulates a process called Thermogenesis, which turns fat calories into heat and which results in a loss of body fat. The side effects start when you use more than 80 micrograms or more. The dosage that's used to treat asthma is around 20 to 40 micrograms twice a day. The muscle building dose would be over 30 micrograms in humans, but scientists consider it to be dangerous and unethical. Another problem is that any effects you get from it wears off rapidly. Bodybuilders who have used this drug shows that it is better for promoting fat burning than for muscle building.

Clinical trials Clinical trials undertake experimental study of human subjects. Trials may attempt to determine whether the finds of basic research are applicable to humans, or to confirm the results of epidemiological research. Studies may be small, with a limited number of participants, or they may be large intervention trials that seek to discover the outcome of treatments on entire populations. The "gold standard" clinical trials are double-blind, placebo-controlled studies which employ random assignment of subjects to experimental and control groups unknown to the subject or the researcher.

Clofibrate Drug used to treat hyperlipidaemia by lowering specifically low density lipoproteins.

Cloning Technique of creating a group of genetically identical cells or DNA molecules from a single ancestor. In horticulture, cloned plants are reproduced asexually from a single parent.

Clostridium botulinum Species of pathogenic bacteria found in some fish, mud and marine sediments. These anaerobic spore-forming bacilli grow only in the absence of

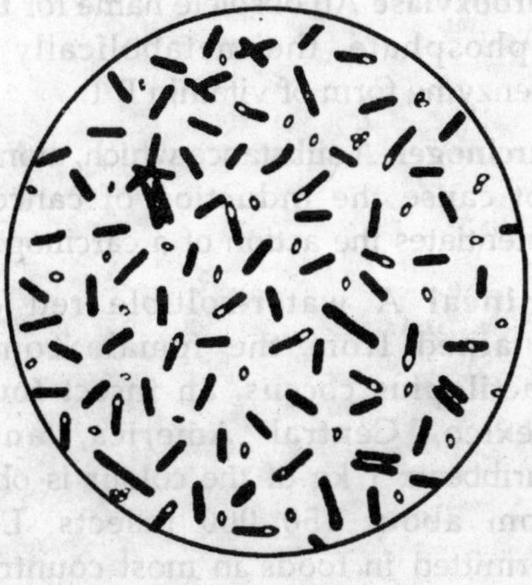

Fig. Clostridium botulinum

oxygen and produce one or more of a series of similar neurotoxins which are readily destroyed by cooking, but the spores of the organism are quite heat resistant. There are seven known types of Clostridium botulinum, referred to as types A to G. Types A, B, E and F consistently produce botulism in humans, and B, E and F are frequently found in the sea. Some authentic cases of botulism have been due to type E, some of which have also been associated with consumption of various seafoods, and so serious attention has been paid to the occurrence of type E in fish and fishery products. Type E and varieties of B and F are found in fish intestines and gills, and in mud from the sea, compared to the other types that are found mostly in soil; they can grow and form toxin at much lower temperatures than the other types; they can grow on fish products at 5°C.

Cloudberry An orange-yellow fruit resembling the raspberry in shape; Rubus chamaemorus, known as avron in Scotland, and baked-apple berry in Canada. It is an extremely rich natural source of benzoic acid and will not ferment; it remains fresh for many months without preservation.

Clove The dried aromatic flower buds of Caryophyllus aromaticus; mother of clove is the ripened fruit, which is inferior in flavour. Used as a flavour in meat products and baked goods.

Club sandwich Americansandwich made with toast, turkey, lettuce, bacon, and mayonnaise.

Clupeotoxin A very potent toxin infrequently present in planktivorous species of fish, such as sardines and sardinellas, in tropical waters. The toxin has also been reported in species of horse mackerel, and trigger fish, and crabs. The toxin, a neurotoxin, is derived from the feed of the fish and causes symptoms similar to those of ciguatera, but with a higher mortality rate, around 50% of cases. The causative chemical has been identified as palytoxin, a toxin found in various species of marine algae. These marine algae are assumed to be the source of the toxin present in fish.

CNS Central nervous system.

Coagulation The process in which colloidal particles come together irreversibly to form larger masses. Coagulation can be brought about by adding ions to change the ionic strength of the solution and thus destabilize the colloid. Ions with a high charge are particularly effective. Alum and iron(III) sulphate are also used for coagulation in sewage treatment. Heating is another way of coagulating certain colloids.

Coating, coated products Fillets and portions may be marketed naked without any coating, or incorporated into other products with sauces. However, there are many products in which a coating of batter and breadcrumbs is applied, and the traditional British battered and deep-fried fish, Japanese tempura, puff-type batter coated fish products, and batter and crumb coated fish stick products from the USA, are classic examples of coated products. Coating technology is a specialist subject and there are many types of specialist batters, crumb textures and applications available. As an example, the manufacture of products with a batter and crumb coating is basically a continuous process known as breading. The frozen stick, or frozen portion or fillet , all of which may be pre-treated to improve adhesion of the batter, is first passed through a falling curtain of batter which coats the whole surface. If necessary, excess batter is blown off with a current of air and the coated product is passed through a bed and curtain of dry crumbs to attain overall coverage. The product may be flash fried to set or fix the coating and then refrozen to –30°C before or after packaging, for storage and distribution. Many coated products are available with a variety of cooking options, such as baking in a conventional oven, grilling, shallow fat frying, deep fat frying, microwave cooking.

Cobalamin A mineral whose main function is in vitamin B 12, although there are some cobalt-dependent enzymes. There is no evidence of cobalt deficiency in human beings, and no evidence on which to base estimates of requirements for inorganic cobalt. 'Pining disease' in cattle and sheep is due to cobalt deficiency and is a growth factor for chicks, turkeys, pigs, and rats. Cobalt salts are toxic in excess, causing degeneration of the heart muscle, and habitual intakes in excess of 300 mg/day are considered undesirable.

Cobamide A derivative of vitamin B 12.

Cobbler Sweetened cold drink made from fruit with wine or liqueur and ice. Meat or fruit dishes topped with scone rounds.

Coburg cakes Small cakes containing syrup and flavoured with spices.

Coca cola Trade name for a cola drink.

Coca leaves From the S. American plant, Erythroxylon coca; they contain the narcotic alkaloid cocaine, and are traditionally chewed by the natives of Peru as a stimulant. Originally the beverage Coca Cola contained coca leaf extract, although this was removed from the formulation many years ago.

Cocarboxylase An obsolete name for thiamin diphosphate, the metabolically active coenzyme form of vitamin B 1.

Cocarcinogen A substance which, alone, does not cause the induction of cancer, but potentiates the action of a carcinogen.

Cochineal A water-soluble red colour obtained from the female conchilla, Dactilopius coccus, an insect found in Mexico, Central America, and the Caribbean: 1 kg of the colour is obtained from about 150 000 insects. Legally permitted in foods in most countries (E-120). Contains carminic acid (E-120). Cochineal red A is an alternative name for Ponceau 4R (E-124), often used to replace cochineal. Carmine is produced from cochineal.

Cockles Several types of marine bivalve molluscs of genus Cardium, often sold preserved in brine or vinegar.

Cocktail Mixed alcoholic drink; there are many recipes based on a wide variety of spirits and liqueurs, with fruit juice, milk, or coconut milk, normally shaken with crushed ice.

Cocoa butter equivalents Vegetable fats with physical and chemical properties similar to those of cocoa butter; useful for the partial or complete replacement of cocoa butter, which is expensive. Examples are illipe and shea nut butter.

Cocoa butter The fat from the cocoa bean, used in chocolate manufacture and in pharmaceuticals; it has a low sharp melting point, between 31 and 35°C, so it melts in the mouth; mostly 2-oleopalmitostearin.

Cocoa nibs Seeds of the fruit of the cocoa plant, Theobroma cacao, are left to ferment, which modifies the bitterness, and their colour darkens. They are then roasted and separated from the husks as two halves of the seed known as cocoa nibs. They contain about 50% fat, part of which is removed in the preparation of chocolate and cocoa for beverages.

Cocoa Originally known as cacao, introduced into Europe from Mexico by the Spaniards in the early sixteenth century. The powder prepared from the seed embedded in the fruit of the cocoa plant, Theobroma cacao, also a milk drink prepared with cocoa powder. Used to prepare chocolate. Contains the alkaloid theobromine; caffiene is trimethylxanthine.

Coconut Fruit of the tropical palm, Cocos nucifera. The dried nut is copra which contains 60-65% coconut oil. The residue after extraction of the oil is used for animal feed. The hollow, unripe nut contains a watery liquid known as coconut milk, which is gradually absorbed as the fruit ripens.

Coconut oil Semi-solid oil extracted from copra, Cocos nucifera; contains 90% saturated fats.

Cocotte, en Describes a dish cooked and served in a small casserole.

Cocoyam New cocoyam is the West African name for tannia; the old cocoyam is the West African name for taro.

Cod A white fish, Gaddus morrhua and other species. The composition of all non-fatty fish, such as cod, hake, haddock, flatfish, is similar.

Cod liver oil The oil from codfish liver; the classic source of vitamins A and D, used for its medicinal properties long before the vitamins were discovered. An average sample contains 120-1200 μg vitamin A and 1-10 μg vitamin D per gram. British Pharmacopoeia standard: minimum 180 μg vitamin A and 2 μg vitamin D per gram.

Cod worm Common name given to a parasitic nematode Pseudoterranova decipiens, found in the flesh of cod and, of other species of fish. Its life cycle is similar to that of Anisakis spp. and seals are commonly the definitive hosts, hence its other common name, 'seal worm'.

Coddle To cook slowly in water kept just below boiling point.

Code of Practice A set of guidelines or instructions for the manufacture of a product or to the carry out an operation. They are elements of a Quality assurance programme within a company, or they can be incorporated in, or form part of, official regulations. They can developed and used entirely within a company, or be issued by trade associations or national or international regulatory bodies for more general application. The degree of detail in a Code of Practice will differ among these various origins.A processing company will have more than one code. Some will apply to operations common to a plant as a whole. Examples, would be Codes of Hygienic Practice and Codes of Sanitary Practice. Some would apply to general operations such as codes for operation of freezing equipment and for chill and cold stores, some for the manufacture of particular products. These will be very specific about where in a factory the operation will be conducted; who in the company personnel structure will make the product or carry out the operation; the particular machine and equipment to be used; the source of raw material; the nature and preparation of materials to be used in the product; and the manufacture, packaging, and storage of the product. The code will specify the persons responsible for supervising the manufacture of a product, or for an operation, and could incorporate instructions for monitoring quality but, usually, quality control would be the subject of separate documents. Trade bodies might issue agreed Codes of Practice to advise members or to act as model codes, or as the basis of quality labelling schemes supported by the trade body. Codes of Practice issued by international agencies, for example, Codex Alimentarius, or by national agencies will often be part of the regulatory mechanism for ensuring the safety and wholesomeness of products, or to control the hygienic and sanitary conditions for the manufacture of products. In such cases, the codes could be enforced by force of law.

Coenzyme Q10 This antioxidant is shown to have heart protective and energy productive properties. COQ10 is involved in cellular energy production. Several studies have reported improved endurance after taking CoQ10. It is considered one of the best antioxidants. It may be an ergogenic aid because it plays a pivital role in a cellular process that leads to ATP synthesis and is known as the electron transport system.

Co-extrusion Is a multi-layer plastic film made by extruding two or more molten plastics and layering them together to produce a film on cooling. They are designed for many special-purpose packaging for flexible wraps, bags and pouches.

Coffee A beverage produced by roasting the beans from the berries of two principal types of shrub: Coffea arabica and C. canephora. Niacin is formed during the roasting process, and the coffee can contain 10-40 mg of niacin per 100 g, depending on the extent of roasting, thus making a significant contribution to average intakes of niacin.

Instant coffee is dried coffee extract which can be used to make a beverage by adding hot water or milk.

Cognac Brandy made in the Charentes region of north-west France, around the town of Cognac, from special varieties of grape grown on shallow soil and claimed to be distilled only in pot, not continuous, stills. Sometimes used as a general name for brandy.

Cointreau Orange-flavoured liqueur.

Cola drinks Carbonated drinks containing extract of cola bean, the seed of the tree Cola acuminata, and a variety of other flavouring ingredients. Cola seed contains caffeine, and the drink contains 10-15 mg caffeine/ 100 mL; decaffeinated or caffeine-free varieties are also made.

Colchicine An alkaloid isolated from the meadow saffron, or autumn crocus. It is an old remedy for gout. It inhibits cell division, and is used in experimental horticulture to produce plants with abnormal numbers of genes.

Cold chain The succession of refrigerated storage and transport links between primary producer and consumer to keep fish chilled or frozen throughout distribution from catching or harvesting to preparation for final consumption.

Cold storage flavour, cold store flavour The unpleasant flavour, with an associated odour, formed in the flesh of lean fish during frozen storage, also called cold store flavour. It is variously described as 'turnipy', 'leathery', 'cardboardy' 'wet dogs'. It tends to be somewhat persistent in the mouth. Some people, probably about 10-15% of individuals are almost completely insensitive to the odour and flavour. Chemical analysis shows that the causative chemical is hept-cis-4-enal formed by the oxidation of phospholipids in the flesh. The flavour, or rather its absence or low intensity, is an important quality attribute of frozen lean fish products, and its development limits their storage life. Packaging to exclude air, or holding in bulk form - whole fish rather than fillets - will reduce the rate of formation of the hept-cis-4-enal and, hence, the flavour.

Cold store, cold storage A refrigerated and insulated room, suite of rooms, or a large refrigerated and insulated enclosed space, on land or in a ship, for the cold storage and preservation of fish and fish products in the frozen state. The generally recommended operating temperature of cold stores for fish and fish products is -30°C, but some species such as tuna are stored at lower temperatures.

Cold-shortening When the temperature of muscle is reduced below 10°C while the pH remains above 6-6.2 the muscle contracts in reaction to cold and, when cooked, the meat is tough.

Colectomy Surgical removal of all or part of the colon, to treat cancer or severe ulcerative colitis.

Coleslaw Salad made from finely shredded white cabbage dressed with cream or mayonnaise; other vegetables may be included.

Coliforms A group of gram negative, lactose fermenting, bacteria characterised operationally by ability to grow in specified culture media under specified conditions rather by any taxonomic considerations.

The number of coliform organisms in a sample is typically estimated by the Most Probable Number (MPN) method. Coliforms include the group of faecal coliforms and high numbers of coliforms in a sample is an indication that faecal coliforms, and perhaps food poisoning organisms of enteric origin might be present. It is more satisfactory for monitoring the hygienic condition of fishery products to test for faecal coliforms, but coliform count is simpler to conduct, and quicker to produce a result.

Colitis Inflammation of the large intestine, with pain, diarrhoea, and weight loss; there may be ulceration of the large intestine.

Collagen An insoluble fibrous protein found extensively in the connective tissue of skin, tendons, and bone. The polypeptide chains of collagen form triple-stranded helical coils that are bound together to form fibrils, which have great strength and limited elasticity. Collagen accounts for over 30% of the total body protein of mammals.

Colon Also known as the large intestine or bowel, consisting of three anatomical regions: the ascending, the transverse, and the descending colon. The colon normally has a considerable population of bacteria, while it is rare to find a significant bacterial population in the small intestine. The colon terminates at the rectum, where faeces are compacted and stored before voiding.

Colorado potato beetle Insect pest that prefers potatoes, but will attack tomatoes and eggplants. Cross protect Make a plant resistant to a severe virus by intentionally infecting it with a mild strain of the same virus.

Colostomy Surgical creation of an artificial conduit on the abdominal wall for voiding of intestinal contents following surgical removal of much of the colon and/or rectum.

Colostrum Colostrum is a fluid produced by the mammary glands of female mammals prior to milk production. It has a high concentration of immunoglobins, antimicrobial agents and hormone factors. Colostrum provides major immune and gastrointestinal system benefits.

Bovine Colostrum usually comes from cows, a dairy product that has similar properties to human colostrum. Normally, adults cannot absorb colotrum's antobodies and growth factors the way a newborn can. But it still has superior nutritional values which may make it a useful supplement.

Colours Widely used in foods to increase their aesthetic appeal; may be natural, nature-identical, or synthetic. Natural colours include carotenoids, some of which are vitamin precursors. Chlorophylls are the green pigments in all leaves and stems. Anthocyanins are the red, blue, and violet pigments in beetroots, raspberries, and red cabbage. Flavones are yellow pigments in most leaves and flowers. There are twenty permitted synthetic colours. Caramel is used for both flavour and as a brown colour made by heating sugar.

In addition to all these there are various ingredients such as paprika, saffron, and turmeric that also provide colour.

Comfrey Wild member of the borage family, Symphytum officinale. Leaves may be cooked like spinach or fried in batter.

Commercial sterility A compromise on the practical impossibility of achieving full sterility in the contents of a hermetically sealed container during commercial heat processing, whereby the initial bacterial load is destroyed through sufficient decimal reductions to reduce the possibility of a single organism surviving to an acceptably low level. This level depends on the organism, usually Clostridium botulinum, which the process is designed to destroy. The time required to reduce the number of spores of this organism by a factor of 10 at a specific reference temperature is the decimal reduction time, or D value, denoted D 0 . The D 0 value for Clostridium botulinum spores can be taken as 0.25 minutes. To achieve a reduction by a factor of 10 12, regarded as an acceptably low level, requires 3 minutes at 121.1°C, and is known as the process value, or F value, designated F 0 so, in this case, F 0 = 3, which is known as a botulinum cook. See Canning and Heat-treated.

Comminuted Finely divided; used with reference to minced meat products and fruit drinks made from crushed whole fruit including the peel.

Comminuted The term is used in a number of ways. It may simply mean minced or fragmented fish flesh. It may refer to flesh mechanically disintegrated to such a degree that the type and previous form of the fish cannot be visually identified, or it may mean similarly disintegrated material from which some protein has been extracted in a salt solution, as in the preparation of surimi. It may refer to fish fillet minced after the removal of bones and skin. See Minced fish.

Complan Trade name for a mixture of dried skim milk, arachis oil, casein, maltodextrins, sugar, and vitamins, used as a nutritional supplement.

Complementation This term is used with respect to proteins when a relative deficiency of an amino acid in one is compensated by a relative surplus from another protein consumed at the same time. The protein quality is not the average of the separate values, but higher.

Composition The constituents of a product and their amounts in unit mass or volume.

Compote Fruit stewed with sugar; a single fruit or a mixture, served hot or cold. Also sometimes used for a stew of small birds such as pigeons.

Con crianza Spanish; aged wines.

Conalbumin One of the proteins of egg-white, comprising 12% of the total solids. It binds iron in a pink-coloured complex; this accounts for the pinkish colour resulting when eggs are stored in rusty containers.

Conde Dessert of creamed rice with fruit and red jam sauce. Also the name of a type of patisserie. condiment Seasoning added to flavour foods, such as salt, or herbs and spices such as mustard, ginger, curry, pepper, etc. Although some are relatively rich in nutrients, they are generally used in such small quantities that they make a negligible contribution to the diet.

Condition The nutritional status of a fish; the amount of flesh on the carcass. Sexually mature fish deplete protein in their flesh to build up their gonads in the approach to the spawning season. The spawning season, in spring, follows a period when food for the fish is not abundant and fish further deplete muscle tissue protein as an energy source. As the protein depletes a fish becomes thinner, and its weight decreases. Fish are then in poor condition. After spawning, in the late spring and early summer, fish feed well, build up muscle tissue, and are then in good condition. Compared with fish in good condition, fish in poor condition have lower fillet yields, softer-textured fillets in both the raw and cooked states, and a tendency to an insipid flavour.

Confectionery Delicacies made with sweet ingredients. Confectioneries sweetened with honey were made in Egypt at least 3,000 years ago. In the Middle Ages the Persians developed confectionery made with refined cane sugar, and during the 18th century in Europe , machinery for confectionery manufacture was first developed. Boiled or hard sweets such as fruit drops and clear mints are made by boiling a flavoured solution of sugar and corn syrup until the sugar concentration reaches a high level. On cooling, a hard, glassy product is formed. Caramels and toffees are manufactured in a similar way, but the mixture includes condensed or evaporated milk. Fondant, the basis for the 'soft centres' of many chocolates, is made by rapidly beating a hot, concentrated sugar mixture so that minute crystals are formed: fudge can be made by similarly beating hot caramel. Agar, pectin, or gelatine is added to sugar syrup to form jellies and Turkish delight, while gums and pastilles are made by dissolving gum arabic in sugar syrup.

Oxford paperback encyclopaedia, Oxford university press 1998

Confit 1. Fruit or vegetables preserved in brandy, vinegar, or sugar.

2. Poultry meat or pork preserved in a vessel and covered with a layer of fat to exclude air.

Congee Chinese soft rice soup or gruel; may be sweet or savoury. Commonly eaten for breakfast.

Congeners Flavour substances in alcoholic spirits that distil over with the alcohol; chemically a mixture of higher alcohols and esters. Said to be responsible for many of the symptoms of hangover after excessive consumption.

Congenital Condition existing at birth, not hereditary.

Congress tart Small pastry case filled with ground almonds, sugar, and egg.

Congris Caribbean; casseroled red beans served with rice.

Conjugated linoleic acid CLA occurs naturally in whole milk and red meat. A collective term used to designate a mixture of positional and geometric isomers of the essential fat linoleic acid. It is actually a fat, derived from linoleic acid. Studies have shown that CLA can increase lean body mass and decrease fat, inhibit the growth of tumors and enhance immune function. CLA is found naturally in beef, cheese and whole milk.

Connective tissue Consists of the protein collagen which in fish is found between the muscle segments; in meat it is spread through the muscle, uniting the muscle fibres into bundles and supporting the blood vessels, and consists of both collagen and elastin.

A high content of connective tissue results in tougher meat.

Collagen is insoluble; it is converted to soluble gelatine by moist heat, so making the food more tender. Tough meat is softened to some extent by stewing, but roasting or frying has little effect. Elastin is unaffected by heating, and remains tough, elastic, and insoluble.

Consomme A clear soup made from meat or meat extract.

Constipation Difficulty in passing stools or infrequent passage of hard stools. In the absence of intestinal disease, frequently a result of a diet low in non-starch polysaccharide, and treated by increasing the intake of fruits, vegetables, and especially wholegrain cereal products.

Consumer Any person who uses goods and services.

Contaminant A substance present in fish or fish products coming from sources outside of the fish or product, and not intentionally added in processing, but having implications for the safety or quality of the products. Contaminants might be derived from the natural environment, or from the environment in which the fish has been handled, stored or processed. Mostly, contaminants are chemicals, but they can be physical material such as pieces of metal or complete or parts of insects, in which case they are referred to a 'foreign bodies'. Contamination of fishery products during handling and processing comes from substances such as cleaning and disinfecting agents, or possibly from packaging materials. These affect the quality of the products, and likely render them unfit for consumption. Environmental contaminants might have implications for human health. Examples of particular concern for fishery products are some heavy metals, particularly mercury and cadmium, and persistent, highly chlorinated

Contamination A process by which harmful or unpleasant substances get into or onto food.

Control chart The term is sometimes expanded to Shewart control chart in recognition of the originator, W. A. Shewart, of this system for quality control. Control charts are used to monitor the performance of repetitive operations and indicate when the process should be adjusted to ensure that the outcome of the operation is within specified limits. They are widely used for quality control in the engineering industry, but there is not much scope for their use in fish processing.There are a few different sorts of control charts, but the one likely to be used in fish processing is control by variables. A variable is a measurable quantity such as the dimensions of a seam in canning, or the weight of a portion in a portioning process. The process is sampled at regular intervals and the variable of interest is measured in all units in the sample. The mean value is plotted on a chart against the time of sampling. In this way it can be seen if the process is drifting, that is, if the mean value of the variable is tending to increase or decrease with time. Also plotted on the chart are permitted upper and lower limits, control limits, of the variable. Appropriate action is taken to adjust the process if the trend of sample mean values approaches either of the limits.

Control group The group of subjects in a study to whom a comparison is made in order to determine whether an observation or treatment has an effect. In an experimental study it is the group that does not receive a treatment. Subjects are as similar as possible to those in the test or treatment group.

Controlled experiment In this type of research, study subjects are selected according to relevant characteristics, and then randomly assigned to either an experimental group, or a control group. Random assignment ensures that factors known as variables, which may affect the outcome of the study, are distributed equally among the groups and therefore could not lead to differences in the effect of the treatment under study. The experimental group is then given a treatment, and the results are compared to the control group, which does not receive treatment. A placebo, or false treatment, may be administered to the control group. With all other variables controlled, differences between the experimental and control groups may be attributed to the treatment under study.

Convenience foods Processed foods in which a considerable amount of the preparation has already been carried out by the manufacturer, e.g. cooked meats, canned foods, baked foods, breakfast cereals, frozen foods.

Conventional cook-chill This is when foods are cooked by conventional methods prior to aseptic packaging and chilling. These foods can be refrigerated for up to five days.

Convicine One of the toxins in broad beans, responsible for the acute haemolytic anaemia of favism.

Coo-coo Caribbean; a cooked side-dish of starchy vegetables such as breadfruit or corn meal with herbs and spices.

Cook-freeze A method of catering involving cooking followed by rapid freezing and storage between -18 and -30°C, giving a storage time of several months.

Cooking Required to make food more palatable, more digestible, and safer. There is breakdown of the connective tissue in meat, softening of the cellulose in plant tissues, and proteins are denatured by heating, so increasing their digestibility.

Cooking, loss of nutrients In general, water-soluble vitamins and minerals are lost in the cooking water, the amount depending on the surface area to volume ratio, i.e. greater losses take place from finely cut or minced foods. Fat-soluble vitamins are little affected except at frying temperatures. Proteins suffer reduction of available lysine when they are heated in the presence of reducing substances, and further loss under extreme conditions of temperature. Dry heat, as in baking, results in some loss of vitamin B 1 and available lysine. The most sensitive nutrient by far is vitamin C, with vitamin B 1 next. Average losses from cereals are: boiling, 40% vitamins B 1, B 2, B 6, niacin, biotin, and pantothenic acid; 50% total folate; baking, 5% niacin, 15% vitamin B 2 ; 25% vitamins B 1, B 6, and pantothenic acid; 50% folate; with biotin being stable. In meat, losses are approximately 20% of all the vitamins for roasting, frying, and grilling and 20-60% for stewing and boiling.

Cook's knife Large sharp knife suitable for slicing and chopping.

Copepod parasites Zoologically, copepods are a subclass of the class Crustacea that includes crabs, shrimps and prawns, but copepods are not used for human food, at least not commercially so. Many species are ectoparasites, that is, live on the outside of fish. Fish lice are copepods. Some species take on bizarre, degenerate forms and are not immediately recognisable as copepods.

Copper Active in the storage and release of iron to form hemoglobin for red blood cells.

Copra Dried coconut 'meat' used for production of coconut oil for margarine and soap manufacture.

Coprophagy Eating of faeces. Since B vitamins are synthesized by intestinal bacteria, animals that eat their faeces can make use of these vitamins, which are not absorbed from the large intestine, the site of bacterial action.

Coquimol Custard made with coconut milk and eggs.

Coral The ovaries of female lobsters, used as the basis for sauces; red-coloured when cooked.

Cordial, fruit Originally a fruit liqueur, and still used in this sense in the USA; in the UK a cordial is now used to mean any fruit drink, usually a concentrate to be diluted.

Coriander Coriandrum sativum; the leaf is used fresh or dried as a herb, and the dried ripe fruit (also called dhanyia) as a spice in meat products, bakery goods, tobacco, gin, and curry powder.

Coring Removal of the pips and central membranes from apples, pears, etc.

Corked Of wines, the development of an unpleasant flavour due to fungal contamination of the cork.

Corm The thickened, underground base of the stem of plants, often called bulbs, as, for example, taro and onion.

Corn oil Extracted from maize germ, Zea mays; 13% saturates, 60% polyunsaturates.

Corn pone Small corn cakes, a speciality of Alabama, USA.

Corn salad Winter salad vegetable, Valeriana olitoria, also known as lamb's lettuce.

Corn Term used in the UK for wheat, in the USA for maize, and sometimes for oats in Scotland and Ireland, originally any grain.

Corn, flour Flour corn is a variety of maize with large, soft grains and very friable endosperm, making it easy to grind to flour.

Corned beef In the UK a canned product made from low-quality beef that has been partially extracted with hot water to make meat extract.

Cornflakes Breakfast cereal made from maize, often enriched with vitamins. A rich source of vitamins B 2, B 6, B 12, niacin, and folate; a good ource of vitamin B 1 and iron; provides 4.4 g of dietary fibre; supplies 145 kcal (610 kJ).

Cornflour Purified starch from maize; in the USA called corn starch; used in custard, blancmange, and baking powders and for thickening sauces and gravies.

Corrinoids The basic chemical structure of vitamin B 12 is the corrin ring; compounds with this structure, whether or not they have vitamin activity, are corrinoids.

Cortisol A catabolic hormone that is released and increases in response to stress when the body is subjected to trauma such as intense exercises, including weight training. Excess cortisol is known to increase catabolism. Cortisol leads to muscle breakdown through promoting a release of muscle amino acids for transport to the liver, where the amino acids are coverted into glucose.

Cos lettuce Long-leafed variety of lettuce.

Cosecha Spanish; vintage of wine.

Cossettes Thin chips of sugar beet shredded for hot-water extraction of the sugar.

Cottage loaf Traditional English loaf consisting of a large round base with a smaller round topknot of bread.

Cottonseed Seed of Gossypium spp.; of double use in the food field since the oil is valuable as cooking oil, or for margarine manufacture when hardened, and the protein residue is a valuable animal feed. The oil is 25% saturated and 50% polyunsaturated.

Coulis Also cullis; originally the juices that run out of meat when it is cooked, now used to mean rich sauce or gravy made from meat juices, pureed shellfish, vegetables, or fruit. Most usually a sauce made from pureed and sieved fruit.

Count The number of items in a fixed weight or volume. Frequently used in the marketing of small or moderately-sized fishery products such as shrimps, squid mantles, small pelagic fish. The products are sorted and packaged by count ranges for marketing. The size grades often have trivial names, like small, large, jumbo, as well as the numeric size ranges. These names are not standardised across commodities, and might not be the same in all markets for a product even though size grades will be harmonised.

Coupe Dessert of ice cream and fruit, a sundae.

Courgette Variety of marrow developed to be harvested when small; also known as Italian marrow, Italian squash, or zucchini.

Courlose Trade name for sodium carboxymethylcellulose.

Court-bouillon Fish stock used in place of water to cook fish; may contain wine, vinegar, or milk.

Couscous North African; millet flour or fine semolina, steamed until fluffy and usually served with mutton stew.

Cow-heel Dish made from heel of ox or cow, stewed to a jelly; also known as neat's-foot.

CPET A type of plastic suitable for microwave packaging. It is heat resistant to 220°C and transparent to microwave energy.

Crab apple Wild varieties of apple, normally very sour; used to make sweet-sour jelly as accompaniment to meat. Commonly grown as an ornamental plant, and to enhance fertilization of cultivated apple trees.

Crab Shellfish; Cancer and Carcinus spp.; king crab is Limulus polyphemus. A rich source of protein, niacin, zinc, copper, and selenium; a good source of iron; a source of vitamins B 2 and B 6.

Crackers Plain, thin biscuits such as water biscuits, cream crackers, and wholemeal crackers, made from wheat flour, fat, and bicarbonate as a raising agent.

Crackling Crisp rind of a joint of pork after baking or roasting.

Cracknel 1. Plain biscuit made with paste which is boiled before baking so that it puffs up.

2. Brittle toffee filling for chocolates.

Cran A traditional measure for herrings containing 37½ gallons (167 L) or about 800 fish.

Cranberry The fleshy, acid fruit of Vaccinium oxycoccus or V. macrocarpon, resembling a cherry; commonly used to make cranberry sauce and for the preparation of juice.

Crapaud Large edible Caribbean frogs, also known as mountain chicken.

Crawfish Crustaceans of the family Palinuridae, also called spiny lobster, rock lobster, sea crayfish.

Cream line index The cream line or layer usually forms about 6% of the total depth of milk. The cream line index is the ratio between the percentage cream layer and the percentage of fat in the milk; in ordinary bulk pasteurized milk it is about 1.7.

Cream of tartar Potassium hydrogen tartrate, used with sodium bicarbonate as baking powder because it acts more slowly than tartaric acid and gives a more prolonged evolution of carbon dioxide. This is tartrate baking powder. Also used to 'invert' sugar in making boiled sweets.

Creatine A muscle fuel that is extracted naturally from meat and fish, or synthesized in the lab. Once it is in the muscles, creatine combines with phosphorous to make Creatine Phosphate (CP), a high powered chemical that rebuilds the muscles ultimate energy source, Adenosine Triphosphate (ATP). CP powers your muscles for high intensity exercise for short periods only, consequently, athletes who compete in power and sprint event will have an advantage if they take supplemental creatine. More CP in the muscle cell translates into a greater resistance to fatigue. Also, CP helps with the transfer of energy in the muscle cells, thus speeding up the action, which may enhance performances that are aerobically taxing. Reports says people who take creatine supplements may recover from intense activity faster and experience less postexercise muscle soreness.

Creatine is a naturally occurring compound in the muscle tissue and when converted in the muscle tissue to phosphocreatine during exercise can provide sudden bursts of energy. Insufficient amounts of phosphocreatine could result in a fatigued feeling in the muscle. The Creatine Monohydrate Powder provides enough energy to delay to onset of fatigue. Creatine Monohydrate is a synthesized metabolite that is the powerful energizer providing instant energy and strength with better endurance and helps to maintain optimal levels of ATP production during intense exercise.

Why Monohydrate? Creatine comes in several forms. Creatine Monohydrate, Creatine Phosphate, and liquid form. Creatine Phosphate is much more expensive to manufacture while it offers no advantage. Liquid creatine has many problems associated with it. When mixing creatine monohydrate with a protein drink, or water, the creatine starts to become unstable. Within 24 hours, the creatine begins to change or 'fallout' into creatinine. Creatinine is a useless substance to the body. Thus, buying a premixed liquid form of creatine is not a legitimate product. The best absorbed form of creatine is the creatine monohydrate. Creatine monohydrate is better absorbed because it is more stable, resulting in higher concentration of available creatine.

Creatine & ATP. ATP is the molecule that releases the energy for contraction of muscles, the breakdown and synthesis of proteins and all other reactions requiring energy. In short, ATP is the energy molecule powering all of our movements. By giving off its energy through its high energy phosphate bond, ATP is reduced to ADP. The problem is that the amount of ATP that is stored in our cells is limited. Depending on the intensity of the activity, ATP supplies can be used up by converting to ADP within seconds. So how do athletes run or workout for long periods. We can do that because there are three way to replenish ATP.

1. You can restore ATP using energy derived from the oxidation of fats and carbohydrates. This is a slow process that occurs in the mitochondria.

2. You can restore ATP through lactic acid, which is utilized to produce energy, which turns ADP back into ATP.

3. Through Creatine Monohydrate, which helps creatine phosphate create more ATP from ADP within seconds. It is a short term, high energy backup for ATP. It does not need carbohydrates, fats or oxygen to recharge ATP.

Creatine Titrated creatine is less expensive than effervescent creatine, but has the same total solubality and absorbability. The result is free-ionized, soluble creatine. Titrated creatine acheives the process by titrating, or changing, the pH of the water when it's stirred in. The altered pH solution enables more than 95 percents of the creatine to go into solution, so you get dissolved creatine.

Creatinine Formed non-enzymically from creatine. Urinary excretion of creatinine is relatively constant from day to day, and reflects mainly the amount of muscle tissue in the body. Therefore the amounts of various components of urine are often expressed as unit/creatinine.

Crescioni Fried pastry squares filled with spinach chopped with shallots and raisins.

Crespolini Pancakes filled with spinach, cream cheese, and chicken liver, baked in bechamel sauce.

Cress Garden cress, pepper grass, Lepidium sativum. The seed leaves can be eaten raw with mustard seed leaves as mustard and cress or salad rape and cress. Can be grown on wet blotting paper.

Creta praeparata Official British Pharmacopoeia name for prepared chalk, made by washing and drying naturally occurring calcium carbonate. The form in which calcium is added to flour.

Cretinism Underactivity of the thyroid gland in children, resulting in poor growth and severe mental retardation. Hypothyroidism in adults is myxoedema. Commonly the result of a dietary deficiency of iodine; may be congenital if the mother's iodine intake was severely inadequate during pregnancy.

Crimping 1 Slashing a large fish at intervals before cooking, to make it easier for the heat to penetrate the flesh. Trimming cucumber or similar foods in such a way that the slices appear to be 'deckled'. Decorating the double edge of a pie or tart or the edge of shortbread by pinching it at regular intervals with the fingers, giving a fluted effect. criolla The traditional cuisine of the Spanish-speaking Caribbean and Latin America; derived from French, Spanish, African, and American cookery.

Crispbread Name given to a flour and water wafer, originally Swedish and made from rye flour, but may be made from wheat flour. It has a much lower water content than bread and some brands are richer in protein because of added wheat gluten. Although popularly believed to be an aid in slimming, crispbreads provide more energy than the same weight of ordinary bread since they contain less water. Three pieces of crispbread (30 g) provide up to 3 g of dietary fibre and supply 100 kcal (420 kJ).

Cristal height A measure of leg length taken from the floor to the summit of the iliac crest. As a proportion of height it increases with age in children, and a reduced rate of increase indicates undernutrition.

Crohn's disease Chronic inflammatory disease of the bowel, of unknown origin, treated with antibiotics to prevent infection and with anti-inflammatory agents. Sufferers may be malnourished as a result of both loss of appetite due to illness and also malabsorption. Also known as regional enteritis, since only some regions of the gut are affected.

Croissant Flaky crescent-shaped rolls traditionally served hot for breakfast, made from a yeast dough with a high fat content.

Cropadeau Scottish; oatmeal dumpling with haddock liver in the middle.

Croque-monsieur Ham and cheese toasted sandwich. When topped with a fried egg it is called croque-madame.

Croquette Finely chopped meat, fish, or vegetables, mixed with a rich sauce or panada, shaped into balls or small cylinders, coated in egg and breadcrumbs and fried. The commonest filling is potato. croustade Small case of fried bread, pastry, or duchesse potato used to serve a savoury mixture.

Croûte, en Literally 'in a crust'; game, meat, or fish served in a pastry case or surrounded by slices of bread.

Croûtons Small diced or shaped pieces of bread, fried in fat, and commonly used to garnish soup and salads.

Crowdie Scottish; soft cheese made from buttermilk or soured milk curd, also a dish of buttermilk and oatmeal.

Cruciferae Family of plants with flowers with four equal petals; most vegetables in this family belong to the genus Brassica.

Crudites Raw vegetables, sliced or cut into small pieces or strips, served with a variety of dressings or dips, as an hors d'œuvre.

Cruller American; deep-fried bun made from baking powder dough. Similar to a doughnut.

Crumb Small particle of bread, cake, or biscuit, as broken off by rubbing. crumble Flour, fat, and sugar baked as a topping over fruit instead of pastry. For savoury crumble the sugar is omitted and cheese and seasoning are added.

Crumb softeners Derivatives of monoglycerides added to bread as emulsifiers to give a softer crumb and retard staling. There are several of these compounds which are combinations of different monoglycerides, E-430-436; also called polysorbates.

Crust The crisp outer part of a loaf of bread; also used for the sediment thrown off as wines mature.

Crustacea Zoological class of hard-shelled marine arthropods (shellfish) including crabs, crayfish, lobster, prawns, scampi, shrimps.

Cryogenic freezing Freezing with extremely cold freezants such as liquid nitrogen or solid carbon dioxide.

Cryoprotectant A substance permitted to be added to products of fish muscle (additive) before freezing to prevent reactions in the products during frozen storage that result in impaired functional properties of the muscle proteins. The cryoprotectants most frequently used in fish technology are sugars and polyalcohols, compounds having many hydroxy groups in the molecule. Examples are sucrose and sorbitol. About 8% of cryoprotectants, singly or in combination, are routinely added to surimi before making into blocks and freezing. Cryptocotyle lingua A digenetic trematode parasite infecting many species of fish. It is seen as numerous small black spots in the skin and in the sclera of the eye. The condition is referred to as 'black spot' disease, but it must not be confused with black spot in crustacea. The definitive host is a seabird, typically a gull, and intermediate hosts are marine molluscs, such as winkles, and fish. It can be transmitted to animals, including humans, which act as definitive hosts following ingestion of affected fish, but does not appear to cause illness.

Cryovac Trade name of thermoplastic resin wrapping film which can be heat-shrunk onto foods.

Cryptoxanthin Yellow carotenoid in a few foods such as yellow maize and the seeds of Physalis, the Cape gooseberry. A hydroxylated derivative of carotene which is a source of vitamin A in the body.

Crystal boiling Chinese method of cooking; food is heated in a pan of boiling water, then removed from the heat and cooking continued by the retained heat.

CSM Corn-soya-milk; a protein-rich baby food made in the USA from 68% pre-cooked maize (corn), 25% defatted soya flour, and 5% skim milk powder, with added vitamins and calcium carbonate.

CTC machine A device consisting of two contra-rotating toothed rollers that rotate at different speeds and provide a crushing, tearing, and curling action: used in breaking up leaves of tea to form small particles.

Cuaranta y tres Spanish; liqueur flavoured with herbs and spices; the name means 'forty-three'.

Cubeb Grey pepper native to South East Asia; pungent flavour akin to camphor.

Cubs Trade name for a breakfast cereal made from wheat.

Cucumber Fruit of Cucumis sativus, a member of the gourd family, eaten as a salad vegetable; it is 95% water.

Cucurbit A term used for vegetables of the family Cucurbitaceae, or gourds.

Cuisson Cooking juices from meat, poultry, or fish.

Culture Cultivate cells or living organisms in a prepared medium under laboratory conditions. "Culture" is both the process and the growing cells.

Cumberland sauce Sauce made from redcurrant jelly, orange, lemon, and port, served with ham, venison, and lamb. cumin Pungent herb, the crescent-shaped seed of Cuminum cyminum; used in curry powder and for flavouring cordials. Black cumin is the seed of Nigella sativa and sweet cumin is anise.

Cup cake Small individual cake or bun, typically baked in a paper case and topped with icing. Also known as fairy cake.

Cup North American and Australian measure for ingredients in cooking; the standard American cup contains 250 mL (8 fl oz).

Curacao A liqueur made from the rind of Seville oranges and brandy or gin; 30% alcohol, 30% sugar; 300 kcal (1260 kJ)/100 mL.

Curd Refers to the unsightly creamy material covering the surface of fish, especially canned salmon produced from previously frozen fish, and sometimes tuna. It consists of coagulated protein, probably denatured during poor cold storage prior to heat processing. Formation is reduced by pre-cooking, by sealing the cut surfaces of the fish with a brine or a tartaric acid dip to prevent the escape of fluid, or by treatment with proteolytic enzymes. See Canning. Curd also refers to the thick slime which is formed on the exterior of spoiled fish. Blood curd refers to the exudate produced during the processing of crabs.

Curd tension A measure of the toughness of the curd formed from milk by the digestive enzymes, and used as an index of the digestibility of the milk.

Curdle To form curds, or to go sour.

Curing Preserving fish by smoking, salting, drying, fermenting, acid curing, or any combination of these.

Currants Made by drying the small seedless black grapes grown in and around Greece and in Australia ; usually dried in bunches on the vine or after removal from the vine on supports. The name is derived from 'raisins of Corauntz' (Corinth).

Curry Name given by the British to an Indian dish of stewed meat or vegetables. It is served with a pungent sauce whose components and pungency vary.

Curry plant An aromatic herb, Murraya koenigii.

Curry powder A mixture of turmeric with several spices including cardamom, cinnamon, cloves, coriander, cumin, and fenugreek, made pungent with mustard, chilli, and pepper.

Cushion The cut nearest the udder in lamb or beef.

Custard apple The fruit of one of a number of species of tropical American trees of the family Anonaceae. Sour sop, Anona muricata, has white fibrous flesh and is less sweet than the others; the fruit may weigh up to 4 kg (8 lb). The sweet sop is also known as the 'true' custard apple, and is especially popular in the West Indies. The bullock's heart has buff-coloured flesh.

Cut surface Fleshy surface exposed by heading, heading and gutting, filleting, splitting or cutting the fish into steaks. The cut surface of a split fish is referred to as the face.

Cutlet Chop cut from the best end of neck of lamb, veal, or pork.

Cutting-in Combining the fat with other ingredients in a mixture by cutting with a knife.

Cyanogen (et) ic glycosides Organic compounds of cyanide found in a variety of plants; chemically cyanhydrins linked by glycoside linkage to one or more sugars. Toxic through liberation of the cyanide when the plants are cut or chewed.

Cyclitols Cyclic sugar alcohols such as inositol, quercitol, and tetritol.

Cymogran Trade name for a protein-rich food which is low in phenylalanine for feeding patients with phenylketonuria.

Cynar Trade name; liqueur made from artichoke hearts.

Cystathionine beta synthase The enzyme that converts homocysteine to cystathionine and then cysteine using vitamin B6 as cofactor.

Cystathioninuria A genetic disease affecting the metabolism of the amino acid methionine and its conversion to cysteine. May result in mental retardation if untreated. Treatment is by feeding a diet low in methionine and supplemented with cysteine, or, in some cases, by administration of high intakes of vitamin B 6.

Cysteine A non-essential amino acid, but nutritionally important since it spares the essential amino acid methionine. In addition to its role in protein synthesis, cysteine is important as the precursor of taurine, in formation of coenzyme A from the vitamin pantothenic acid and in formation of the tripeptide glutathione. It is used as a dough 'improver' in baking.

Cystic fibrosis Disease of mucous glands throughout the body that usually develops during childhood, and makes breathing increasingly difficult. If a child receives two copies of the defective gene called the CF gene - one copy from each parent - then the child will develop the disease.

Cysticercosis Infection by the larval stage of tapeworms caused by ingestion of their eggs in food and water contaminated by human faeces. Normally the larval form develops in the animal host, and human beings are infected with the adult form by eating undercooked infected meat.

Cystine The dimer of cysteine produced when its sulphydryl group is oxidized forming a disulphide bridge. Such disulphide bridges are especially important in maintaining the structure of proteins, and also in the role of the tripeptide glutathione as an antioxidant. Hair protein (keratin) is especially rich in cystine, which accounts for about 12% of its total amino acid content.

Cystinuria A genetic disease in which there is abnormally high excretion of the amino acids cysteine and cystine, resulting in the formation of kidney stones. Treatment is by feeding a diet low in the sulphur amino acids methionine, cysteine, and cystine.

D

Dairy products Foods and other products derived from the processing of milk. Separation of milk by centrifugation yields skimmed milk and cream. Churning the cream disrupts the fat globules, removes water, and produces butter, containing over 80% fat, and buttermilk. Cream is retailed in various stages of concentration, for example single cream and double cream; other less concentrated forms include evaporated milk and condensed milk. Yogurt is produced by inoculating whole milk with bacteria, principally of the genera Streptococcus and Lactobacillus. The acid they produce during incubation at about 43°C for four to five hours coagulates the milk, to which sweetening and flavouring may be added. Whey is a by-product of cheese manufacture and, as with skimmed milk, may be dried to a powder form for use in the food industry or fed to farm animals in the fresh liquid state.

Daltose Trade name for a carbohydrate preparation consisting of maltose, glucose and dextrin for infant feeding.

Damascene Original name for damson.

Damson Small dark purple plum; very acid and mainly used to make jam.

Dandelion The leaves of the weed Taraxacum officinale may be eaten as a salad or cooked. In France dandelion greens are known as pis- en-lit because of their diuretic action. A good source of vitamins C and A; a source of calcium and iron. The root can be cooked as a vegetable, or may be roasted and used as a substitute for coffee.

Danish pastry Rich, sweet, yeast pastry confection filled with fruit or nuts. Originally Viennese, not Danish; indeed in Danish it is known as Wienerbrod.

Dariole Small narrow mould with sloping sides used to make individual sweets and savouries; originally the name of a small cake.

Dark adaptation In the eye, the visual pigment rhodopsin is formed by reaction between vitamin A aldehyde and the protein opsin, and is bleached by exposure to light, stimulating a nerve impulse. At an early stage of vitamin A deficiency it takes considerably longer than normal to adapt to see in dim light after exposure to normal bright light, because of the limitation of the amount of rhodopsin that can be reformed. Measuring the time taken to adapt to dim light thus provides a sensitive index of early vitamin A deficiency. More severe vitamin A deficiency results in night blindness, and eventually complete blindness.

Darne Thick cut from the middle of a fish.

Dartois (d'Artois) Small light pastry, filled and flavoured served as an hors d'oeuvre or dessert.

Date Fruit of date palm, Phoenix dactylifera, known as far back as 3000 BC. There are three types: 'soft'; semi-dry; and dry. A good source of vitamin C.

Defect, Defective Terms with particular meaning in quality control, especially in the sampling and testing of products for compliance with specifications. A defect is a failure in an item to comply with one or more criteria in a specification. For example, a specification for boneless, skinless fillets might require the weight of the fillets to be within a stated weight range. If a particular item in a sample of a batch of fillets is outside of this range then there is a defect in that item. An item can have more than one defect. Using the fillets specification as an example, a fillet might fail on criteria for weight, presence of bones, blood spots and presence of skin. One or more defects in an item will render that item defective, that is, not complying with the specification. The decision rules of a sampling plan will define the number of defectives that may be allowed in a sample for the sampled batch to be declared unacceptable.

The specification, or the associated sampling plan, can weight the seriousness of defects. Some defects might be declared critical, and a defect of this sort will render the item defective no matter what other defects are present, or not. Presence of pathogenic bacteria, substances that present a risk to health, or non-compliance with regulatory controls, are examples of critical defects. A few bones, or small amounts of skin in the boneless, skinless fillet example could be declared minor defects, but more than this would be declared major defects. The specification or the sampling plan would specify the maximum number, if any, of each type of defect that may be allowed in a non-defective item.

Defrosting The term is used in a number of contexts. It refers to the removal of frost from a cooler to improve heat transfer, e.g. from a cooling coil. It is used to describe the heating of plates in a contact freezer to release frozen produce, ease reloading and improve heat transfer. It also refers to the thawing of frozen fish or fish products.

Degradability The ability of materials to be broken down.

Dehydration Removal of water from the tissues of an organism.

Demerit points The essential element of a system for rating the overall quality of an item. The attributes of the product that contribute to its quality are defined. The degree of change in each attribute as quality decreases is rated on a numeric scale starting at a low value, usually 0, for excellent quality of that attribute. The scores on these scales are demerit points and the sum of the demerit points allocated to a sample unit grades the quality of the item. The grading might be just a simple pass/fail. The demerit points system is analogous to the weighting of defects for seriousness, but allows for more levels of seriousness. The scales need not be of the same length and this allows for a weighting of the seriousness of deviations from excellent quality in the various attributes. Defects rated on longer scales will be weighted more in the summation of the demerit points than defects rated on shorter scales (2).

Demersal fish Those fish found living on or near the bottom of the sea, including cod, haddock, whiting, and halibut. They contain little oil (1-4%).

Demographic Vital and social statistics concerning the population including distribution, ethnic background, marriage rates, birth rates, etc.

Denaturation, protein A change in the structure of protein by heat, acid, alkali, or other agents which results in loss of solubility and coagulation. It is normally irreversible. Denatured proteins lose their biological activity, but not their nutritional value. Indeed, their digestibility is improved compared with the original structures, which are relatively resistant to enzymic hydrolysis.

Denature, denaturation 1. One meaning of denature in food science is to render a foodstuff unsuitable for consumption because it does not comply with food laws or regulations. An example in fish technology would be where a public health inspector or veterinarian had considered fish unfit for human consumption, or where fish had been withdrawn from the market as a consequence of the EC marketing regulations. The batch of condemned fish might then be sprayed with a dye to indicate it unsuitable or not to be used for human food, and to ensure it is not illegally returned to the market.

2. In protein chemistry, denaturation is the phenomenon of an alteration in the tertiary structure of a protein. Covalent bonds such as disulphide bonds might be broken during denaturation, but there is no breaking the polypeptide chain of the molecule. Denaturation almost invariably results in loss of biological activity of the protein, for example, enzymic activity, and in loss of solubility in water or dilute salt solutions. Susceptibility to denaturation differs among proteins, and can be brought about by a variety of treatments, chemical and physical. The processes most frequently encountered in fish technology are heating - cooking, pasteurising, retorting - and treatment with acids - marinading. The main muscle proteins in fish, the actomyosin complex, become more difficult to solubilise in dilute salt solution as storage proceeds, and this process is often referred to as 'denaturation' in descriptions of changes occurring during frozen storage of fish and fishery products. However, it is not clear that denaturation of the proteins has occurred; other mechanisms such as cross-linking of actin and myosin molecules, with themselves or other proteins, or with structural elements of the muscle fibre could lead to loss of solubility with unfolding of tertiary structure.

Denatured alcohol The addition of dyestuffs and unpleasant flavours to prevent the consumption of alcohol which is intended for domestic or industrial purposes; up to 15% methyl alcohol is added.

Dendritic salt A form of ordinary table salt, sodium, chloride, with the crystals branched or star-like instead of the normal cubes. This is claimed to have a number of advantages: lower bulk density, more rapid solution, and an unusually high capacity to absorb moisture before becoming wet.

Dental caries Popularly known as cavities, dental caries occur when bacteria in the mouth feed on fermentable carbohydrates and produce acids that dissolve tooth enamel. Various conditions affect this process, such as heredity and the composition and flow of saliva. Any fermentable carbohydrate can serve as food for cavity-causing bacteria. The amount of carbohydrate is not as important as how often these foods are eaten and how long they stay in the mouth. Widespread use of fluoride in water supplies and oral health products is credited with the dramatic decline in dental caries among children and adults alike over the past 20 years.

Deodorization The removal of an undesirable flavour or odour. Fats are deodorized during refining by bubbling superheated steam through the hot oil under vacuum, when most of the flavoured substances are distilled off.

Depectinization The removal of pectins from fruit juice to produce a clear, thin juice instead of a viscous, cloudy liquid, by the use of enzymes which hydrolyse pectins to smaller, soluble compounds.

Dermatitis A lesion or inflammation of the skin; many nutritional deficiency diseases include more or less specific skin lesions, but most cases of dermatitis are not associated with nutritional deficiency, and do not respond to nutritional supplements.

Desmosine The compound that forms the cross-linkage between chains of the connective tissue protein elastin.

Dessication See Freezer burn.

Detoxication The metabolism of toxic compounds to yield less toxic derivatives which are more soluble in water and can be excreted in the urine or bile. A wide variety of 'foreign compounds', sometimes referred to as xenobiotics, and some hormones and other normal body metabolites, are metabolized in the same way.

Deveining Refers to the removal of the vein from the tail of crustaceans to yield deveined tail meat. The vein runs down the back of the animal between the flesh and the shell to the vent at the posterior end.

Developmental apraxia Apraxia in children.

Devilled Food grilled or fried after coating with condiments or breadcrumbs.

Dewatering press Minced fish recovered from a bone separator often contains an undesirable amount of free water, probably from the water used to lubricate the drum of the bone separator. This can be removed in a dewatering press in which the minced fish is forced by an endless screw along a dewatering drum, with fine perforations, such that the pressure of the screw in the drum may be varied to control the amount of water removed.

Dewberry A hybrid fruit, a large variety of blackberry; rather than climbing, the plant trails on the ground.

Dexedrine Anorectic drug used in the treatment of obesity.

Dexfenfluramine Anorectic drug used in the treatment of obesity.

Dextran A polysaccharide composed of linked fructose units, produced by the action of Betacoccus arabinosus on sugar. It can cause problems in sugar factories, but is clinically useful as a plasma extender for transfusion.

Dextrin An intermediate polysaccharide compound resulting from the hydrolysis of starch to maltose by amylase enzymes.

Dextrose Alternative name for glucose. Commercially the term 'glucose' is often used to mean corn syrup and pure glucose is called dextrose.

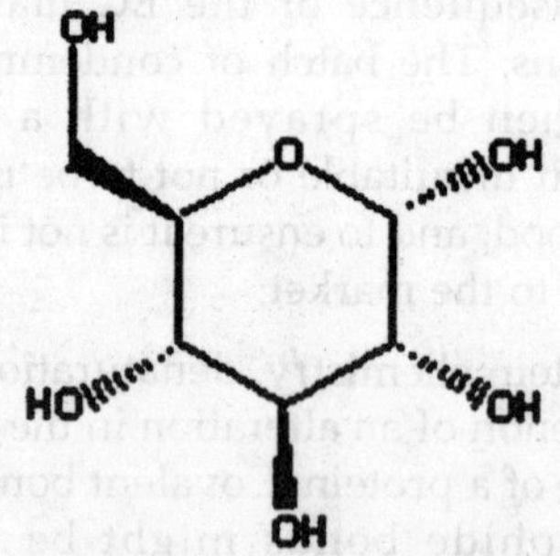

Fig. Dextrose

Dextrose equivalent value A term used to indicate the degree of hydrolysis of starch into glucose syrup. It is the percentage of the total solids that have been converted to reducing sugars: the higher the DE, the more sugars and less dextrins are present.

Liquid glucoses are commercially available ranging from 2 DE to 65 DE. A complete acid hydrolysis converts all the starch into glucose but produces bitter degradation products. Glucose syrups above 55 DE are termed 'high conversion'; of 35-55 DE, regular conversion; below 20 DE the products of hydrolysis are maltins or maltodextrins

Diabetes Diabetes is the name for a group of medical disorders characterized by high blood sugar levels. Normally when people eat, food is digested and much of it is converted to glucose – a simple sugar – which the body uses for energy. The blood carries the glucose to cells where it is absorbed with the help of the hormone insulin. For those with diabetes, however, the body does not make enough insulin, or cannot properly use the insulin it does make. Without insulin, glucose accumulates in the blood rather than moving into the cells. High blood sugar levels result.

Diacetyl The main flavour and aroma agent in butter, formed during the ripening stage by the organism Streptococcus lactis cremoris. Synthetic diacetyl is added to margarine as 'butter flavour'. Chemically it is CH_3.CO.CO.CH_3.

Diagnosis Hemorrhagic colitis is diagnosed by isolation of E. coli of serotype O157:H7 or other verotoxin-producing E. coli from diarrhoeal stools. Alternatively, the stools can be tested directly for the presence of verotoxin. Confirmation can be obtained by isolation of E. coli of the same serotype from the incriminated food.

Diarrhoea Frequent passage of loose watery stools, commonly the result of intestinal infection; rarely as a result of adverse reaction to foods or disaccharide intolerance. Severe diarrhoea in children can lead to dehydration and death; it is treated by feeding a solution of salt and sugar to replace fluid and electrolyte losses.

Osmotic diarrhoea is diarrhoea associated with retention of water in the bowel as a result of an accumulation of non-absorbable watersoluble compounds; especially associated with excessive intake of sorbitol and mannitol. Also occurs in disaccharide intolerance.

Diarrhoea Frequent passage of loose watery stools, commonly the result of intestinal infection; rarely as a result of adverse reaction to foods or disaccharide intolerance. Severe diarrhoea in children can lead to dehydration and death; it is treated by feeding a solution of salt and sugar to replace fluid and electrolyte losses.

Osmotic diarrhoea is diarrhoea associated with retention of water in the bowel as a result of an accumulation of non-absorbable water-soluble compounds; especially associated with excessive intake of sorbitol and mannitol. Also occurs in disaccharide intolerance.

Diastatic activity of flour A measure of the ability of flour to produce maltose from its own starch by the action of its own maltase. This sugar is needed for the growth of the yeast during fermentation.

Dice To cut into small cude-shaped pieces.

Dieppoise, a la Sea fish garnished with crayfish tails and mussels, served with a white wine sauce.

Diet The types and amounts of food an individual typically consumes.

Dietary fibre Material mostly derived from plant cell walls which is not digested by human digestive enzymes but is partially broken down by intestinal bacteria to volatile fatty acids that can be used as a source of energy. A large proportion consists of non-starch polysaccharides (NSP); these include soluble fibre that reduces levels of blood cholesterol and increases the viscosity of the intestinal contents: and insoluble fibre that acts as a laxative. Earlier known as roughage or bulk.

Dietary reference values A UK set of standards of the amounts of each nutrient needed to maintain good health. People differ in the daily amounts of nutrients they need; for most nutrients the measured average need plus 20% takes care of the needs of nearly everyone and in the UK this is termed Reference Nutrient Intake, elsewhere known as Recommended Daily

Allowances or Intakes (RDA or RDI), or Population Reference Intake (PRI). This figure is used to calculate the needs of large groups of people in institutional or community planning. Obviously some people require less than the average. This lower level is termed the Lower Reference Nutrient Intake, LRNI. This is an intake at or below which it is unlikely that normal health any nutrient at or below LRNI then detailed investigation of his/her nutritional status would be recommended.

For energy intake only a single Dietary Reference Value is used, the average, because there is potential harm from ingesting too much.

Dietetic foods Foods prepared to meet the particular nutritional needs of people whose assimilation and metabolism of foods are modified, or for whom a particular effect is obtained by a controlled intake of foods or individual nutrients. They may be formulated for people suffering from physiological disorders or for healthy people with additional needs.

Dietetics The study or prescription of diets under special circumstances and for special physiological needs such as pregnancy, growth, weight reduction.

Diethylpropion Anorectic drug used in the treatment of obesity.

Diet-induced thermogenesis The increase in heat production by the body after eating. It is due to both the metabolic energy cost of digestion and the energy cost of forming tissue reserves of fat, glycogen, and protein. It can be up to 10-15% of the energy intake. Also known as the specific dynamic action (SDA) or thermic effect of foods and luxus konsumption.

Dietitian, dietician According to the US Department of Labor, Dictionary of Occupational Titles, one who applies the principles of nutrition to the feeding of individuals and groups; plans menus and special diets; supervises the preparation and serving of meals; instructs in the principles of nutrition as applied to selection of foods. In the UK the training and state registration of dietitians is controlled by law.

Digestibility The proportion of a foodstuff absorbed from the digestive tract into the bloodstream, normally 90-95%. It is measured as the difference between intake and faecal output, with allowance being made for that part of the faeces that is not derived from undigested food residues. Digestibility measured in this way is referred to as 'true digestibility', as distinct from the approximate measure, 'apparent digestibility', which is simply the difference between intake and output.

Digestif French term for a liqueur or spirit drunk after dinner, supposedly to aid digestion.

Digestion The breakdown of a complex compound into its constituent parts, achieved either chemically or enzymically. Most frequently refers to the digestion of food, which means breakdown by the digestive enzymes of proteins to amino acids, starch to glucose, fats to glycerol and fatty acids. These breakdown products are then absorbed into the bloodstream.

Digestive juices The secretions of the gastro-intestinal tract which are involved in the digestion of foods: bile, gastric secretion, intestinal juice, pancreatic juice, saliva.

Dill The aromatic herb Anethum graveolens. The dried ripe seeds are used in pickles, sauces, etc. The young leaves are also used, fresh, dried, or frozen (dill weed) to flavour fish and other dishes. Dill pepper is a mixture of dill seed, dill weed, and ground black pepper, used as a condiment, especially with salmon.

Dimethylamine An amine of formula $(CH_3)_2NH$. Formed in the muscle tissue of some species of fish by the action of enzymes on trimethylamine oxide.

Dimethylsulphide A compound with the formula $(CH_3)S$. In common with other sulphides, it has a foul smell. It is formed in spoiling fish by the action of bacteria on sulphur-containing amino acids and very likely contributes to the unpleasant smell of spoiled fish. DMS is present in small amounts in the marine environment, derived from the decomposition of a precursor, dimethyl-a-propiothetin, present in some species of marine algae. Some species of planktonic invertebrates graze on planktonic algae containing dimethyl-a-propiothetin, and in turn the invertebrates are eaten by fish. The dimethyl-a-propiothetin is degraded to DMA by the fish which then has the foul smell of the sulphide. This odour, which can render the fish commercially unacceptable if it is strong enough, is variously described as 'blackberry' or 'weedy'.

Dim-sum Steamed dumplings and other delicacies.

Dioxins A shortened form of the name for polychlorinated dibenzo-p-dioxins, a group of compounds with several, typically 4 to 8, chlorine atoms in the molecule. Typically used as a collective term to include the furans. Small amounts are formed from natural processes such as forest fires, but by far the major contributor of dioxins in the environment is human activities. They are not manufactured for any commercial purpose, but are present as minor impurities in preparations of chlorinated chemicals, and are formed in the combustion of wastes containing chlorinated products. They are very persistent in the environment, and atmospheric transport distributes them around the world so that they are present even in sites remote from industrial activities. They can enter the aqueous environment directly by discharges from terrestrial sources, and concentrations can be relatively high in some bodies of waters affected by industrial discharges, but deposition from the atmosphere is an important additional source so that dioxins are present throughout the aqueous environment. They are not taken up directly from the water by fish, but fish absorb them from contaminated feed. They are powerful chronic toxins producing a variety of effects. Analysis of dioxins in foods, including fishery products show the presence of several varying in number of chorine atoms in the molecule. The various dioxins are not equally toxic and when expressing summary dioxin concentrations it is usual to weight the concentration of each dioxin for its toxicity relative to the 2,3,7,8-tetrachloro isomer which is considered the most toxic. The WHO has issued toxic equivalency factors for dioxins and summarised dioxin concentrations are often expressed as Toxic Equivalents (TEQ) concentrations based on these factors. The WHO has established Tolerable Daily Intakes (TDIs) of dioxins at 1-4 TEQ picograms/kg of body weight. Data on dioxin concentrations in fish, and the contribution of fish to dietary intakes of dioxins are summarised in chapters in Moffat & Whittle (1999).

Dipeptide A peptide consisting of two amino acids.

Diphenyl Also known as biphenyl (E-230), one of two compounds used for the treatment of fruit after harvesting to prevent the growth of mould. For making marmalade, citrus fruits that have not been treated with diphenyl are available.

Diphosphopyridine nucleotide Obsolete name for nicotinamide adenine dinucleotide, NAD.

Diphyllobothrium The definitive hosts of Diphyllobothrium are warm-blooded animals including man, and intermediate hosts are firstly aquatic crustacea then fish. Human are infected if they eat raw or lightly processed fish containing the intermediate form of the parasite.

Dipsa Foods that cause thirst.

Dipsesis Extreme thirst, a craving for abnormal kinds of drinks.

Dipsogen A thirst-provoking agent.

Dipsomania A morbid craving for alcoholic drinks.

Disaccharidases Enzymes that hydrolyse disaccharides to their constituent monosaccharides in the intestinal mucosa: sucrase acts on sucrose, lactase on lactose, and maltase on maltose.

Disaccharide intolerance Impaired ability to digest lactose, maltose, or sucrose, due to lack of lactase, maltase, or sucrase in the small intestinal mucosa. The undigested sugars remain in the intestinal contents, and are fermented by bacteria in the large intestine, resulting in painful, explosive, watery diarrhoea. Treatment is by omitting the offending sugar from the diet.

Lack of all three enzymes is generally caused by intestinal infections, and the enzymes gradually recover after the infection has been cured. Lack of just one of the enzymes, and hence intolerance of just one of the disaccharides, is normally an inherited condition.

Lactose intolerance due to loss of lactase is normal in most ethnic groups after puberty; it is only among people of northern ethnic origin that lactose persists into adult life.

Disc mill One or more revolving circular plates between which substances are ground; the discs are separated by projecting teeth or pins. Used to grind grain, fruit, sugar, chocolate, pastes, etc.

Discoloration An abnormal coloration of a product affecting the quality of the product and usually considered a defect.

Discrimination test A type of testing procedure used in the sensory evaluation of foods in which two or more samples are compared and some judgement made on the difference between them. Difference test is the preferred term.

Disease symptoms Hemorrhagic colitis is the name of the acute disease caused by E. coli O157:H7.

The illness is characterized by severe cramping and diarrhoea which is initially watery but becomes grossly bloody. Occasionally vomiting occurs. Fever is either low-grade or absent. The illness is usually self-limited and lasts for an average of 8 days. Some individuals exhibit watery diarrhoea only.

Infective dose — Unknown, but from a compilation of outbreak data, including the organism's ability to be passed person-to-person in the day-care setting and nursing homes, the dose may be similar to that of Shigella spp.

Disinfectant, disinfection A disinfectant is a chemical that kills microorganisms; disinfection is the process of using disinfectants to kill microorganisms. Corresponding equivalent terms often used in the context of food premises are sanitizer and sanitation. Many chemicals act as disinfectants, but only a selected range are suitable for use in food premises. Chlorine in various forms is frequently used as a disinfectant in fish processing plants. It is available as a gas, in a solution as sodium hypochlorite, and in a solid form as calcium hypochlorite. Iodophors are iodine-based disinfectants in solution with surfactants, phosphoric acid, and possibly other compounds. They are often used in hand washes. Quaternary ammonium salts (QUATS) are not as effective as chlorine and iodine, but have advantages of being detergents and of being less corrosive of metal structures. Disinfectants based on phenol or compounds of phenol should not be used in, or in proximity to, fish processing as the they have very strong odours and can taint products. (Shapton & Shapton, 1994).

Disjunction Separation of the members of a chromosome pair during cell division.

Disodium hydrogen phosphate A chemical of formula Na_2HPO_4, also called disodium hydrogen orthophosphate and sodium phosphate, dibasic. A permitted food additive used especially to modulate pH as it has an alkaline reaction in water.

Disordered eating Patterns of food consumption temporarily adopted to improve body image or physical performance. Disordered eating relates to misinformation and food fads, which does not allow an individual to meet basic nutritional needs.

Diuresis Increased formation and excretion of urine; it occurs in diseases such as diabetes, and also in response to diuretics.

Diuretic Any agent or compound that increases the flow of urine from the body. They can range from herbal teas to powerful drugs that flush out electrolytes and water. They are classed based on the location and mechanism of action in the kidneys. Athletes use diuretics to eliminate water weight to further emphasize their muscular definitions. Most bodybuilding and fitness federations have banned the use of diurectics and test for them.

Diverticular disease Diverticulosis is the presence of pouch-like hernias through the muscle layer of the colon, associated with a low intake of dietary fibre and high intestinal pressure due to straining during defecation. Faecal matter can be trapped in these diverticula, causing them to become inflamed, causing pain and diarrhoea, the condition of diverticulitis.

Djenkolic acid A sulphur-containing amino acid found in the djenkol bean, Pithecolobium lobatum, which grows in parts of Sumatra. It is a derivative of cysteine, and is metabolized but, being relatively insoluble, any djenkolic acid that escapes metabolism can crystallize in the kidney tubules and cause damage.

DMAE Supplement reported to minimize buildup of lipofuscin (age spots) in the brain. Plays a participatory role in acetycholine synthesis. DMAE has been shown to stimulate vivid, lucid dreams, suggesting possible sleep pattern enhancement.

DNA Deoxyribonucleic Acid. The genetic material of living organisms; the substance of heredity. It is a large, double-stranded, helical molecule that contains genetic instructions for growth, development, and replication. The rungs of this double helix are made of base pairs.

DNA DNA is present in all cellular organisms and in viruses. It contains the genetic information of the organism. DNA, the abbreviation for Deoxyribonucleic Acid, consists of two antiparallel chains consisting of sugar, a phosphate-unit and 4 different bases. The two chains are held together by hydrogen bonds between base pairs, while the sugar-phosphate backbone is on the outside. The sugar-phosphate backbone stays the same within a strand, whereas the bases can differ between Adenine, Thymine, Guanine and Cytosine. The order of the bases in the double-chain determines the genetic code, that means the order and the distribution of the bases is responsible for all heritable traits of an organism. It is a very precise code, and a mistake in even a single letter of the DNA code can have disastrous consequences.

A unit of the double-helical chain, responsible for a certain trait, is called a gene.

DNA fingerprinting Detecting patterns in DNA that indicate the presence of a gene for a trait. The pattern resembles a bar code printed on a commercial product so computers can scan the price. Forensics experts can use this distinct pattern to link or clear an Individual suspected of being involved in a crime, like they compare fingerprints. Breeders can use these patterns to find and select breeding stock with traits such as disease resistance.

DNP The first weight loss drug ever offered to the public, around 1933. DNP is a yellow crystalline sold that's slightly soluable in water. After being injected, it increases the metabolism an average of 30% over baseline in less than a minute, returning back to normal anywhere from 6 to 48 hours. However, DNP has way too many side effects, including blinding people, and if you take too much of it, it can cook you to death from the inside. You can lose weight with it, but it is highly dangerous. Street Names for DNP include Hexalon. DNP is now classified as a poison.

Dockage Name given to foreign material in wheat which can be removed readily by a simple cleaning procedure.

Docosahexaenoic acid A long-chain polyunsaturated fatty acid.

Docosanoids Long-chain polyunsaturated fatty acids with 22 carbon atoms.

Dogfish A cartilaginous fish, Scilliorinus caniculum, or Squalis acanthias, related to the sharks; sometimes called rock salmon or rock eel.

Doh peeazah Indian dish; a variant of korma. The name means 'double onion'.

Dolce Sweet wines.

Dolma Greek, Turkish; stuffed vegetables, especially vine leaves stuffed with rice and minced meat.

Dolomite Calcium magnesium carbonate.

Do-maker process For continuous bread making. Ingredients are automatically fed into a continuous dough mixer, the yeast suspension being added in a very active state.

DOMS The pain and soreness you feel a few days after a heavy workout.

Döner kebab Middle-Eastern, Greek, and Turkish. Slices of lamb, highly flavoured with herbs and spices, wound around a revolving spit, cooked in front of a vertical charcoal fire.

Dopamine A hormone-like substance that acts as a neurotransmitter, transmitting nerve signals from neuron to neuron.

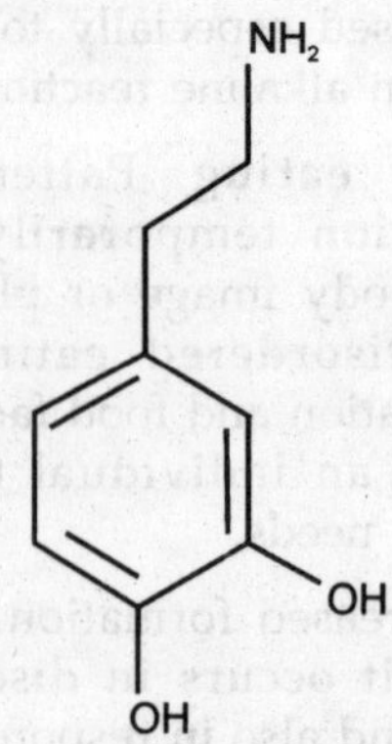

Fig. Dopamine

Double-blind, placebo-controlled study In a double-blind, placebo controlled study, neither the researchers nor the participants in the study are aware of which subjects receive the treatment under study and which subjects receive the placebo until after the study is completed. The study design is intended to remove bias on the part of both researcher and study subject.

Dough cakes A general term to include crumpets, muffins, and pikelets, all made from flour, water and milk. The batter is raised with yeast and baked on a hot plate or griddle. Crumpets have sodium bicarbonate added to the batter; muffins are thick and well aerated, less tough than crumpets; pikelets are made from crumpet batter that has been thinned down.

Dough Mixture of flour and liquid used to make bread and pastry. May contain yeast or baking powder as leavening agent.

Doughnut Cake made from fried, sweetened dough leavened with yeast or baking powder; may be filled with jam or cream.

Douglas bag An inflatable bag for collecting expired air to measure the consumption of oxygen and production of carbon dioxide, for the measurement of energy expenditure by indirect calorimetry.

Down syndrome The most frequent genetic cause of mild to moderate mental retardation and associated medical problems and occurs in one out of 800 live births, in all races and economic groups. Down syndrome is a chromosomal disorder caused by an error in cell division, when a fertilised ovum, which will develop into the foetus, contains extra material from chromosome number 21. Three genetic variations can cause Down syndrome: · In most cases, approximately 92% of the time, Down syndrome is caused by the presence of an extra chromosome 21 in all cells of the individual. In such cases, the extra chromosome originates in the development of either the egg or the sperm.

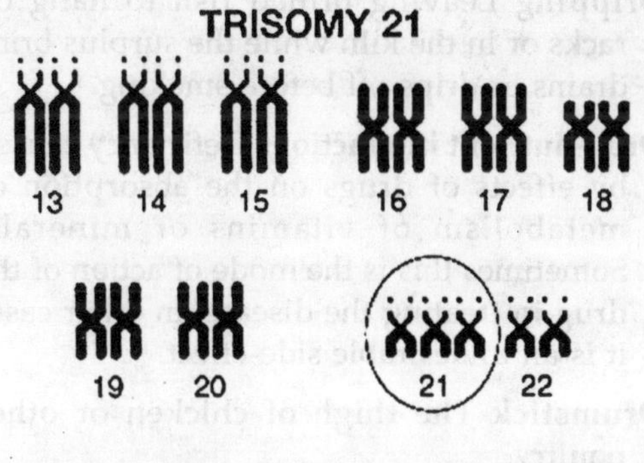

This condition, in which three copies of chromosome 21 are present in all cells of the individual, is called trisomy 21. · In approximately 2-4% of cases, Down syndrome is due to mosaic trisomy 21. This situation is similar to simple trisomy 21, but, in this instance, the extra chromosome 21 is present in some, but not all, cells of the individual. This may be due to an error in chromosome division early in embryonic development. In this situation, the range of the physical problems may vary, depending on the proportion of cells that carry the additional chromosome 21. Approximately 3-4% of individuals with Down syndrome have cells containing 46 chromosomes, but still have the features associated with Down syndrome. In such cases, material from one chromosome 21 gets stuck or translocated onto another chromosome, either prior to or at conception. In such situations, the individual with Down syndrome is said to have translocation trisomy 21. For parents of a child with Down syndrome due to translocation trisomy 21, there may be an increased likelihood of Down syndrome in future pregnancies. This is because one of the two parents may be a balanced carrier of the translocation.

DPN Obsolete name for nicotinamide adenine dinucleotide, NAD.

Dragee whole nuts with hard sugar or sugared chocolate coating. Silver dragees are coated with silver leaf.

Drained weight The weight of the contents of a container of fish after the liquid has been drained off. Also called the fill. Food labelling regulations in most countries, and a requirement in Codex Alimentarius standards for labelling of canned fish products, is that the drained weight of the pack be declared.

Drained weight is determined by emptying the contents of the can onto a sieve and allowing the liquids to drain off. The remaining material is weighed. The specifications or regulations will usually specify the conditions, for example, size of mesh of the sieve, draining time, of test. Though originally used for canned products, the principle of drained weight can be applied to frozen products, especially to glazed frozen products. Here, the drained weight is the weight of product remaining after it has been allowed to thaw and drain under specified conditions. The following is the procedure for determining drained weight included in the Codex Alimentarius Standard for canned salmon, codexstan 3-1991, Rev. 1-1995 '7.4.

Determination of drained weight for products packed with edible oils other than salmon oil The drained weight of all sample units shall be determined by the following procedure

(i) Maintain the container at a temperature between 20ºC and 30ºC for a minimum of 12 hours prior to examination.

(ii) Open and tilt the container to distribute the contents on a pre-weighed circular sieve which consists of wire mesh with square openings of 2.8 mm x 2.8 mm.

(iii) Incline the sieve at an angle of approximately 17-20°.... and allow the fish to drain for two minutes, measured from the time the product is poured into the sieve.

(iv) Weigh the sieve containing the drained fish.

(v) The weight of drained fish is obtained by subtracting the weight of the sieve from the weight of the sieve and drained product.'

Drambuie Scottish liqueur based on malt whisky, sweetened with heather honey, and flavoured with herbs.

Dredging Sprinkling food with flour, sugar, etc. Fish and meat are often dredged with flour before frying, while cakes and biscuits are dredged with fine sugar as a decoration.

Dressing food To prepare food for cooking or serving in such a way that it looks as attractive as possible. Sometimes denotes a special method of preparation, as in dressed crab.

DRI Dietary Reference Intakes. An umbrella term for groups of values that specify recommended dosages. Proposed by the Food and Nutrition Board of the Institute of Medicine of the National Academy of Sciences in partnership with Health Canada.

Dried fruits Dried currants, dates, figs, prunes, raisins, and sultanas all have similar analyses.

Dried shrimp In many parts of the world, this product is prepared as a condiment, especially for soups, from shrimp regarded as too small for the market. Typically the shrimp are boiled in sea water, and sun-dried or smoked and dried.

Dried, milk Milk that has been evaporated to dryness, usually by spray- or roller-drying. May be whole milk, three-quarter cream, half-cream, quarter-cream, or skim milk.

Drip, drip loss Liquid that exudes from fish flesh, especially when frozen fish is thawed, when it is also called thaw drip. The drip loss is the amount of liquid that is lost and is usually expressed as a percentage of the original weight.

Dripping Leaving brined fish to hang on racks or in the kiln while the surplus brine drains or drips off before smoking.

Drug-nutrient interactions Deficiency caused by effects of drugs on the absorption or metabolism of vitamins or minerals. Sometimes this is the mode of action of the drug in treating the disease; in other cases it is an undesirable side-effect.

Drumstick The thigh of chicken or other poultry.

Drunken foods meat or fish is highly seasoned and marinated, then steamed or lightly simmered. After draining it is steeped in wine for several days before serving.

Drupe Botanical term for a fleshy fruit with a single stone enclosing the seed that does not split along defined lines to liberate the seed, e.g. apricot, cherry, date, mango, olive, peach, plum.

Dry frying Frying without the use of fat by using an anti-stick agent of silicone or a vegetable extract.

Dry ice Solid carbon dioxide, which has a temperature of -79°C; used to refrigerate foodstuffs in transit and for carbonation of liquids. It sublimes from the solid to a gas without liquefying.

Dry salting Embedding whole fish, gutted fish, gibbed fish, split fish, fillets or sides in solid salt for salt curing, or as an alternative to brining before smoking. See Salting.

Dry-blanch-dry process A method of drying fruit so as to retain the colour and flavour; it is faster than drying in the sun and preserves flavour and colour better than hot air drying. The material is dried to 50% water at about 82°C, blanched for a few minutes, then dried at 68°C over a period of 6-24 h to 15-20% water content.

Drying Removal of moisture by evaporation from fish, originally as a means of curing for preservation. Natural drying in the sun and wind, or drying over wood fires, were the principal traditional means of drying and are still used today in some parts of the world. Drying may take several or more weeks to complete depending on climate and on the size and thickness of fish. Insect infestation is a problem in all climates differing only in species and degree of damage or loss. Removal of moisture reduces the water activity (aw) of the product. The aim for some dried fish products is to reduce the water content to about 15%, corresponding to a w 0.6 to prevent the growth of bacteria and moulds, such that the product remains edible for several years.

In addition to production of dried fish, drying is used in combination with salting and smoking, and in the production of fish meal. It has been an important and integral part of the smoking process to give the required texture to the product, and still is for some products. But, in modern smoking processes drying may only play a small role because a succulent product similar to raw fish may be required. Fish will lose water by drying only if the water vapour pressure of the surrounding air is less than the vapour pressure of the moisture in the fish. The relative humidity of the air indicates the degree of saturation of air with water vapour. The fish cools as water is evaporated, termed evaporative cooling. Heat is supplied from the surrounding air, or from a heated surface, to balance the heat loss by evaporative cooling and, if necessary, increase the temperature of the fish to increase the rate of evaporation. Drying occurs in two stages. During the first stage, whilst the surface of the fish is wet, the drying rate depends on the airspeed, and the moisture content and temperature of the surrounding air. If these conditions remain constant, the drying rate remains constant and so this stage is called constant rate drying. The second stage begins when most of the moisture on the surface of the fish has evaporated because water can then only evaporate as quickly as it reaches the surface from within by diffusion. When the movement of water through the fish can no longer match the evaporation rate from the surface, the surface becomes dry. This transition point in the drying sequence is termed the critical moisture (water) content. As the amount of water within the fish diminishes, drying becomes progressively slower because the water from deep within the fish takes longer to diffuse to the surface, and so this stage is known as falling rate drying.

Three types of process can be used to dry fish to provide faster drying rates than natural drying. However, too rapid drying must be avoided to prevent occurrence of case hardening. In air drying or contact drying, heat is transferred to the fish from heated air or from a heated surface and increased air movement above the fish is used to carry the moisture away. Suitable conditions for air drying in the dryer are air speed over the fish, 60 to 90 m/minute;

air temperature about 24°C, but within the range 15 to 27°C; relative humidity of the air in the dryer, 50 to 55%. In vacuum drying, the evaporation rate of water from the surface of the fish is increased as the atmospheric pressure is reduced. Heat is supplied to evaporate the water by conduction by contact with a heated surface, or by radiation, and the water vapour is removed by means of a vacuum pump. In freeze drying, highly efficient vacuum pumps attain very low pressures in a sealed chamber that contains the fish in contact with refrigerated plates to freeze the fish. Ice sublimes to water vapour at pressures below 0.64 kPa and the water vapour is removed by the vacuum pump. See Stockfish, Salting, Smoking, and Freezer burn.

Drying oil Any highly unsaturated oil that absorbs oxygen and, when in thin films, polymerizes to form a skin. Linseed and tung oil are examples of drying oils used in paints and in the manufacture of linoleum, etc. Nutritionally they are similar to edible oils, but when polymerized, are toxic.

Du bois formula A formula for calculating body surface area.

Dubarry, a la A rich cauliflower soup or a cauliflower garnish.

Duchesse potatoes Sieved boiled potato mixed with cream and egg, glazed with beaten egg, and baked until browned.

Duck Wild duck or wildfowl; mallard (Anas platyrhynchos), teal, small dabbling ducks of genus Anas.

Duglere Method of cooking white fish in white wine and water, adding cream and a veloute sauce.

Dulce Spanish; sweet wines.

Dulcin A synthetic material which is 250 times as sweet as sugar but is not permitted in foods. Also called sucrol and valzin.

Dulcitol A six-carbon sugar alcohol which occurs in some plants and is formed by the reduction of galactose; also known as melampyrin, dulcite, or galacticol.

Dulse Edible purplish-brown seaweeds, Rhodymenia palmata and Dilsea carnosa, used in soups and jellies.

Dumpling A ball of dough, usually boiled, but may be baked. Generally served with soups and stews.

Dun The brownish discoloration of dried salted fish caused by caused by moulds, especially Sporendonema spp.

Dunaliella bardawil A red marine alga discovered in 1980 in Israel, which is extremely rich in U+0B2-carotene, containing 100 times more than most other natural sources.

Dundee cake Rich fruit cake decorated with split almonds.

Dunst Very fine semolina approaching the fineness of flour. Also called break middlings.

Duodenum First part of the small intestine, between the stomach and the jejunum; the major site of digestion. Pancreatic juice and bile are secreted into the duodenum. So called because it is about twelve fingerbreadths in length.

Durian Fruit of tree Durio zibethinum, grown in Malaysia and Indonesia. Each fruit weighs 2-3 kg and has a soft, cream-coloured pulp, with a smell considered disgusting by the uninitiated.

Durum wheat A hard type of wheat of the species Triticum durum; largely used for the production of semolina intended for the preparation of pasta.

Dutch oven A semicircular metal shield which may be placed close to an open fire; fitted with shelves on which food is roasted. It may also be clamped to the fire bars.

Duxelles Mince of mushrooms with chopped shallot and herbs, used to flavour stuffing, soups, and sauces.

DV Daily Value. New dietary supplement regulations went into effect in March of 1999 that require labeling of dietary reference intakes (DRI) and which appear as "DV" on product labels.

Dye Permitted colours satisfying the criteria for purity for food use, that can be used only as specified in the food additives legislation of different countries. In the case of some smoked fish, the permitted colour, mainly yellow, orange or brown, is usually applied as a component of the brine used for brining to intensify the colour obtained during smoking. For example, in European Community legislation, annatto (anatto) is a yellow vegetable dye sometimes used to colour fish before smoking, such that the maximum level in the smoked fish is 10 mg/kg; it is identified as E 160b, and is also known as bixin, bixine, and norbixin. Brown FK is sometimes used in the brine to colour herring for kippering, such that the maximum level in the kipper is 20 mg/kg; FK is simply an abbreviation of 'for kippers'. It is identified as E 154 and also known as kipper dye. Amaranth, a red colour identified as E 123, is permitted in fish roe up to 30 mg/kg. Other specific colours are also permitted in EC legislation for inclusion in other fish products such as pastes, pre-cooked crustacean meats, salmon substitutes, surimi, fish roe, soups and smoked fish.

Dyox Trade name for chlorine dioxide used to treat flour.

Dyspepsia Any pain or discomfort associated with eating. Dyspepsia may be a symptom of gastritis, peptic ulcer, gall-bladder disease, etc., or, if there is no structural change in the intestinal tract, it is called 'functional dyspepsia'. Treatment includes a bland diet.

Dysphagia Difficulty in swallowing, commonly associated with disorders of the oesophagus. Inability to swallow is aphagia.

E

E. coli Escherichia coli, a group of bacteria including both harmless ones that inhabit human intestines and some types that can cause food poisoning.

Earthy/muddy odours/flavours Taints sometimes present in freshwater and brackish water fish and crustacean shellfish in both natural and farmed environments. The causative chemicals have been shown to be geosmin and 2-methylisoborneol released by microorganisms in the water. The chemicals are taken up by the fish through their gills to induce the taint in the flesh. The two chemicals have extremely low detection thresholds, in the order of 1 nanog/l. The character of the odours of the chemicals are similar and give rise to the same descriptions of earthy or muddy, though experienced assessors can differentiate between them. The presence and intensity of the taint varies seasonally and is associated with blooms of cyanobacteria in the phytoplankton in the warmer months. In aquaculture, pond conditions affect both the intensity and the composition of the phytoplankton flora; nutrient levels influence the growth of the phytoplankton, and factors like hardness and pH affect the composition. Geosmin and 2-methyliso-borneol are produced by different species of organisms. For example Anabaena species produce geosmin, Oscillatoria species 2-methylisoborneol, and generally a pond will support one genus or the other. The responsible cyanobacteria do not grow in seawater - they are inhibited above about 2% salt content - and marine fish are not susceptible to the taint. The taint is taken up very rapidly when the chemicals are formed, but takes several days to depurate when affected fish are transferred to uncontaminated water.

Easter soup Greek soup traditionally prepared to celebrate the end of the Lenten fast, made from the pluck and feet of a lamb.

Eating disorders Extreme expressions of weight and food issues experienced by both men and women. They include anorexia nervosa, bulimia nervosa, and compulsive overeating. All are behavioral disorders resulting from serious emotional problems.

Eau de vie Spirit distilled from fermented grape juice; may be flavoured with fruits, etc.

EC grading A system of grading batches of fish, including crustacean shellfish and cephalopods, at port markets for freshness and size. The grades are specified in European Community Council Regulation No. 2406/96 and apply to almost all of the fish landed in the European Community.

The fish must first be sorted into batches of similar size and freshness, and grades are allocated to the batches by inspectors. Four freshness grades of vertebrate fish are defined - Extra, A, B, C, and unfit; three, Extra, A, and B for Norway lobsters and cephalopods; and two, Extra and A for shrimps. Only fish in grades Extra, A, B, and C may be marketed for human consumption. Two to five size grades are defined depending on the species of fish. Size grades for vertebrate fish and for Norway lobsters are defined in the Regulation by weight, and shrimp and crabs by width of shell. The freshness grades are determined by sensory assess-ment of the raw fish, predominantly by appearance, and the Regulation gives tables for the grading in five broad groups of species - whitefish, bluefish, elasmo-branchs, cephalopods, and crustacea. The grading scheme is intended to help the sale and marketing of fish at the port markets, and it is a component of the price support mechanism operated by Producer's Organisations.

Eccles cake Pastry filled with dried fruit, melted butter, and sugar. Named originally for the town of Eccles in greater Manchester.

Eclair Small finger-shaped cake prepared from choux pastry, filled with cream or confectioner's custard, coated with chocolate or coffeeflavoured icing.

Ecological To do with a branch of science which investigates the relationships among living things and their surroundings.

Ectomorph Description given to a tall, thin person, possibly with underdeveloped muscles.

Ecuelle A device for obtaining peel oil from citrus fruit. It consists of a shallow funnel lined with spikes on which the fruit is rolled by hand. As the oil glands are pierced, the oil and cell sap collect in the bottom of the funnel.

Edam Pale yellow, semi-hard Dutch cheese usually round in shape with red wax coating.

Edosol Trade name for a low-sodium milk substitute, containing 43 mg of sodium/100 g, compared with dried milk at 400 mg.

Effluent Liquid industrial waste.

Egg nog Hot, sweetened milk with an egg and brandy or sherry mixed in.

Egg substitute Name formerly used for golden raising powder, a type of baking powder.

Egg white The white of an egg is in three layers: an outer layer of thin white, a layer of thick white, richer in ovomucin, and an inner layer of thin white surrounding the yolk. The ratio of thick to thin white varies, depending on the individual hen. A higher proportion of thick white is desirable for frying and poaching, since it helps the egg to coagulate into a small firm mass instead of spreading; thin white produces a larger volume of froth when beaten than does thick.

EI Specialized services provided to infants and young children to minimize delays in early development. These services include physical therapy, occupational therapy, speech therapy/oral motor therapy, and special education. They are provided by the Birth to Three programs.

Eicosanoids Compounds formed in the body from long-chain polyunsaturated fatty acids, including the prostaglandins, prostacyclins, thromboxanes, and leukotrienes, all of which act as local hormones and are involved in wound healing, inflammation, platelet aggregation, and a variety of other functions.

Eicosenoic acids Long-chain polyunsaturated fatty acids; with 20 carbon atoms.

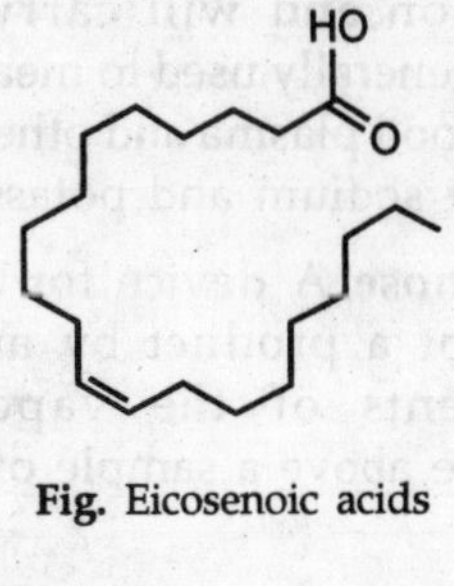

Fig. Eicosenoic acids

Einkorn A type of wheat, the wild form of which, Triticum boeoticum, was probably one of the ancestors of all cultivated wheats. Still grown in some parts of southern Europe and the Middle East, usually for animal feed. The name means 'one seed', from the single seed found in each spikelet.

Eiswein Wine made from grapes that have frozen on the vine, picked and processed while still frozen, so that the juice is highly concentrated and very sweet.

Elasticity A textural property. The ability of a material to return to its original shape when a deforming force is removed. Synonyms used in food quality are rubberiness and springiness. When very fresh fish is pressed with a finger, the finger indentation restores rapidly when the pressure is removed, that is, it is elastic; in a stale fish the indentation remains, that is, it is plastic, capable of flowing under pressure.

Elastin A fibrous protein that is the major constituent of the yellow elastic fibres of connective tissue. It is rich in glycine, alanine, proline, and other nonpolar amino acids that are cross-linked, making the protein relatively insoluble. Elastic fibres can stretch to several times their length and then return to their original size. Elastin is particularly abundant in elastic cartilage, blood-vessel walls, ligaments, and the heart.

Elder A common hedgerow bush; the flowers are used to flavour cordials, syrups, fruit jellies, and elderflower wine. The fruit is used for making jelly and wine.

Electrolytes Chemically salts that dissociate in solution and will carry an electric current; generally used to mean the mineral salts of blood plasma and other body fluids, especially sodium and potassium.

Electronic nose A device for assessing the quality of a product by analysing the constituents of the vapours in the headspace above a sample of the product. The instrument consists of an array of detectors of different sensitivities to chemicals likely to be in the headspace coupled to a data collection and processing system. There are analogies to odour detection and recognition - the sensor array mimics the olfactory epithelium, and the data recognition and processing system the brain. Hence the term. A variety of types of sensors are available, and usually 4-30 sensors are assembled in the detector unit. The sensors have different outputs when exposed to a mix of gases and the pattern of responses is analysed by the data processing system. For use, the instrument has to be trained with known mixes of gases or, more practically, examples of a particular food commodity of different, and defined, qualities. An example from fish technology would be iced fish of a particular species differing in freshness; the freshnesses must be known or be measured by a reference method. The patterns of responses of the members of the training set are related to the corresponding quality parameter by multivariate statistical methods, for example, discriminant analysis, or by artificial intelligence, neural network, procedures. The measured response pattern from a test sample is referred to the calibration parameters and the most likely value of the quality parameter is estimated. The use of artificial noses for measurement of the quality of fish products has been studied in research laboratories, but it is too soon to evaluate the applicability of artificial noses for quality control in industry.

Electrophoresis The motion of electrically charged molecules or particles in a fluid medium under the influence of an electric field. Particles with a positive charge move to the cathode, those with a negative charge to the anode, and the larger the charge on the particle, the faster the motion. Proteins and deoxyribonucleic acid (DNA) molecules are charged, and electrophoresis forms the basis of procedures for separating these compounds. For various reasons, the separation is usually effected in weak gels of agar or polyacrylamide rather than in free solution. The gels are cast in tubes to form a rod, or as slabs. Electrophoresis is used to identify the species of fish of unknown provenance or which might be mislabelled, accidentally or fraudulently. An aqueous extract of the sample is prepared and applied to one end of the rod or slab of gel. An electric potential is applied to the ends of the gel preparation in a suitable apparatus, and the proteins separate as bands in the gel under the applied potential. After a suitable time the gel is stained to show up the protein bands. A pattern of bands is peculiar to a species and allows identification of the sample.

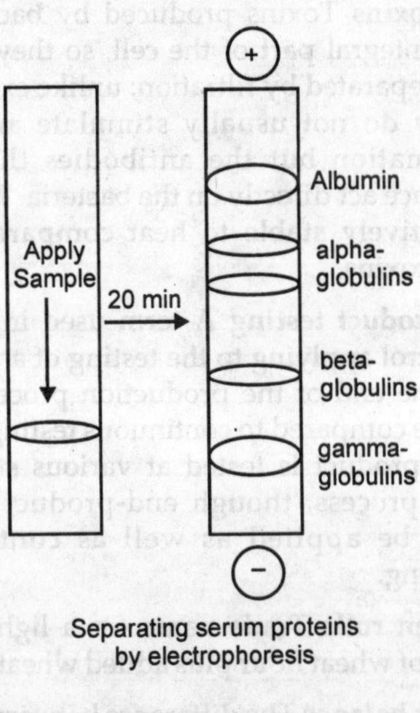

Fig. Electrophoresis

Electropure process A method of pasteurizing milk by passing through it low-frequency, alternating current.

ELISA An acronym for Enzyme Linked Immunosorbant Assay. The basis of a test which is widely used in food microbiology to identify, and quantify the numbers of, microorganisms. The test is based on the reaction of an antigen in the microorganism with an antibody specific to that antigen in the presence of an enzyme which is usually coupled with the antibody. The enzyme is involved in a reaction which has a coloured end-product, and the development of the colour, and its intensity, indicates the presence of the organism and its numbers. A wide variety of commercially produced kits are available to test for microorganisms and toxins in fish products.

Elixir Alcoholic extract of a naturally occurring substance; originally devised by medieval alchemists, now used for a variety of medicines, liqueurs, and bitters.

Elute To wash off or remove. Rather specifically applied to removal of adsorbed chemicals from the substance that adsorbed them, as in chromatography.

Elver Young eel, about 5 cm in length.

Emaciation Extreme thinness and wasting, caused by disease or undernutrition.

Embargoes Bans placed on the importation of a particular product.

Emblic Berry of the South-East Asian malacca tree, Emblica officinalis, similar in appearance to the gooseberry. Also known as the Indian gooseberry. An exceptionally rich source of vitamin C: 600 mg per 100 g.

Embolism Blockage of a blood vessel caused by a foreign object such as a quantity of air or gas, a piece of tissue or tumour, a blood clot, or fatty tissue derived from atheroma, in the circulation.

Embryo Early stages of an animal's development that results when a sperm fertilizes an egg. Embryo cloning produces many calves from one embryo, for example.

Emetic Substance that causes vomiting.

Emince A dish of meat cut into thin slices, covered with sauce, and baked in an earthenware dish.

Emissions Discharge released into the atmosphere from processing.

Emmental Semi-hard cow's milk cheese originally from Switzerland; has large round holes formed by gases from bacterial fermentation during ripening.

Emmer A type of wheat known to have been used more than 8000 years ago. Wild emmer is Triticum dicoccoides and true emmer is T. dicoccum. Nowadays grown mainly for animal feed.

Empanadas Spanish; pies and pasties with a variety of savoury fillings.

Emprote Trade name for a dried milk and cereal preparation consumed as a beverage, containing 33% protein.

Emulsifiers A food additive consisting of a substance that enables nonmiscible liquids to maintain a stable emulsion, as in mayonnaise, margarine, and peanut butter. Natural emulsifying agents, such as egg yolk, agar, and lecithin, have now been largely replaced by synthetic chemical emulsifiers.

Emulsin A mixture of enzymes in bitter almond which hydrolyse the glucoside amygdalin to benzaldehyde, glucose, and cyanide.

En papillote French method of cooking in a closed container, a parchment paper or aluminium foil case.

Enchilada Mexican; tortilla fried in oil, filled with meat, cheese, or vegetables and served with chilli sauce.

Endive Curly serrated green leaves of Cichorium endivia. Called chicory in the USA and chicoree frisee in France.

Endocrine glands Those glands that produce and secrete hormones, including the thyroid gland, pancreas (insulin and glucagon), adrenal glands, ovary and testes. Some endocrine glands respond directly to chemical changes in the bloodstream; others are controlled by hormones secreted by the pituitary gland, under the control of the hypothalamus.

Endomorph In relation to body build, means short and stocky.

Endopeptidases Enzymes that hydrolyse proteins, by cleaving peptide bonds inside protein molecules, as opposed to exopeptidases, which remove amino acids from the end of the protein chain. The main endopeptidases in digestion are chymotrypsin, elastase, pepsin, and trypsin.

Endosperm The inner part of cereal grains; in wheat it comprises about 83% of the grain. Mainly starch, it is the source of semolina. Contains only about 10% of the vitamin B 1, 35% of the vitamin B 2, 40% of the niacin, and 50% of the vitamin B 6 and pantothenic acid of the whole grain.

Endotoxins Toxins produced by bacteria as an integral part of the cell, so they cannot be separated by filtration; unlike exotoxins, they do not usually stimulate antitoxin formation but the antibodies that they induce act directly on the bacteria. They are relatively stable to heat compared with exotoxins.

End-product testing A term used in quality control applying to the testing of a product at the end of the production process. It is to be compared to continuous testing where the product is tested at various stages in the process, though end-product testing can be applied as well as continuous testing.

Energen rolls Trade name for a light bread roll of wheat flour plus added wheat gluten.

Energy balance The difference between intake of energy from foods and expenditure on basal metabolism and physical activity. Positive energy balance leads to increased body tissue, the normal process of growth. In adults positive energy balance leads to creation of body reserves of fat, resulting in overweight and obesity. Negative energy balance leads to utilization of body reserves of fat and protein, resulting in wasting and undernutrition.

Energy conversion Factors Various factors are used to calculate the energy yields of foodstuffs. The complete heats of combustion as determined by calorimetry are: protein, 5.7 kcal (23.9 kJ); fat, 9.4 kcal (39.5 kJ); carbohydrate, 4.1 kcal (17.2 kJ)/gram. The Rubner conversion factors for metabolic energy yield are: protein, 4.1 kcal (17 kJ); fat, 9.3 kcal (39 kJ); carbohydrate, 4.1 kcal (17 kJ)/gram The Atwater factors also allow for losses in digestion and incomplete oxidation of the nitrogen of proteins: protein, 4 kcal (16.8 kJ); fat, 9 kcal (37.8 kJ); carbohydrate, 4 kcal (16.8 kJ)/gram. The following factors are used in general practice: carbohydrate 4 kcal (17 kJ); fat, 9 kcal (38 kJ); carbohydrate (as monosaccharides), 4 kcal (1.7 kJ); alcohol, 7 kcal (29 kJ); sugar alcohols, 2.4 kcal (10 kJ); organic acids, 3 kcal (13 kJ).

Energy expenditure The total energy cost of maintaining constant conditions in the body, i.e. homeostasis plus the energy cost of physical activities. The average total energy expenditure in Western countries is about 1.4 times BMR; a desirable level of physical activity should be about 1.7 times BMR.

Energy requirements Energy requirements are calculated from estimated basal metabolic rate and physical activity. Average energy requirements for adults are 1900 kcal (8 MJ)/day for women and 2400 kcal (10 Mj)/day for men.

Energy-rich bonds An outdated and chemically incorrect concept in energy metabolism, which suggested that the bond between ADP and phosphate in ATP, and between creatine and phosphate in creatine phosphate, which have a high chemical free energy of hydrolysis, somehow differs from 'ordinary' chemical bonds.

Enfleurage A method of extracting essential oils from flowers by placing them on glass trays covered with purified lard or other fat, which eventually becomes saturated with the oil.

Enocianina Desugared grape extract used to colour fruit flavours. Prepared by acid extraction of the skins of black grapes; it is blue in neutral conditions and turns red in acid.

Enriched food Food that has had nutrients replaced, which were originally present, but were lost during food processing.

Enrichment The addition of nutrients to foods. Although often used interchangeably, the term fortification is used of legally imposed additions, and enrichment means the addition of nutrients beyond the levels originally present.

Enteral nutrition Tube feeding with a liquid diet directly into the stomach or small intestine.

Enteritis Inflammation of the mucosal lining of the small intestine, usually resulting from infection. Regional enteritis is Crohn's disease.

Enterocolitis Inflammation of the mucosal lining of the small and large intestine, usually resulting from infection.

Enterogastrone Hormone secreted by the small intestine which inhibits the activity of the stomach. Its secretion is stimulated by fat; hence, fat in the diet inhibits gastric activity.

Enterokinase Obsolete name for the intestinal enzyme enteropeptidase.

Enteropathy Any disease or disorder of the intestinal tract.

Enteropeptidase An enzyme secreted by the small intestinal mucosa which activates trypsinogen to the active proteolytic enzyme trypsin.

Enterotoxin Substances more or less specifically toxic to the cells of the intestinal mucosa, normally produced by bacteria.

Enteroviruses Viruses that multiply mainly in the intestinal tract.

Entoleter A machine used to disinfest cereals and other foods. The material is fed to the centre of a high-speed rotating disc carrying studs so that it is thrown against the studs; the impact kills any insects and destroys their eggs.

Entrecote Steak cut from the middle part of the sirloin of beef; in France a steak taken from between two ribs.

Entree A dressed savoury dish, served hot or cold, complete in its dish with the sauce. The term is used both for the main part of a meal and also to describe courses intermediate between hors d'oeuvre and main course.

Entremeses Spanish; mixed hors d'oeuvre dishes served as a prelude to a meal.

Entremets Originally a dish served between courses, now the dessert course.

E-numbers Within the EU food additives may be listed on labels either by name or by their number in the EU list of permitted additives.

Enzyme A protein substance found in living cells that brings about chemical changes; necessary for digestion of food.

Epidemiology The study of distribution and determinants of diseases or other health outcomes in human populations. It seeks to expose potential associations between aspects of health and diet, lifestyle, habits or other factors within populations. Epidemiological studies may suggest relationships between two factors, but do not provide the basis for conclusions about cause and effect. Possible associations inferred from epidemiological research can turn out to be coincidental.

Epigramme Small pieces of neck or breast of lamb, cooked, then dipped in egg and breadcrumbs and grilled or fried.

Epsom salts Magnesium sulphate, originally found in a mineral spring in Epsom, Surrey, England; acts as a purgative because the osmotic pressure of the solution causes it to retain water in the intestine and so increase the bulk and moisture content of the faeces.

Equal Trade name for aspartame.

Equilibrium humidity The relative humidity of the atmosphere with which the substance under consideration is in equilibrium.

Erdbeergeist German spirit distilled from strawberries with added alcohol.

Erepsin Name given to a mixture of enzymes contained in intestinal juice, including aminopeptidases and dipeptidases.

Ergosterol A sterol isolated from yeast; when treated with ultra-violet light, it is converted to ercalciol. This is the main source of manufactured vitamin D.

Ergot A fungus that grows on grasses and cereal grains; the ergot of medical importance is Claviceps purpurea, which grows on rye. The consumption of infected rye is harmful, causing the disease known as St Anthony's fire, and can be fatal.

The active principles in ergot are alkaloids, which yield lysergic acid on hydrolysis. This is believed to be the active component. Its effect is to increase the tone and contraction of smooth muscle, particularly of the pregnant uterus. For this reason ergot has been used in obstetrics, but pure ergonovine maleate and ergotonine tartrate are preferable.

Ergotism Poisoning due to an ergot infection of rye which occurs from time to time among people eating rye bread. The last outbreak in the UK was in Manchester in 1925, when there were 200 cases. Symptoms appear when as little as 1% of ergot-infected rye is included in the flour.

Eriodictin A flavonoid found in citrus pith, a constituent of what is sometimes called vitamin P.

Erucic acid A mono-unsaturated fatty acid, cis-13-docosenoic acid found in rape seed and mustard seed oils; it may constitute 30-50% of the oil in some varieties. It causes fatty infiltration of heart muscle in experimental animals, and the amount of ordinary rape seed oil used in margarines is consequently limited. Low erucic acid varieties of rape seed have been developed for food use.

Eructation The act of bringing up air from the stomach, with a characteristic sound. Also known as belching.

Erythorbic acid The D-isomer of ascorbic acid, also called Daraboascorbic acid and iso-ascorbic acid, with only slight vitamin C activity. It is as powerful an antioxidant as vitamin C and is used in foods for that purpose.

Erythroamylose An old name for amylopectin.

Erythropoiesis The formation and development of the red blood cells in the bone marrow.

Erythrosine BS Red colour permitted in foods in most countries. Used in preserved cherries, sausages, and meat and fish pastes; it is unstable to light and heat. Chemically the disodium or potassium salt of 2,4,5,7- tetraiodofluorescein.

Escabeche Caribbean; fish, poultry, or game cooked in oil and vinegar or cooked and then pickled in oil and vinegar marinade. From the Spanish for pickled. Fish prepared in this way is also known as escovitch or caveached fish.

Escalope Thin slice of meat, round or oval shaped, cut from the top of the leg or fillet. Also a thin slice of fish.

Escherichia coli The most common aerobic organism in the gut of warm blooded animals and man, and therefore found in sewage polluted waters, but also found in unpolluted warm tropical waters. Most strains are harmless but, within the species, there are at least 4 types of strains pathogenic to man. Generally, the presence of E.coli in foods is undesirable because this indicates poor hygiene, but shellfish, particularly bivalve molluscs, from sewage contaminated inshore waters will carry high loads of E. coli. Many countries prohibit the sale of live bivalve molluscs and other shellfish unless contamination with E.coli or faecal coliforms is below a specified threshold value. For example, in the European Community live bivalve molluscs for immediate human consumption must contain less than 300 faecal coliforms or less than 230 E.coli per 100g flesh and intravalvular fluid.

Espagnole sauce Spanish sauce; basically a thick, brown roux, diluted with stock made from meat or bones, with onion, carrot, garlic, herbs, tomato pulp, and sherry.

Espresso Strong black coffee made by forcing steam through finely ground coffee.

Espumante Portuguese; sparkling wines.

Espumoso Spanish; sparkling wines.

Essential oils Volatile, aromatic, or odoriferous oils found in plants and used for flavouring foods. Chemically they are quite distinct from the, edible oils, since they are not glycerol esters.

Esterases Enzymes that hydrolyse esters, i.e. cleave the ester linkage to form free acid and alcohol. Those that hydrolyse the ester linkages of fats are generally known as lipases, and those that hydrolyse phospholipids as phospholipases.

Esters The chemical name for the compounds formed by condensation between an acid and an alcohol, e.g. ethyl alcohol and acetic acid yield the ester ethyl acetate. Fats are esters of the alcohol glycerol, and long-chain fatty acids. Many esters are used as synthetic flavours.

Estouffade Meat cooked very slowly in very little liquid; braised or casseroled.

Ethanol Alcohol; yeasts ferment corn starch to yield ethanol for gasohol.

Ethanolamine One of the water-soluble bases of phospholipids, chemically it is 2-aminoethanol. Used as softening agent for hides, as dispersing agent for agricultural chemicals, and to peel fruits and vegetables.

Ethoxyquin The trivial name for the chemical 6-ethoxy-1,2-dihydro-2,2,4-trimethylquinoline. It is used as an antioxidant in fish meal, especially meal from fatty fish, and in animal feeds. Generally not approved for use in foods for humans other than in minor ingredients such as spices.

Ethyl formate Used as a fumigant against raisin moth, dried fruit beetle, fig moth, etc., and as a flavour; an ingredient of lemon, strawberry, and artificial rum flavours.

Ethyl vanillin A synthetic compound, chemically the ethyl analogue of vanillin, the major flavouring principle of vanilla.

Ethylene diamine tetra-acetic acid A generally permitted food additive used especially in the fish canning industry. See chelating agents.

Ethylene Ethene; a gas of the formula $CH_2{=}CH_2$, produced by fruit as a hormone to speed ripening. This explains why some fruits ripen faster if they are stored in a plastic bag. It is used commercially in very small amounts to speed fruit ripening after harvesting.

Eutectic ice The solid formed when a mixture of 76.7% water and 23.3% salt is frozen. It melts at -21°C. It has about three times the refrigerant effect of solid carbon dioxide, and is especially useful for icing fish on board trawlers.

Eutrophia Normal nutrition.

Evaporated, condensed, milk Full-fat, skimmed, or partly skimmed milk, sweetened or unsweetened, that has been concentrated by partial evaporation; fat and total solids for each type defined by law.

Evaporation The loss of molecules from a liquid or solution as vapour.

Evaporation, flash A short, rapid application of heat so that a small volume is quickly distilled off, carrying with it the greater part of the volatile components. The flash distillate is collected separately from the later distillate and is added back to the concentrate to restore the flavour; applied to the concentration of products such as fruit juices.

Evening primrose Oenothera biennis, the oil from the seeds is a rich source of ?-linolenic acid, which may account for 8% of total fatty acids. It is used as a dietary supplement and may have beneficial effects in a number of conditions, although there is little firm scientific evidence of its efficacy.

Evian water A still mineral water from a source in Evian in southern France.

Exchange list List of portions of foods in which energy yield, fat, carbohydrate, and/or protein content are equivalent, so simplifying meal and diet planning for people with special needs.

Exclusion diet A limited diet excluding foods known possibly to cause food intolerance, to which foods are added in turn to test for intolerance.

Exergonic Chemical reactions that proceed with the output of energy, usually as heat or light. The reactions involved in the oxidation of food-stuffs are generally exergonic.

Exopeptidases Proteolytic enzymes that hydrolyse the peptide bonds of the terminal amino acids of proteins or peptides, as opposed to endopeptidases, which cleave at sites in the middle of a peptide chain. There are two groups: aminopeptidases which remove the amino acid at the amino terminal of the protein, and carboxypeptidases, which remove the amino acid at the carboxyl terminal.

Exotaxins Toxic substances produced by bacteria which diffuse out of the cells and stimulate the production of antibodies which specifically neutralize them, They are generally heat-labile and inactivated in about 1 hour at 60°C. Exotoxins include those produced by the organisms responsible for botulism, tetanus, and diphtheria.

Expansion rings In relation to cans, the concentric rings stamped into the ends of the can to allow bulging during heat processing without straining the seams unduly.

Expeller cake The residue from oilseeds after most of the oil has been removed by pressing; it is a valuable source of protein, especially for animal feeding.

Experimental group The group of subjects in an experimental study which receives a treatment.

Expression In genetics, manifestation of a characteristic specified by a gene. In industrial biotechnology, production of a specific protein by inserting a gene into a new host organism.

Extensograph An instrument for measuring the stretching strength of dough as an index of its baking quality.

Extremophiles Micro-organisms that can grow under extreme conditions of heat and extreme thermophiles, some of which live in hot-springs at 100°C), or cold (pshyrophiles), in high concentrations of salt (halophiles), high pressure, or extremes of acid or alkali.

Extrusion Forcing a viscous solution through a spinneret-like machine.

F

Factory ship A vessel, that may also be a fishing vessel, equipped to process fish on board in one or more ways, such as canning, freezing, production of frozen fillet blocks, production of fish meal.

FAD Flavin adenine dinucleotide, one of the coenzymes formed from vitamin B 2.

Faecal coliforms A subset of the coliform group characterised by ability to grow in selective culture media at a temperature of 44.5°C. The number of faecal coliform organisms in a sample is typically estimated by the Most Probable Number (MPN) method. The presence of faecal coliforms in a sample of a food is almost certainly evidence of contamination of the food with organisms of enteric origin. The organism Escherichia coli makes up most of the faecal coliforms, but further tests are required to identify the organism.

Faeces Body waste, composed of undigested food residues, remains of digestive secretions that have not been reabsorbed, bacteria from the intestinal tract, cells, cell debris and mucus from the intestinal lining, and substances excreted into the intestinal tract. The average amount is about 100 g/day, but varies widely depending on the intake of dietary fibre.

Faggot 1. Traditional British meatball made from pig offals and meat.

2. Bundle of herbs, bouquet garni.

Fair maids Cornish name for pilchards.

Farinograph An instrument for measuring the physical properties of a dough.

Fast food General term used for a limited menu of foods that lend themselves to production-line techniques; suppliers tend to specialize in products such as hamburgers, pizzas, chicken, or sandwiches.

Fast foods General term used for a limited menu of foods that lend themselves to production-line techniques; suppliers tend to specialize in products such as hamburgers, pizzas, chicken, or sandwiches.

Fasting Going without food. The metabolic fasting state begins some 4 hours after a meal, when the digestion and absorption of food are complete and body reserves of fat and glycogen begin to be mobilized. In more prolonged fasting the blood concentration of ketone bodies rises, as they are exported from the liver for use by muscle and other tissues as a metabolic fuel.

Fat A component of foods. Fats are mixtures of components and are operationally defined as the fraction of the food that is soluble in organic solvents such as ether, hexane, chloroform. The amount of fat extracted depends on the solvent used and the extraction conditions used, so the reported fat content of a sample can vary slightly with the analytical procedure. The fats extracted from fish are liquid at typical room temperatures and are often referred to as oils. Fats are also called lipids. The main components of depot fats are triglycerides, fatty acid esters of glycerol.

Fat free A food containing less than 0.5 gram of fat per serving.

Fat-soluble vitamin Vitamins that are present only in fat. These vitamins include A, D, E, and K.

Fatty acids Chemically, a class of aliphatic compounds containing a carboxyl group. The fatty acids in foods or derived from fats are monocarboxylic acids of the general formula R-COOH where R is an unbranched hydrocarbon chain. The higher carbon number members of the series with a carbon chain length above 8, are odourless and flavourless when pure. The lower members are flavoursome and are produced during microbiological spoilage of fish imparting sour and cheesy flavours to spoiled fish, especially fatty fish. The higher members are components of fats, hence the origin of the term. A wide range of fatty acids have been identified in fats, but the most frequently present have even numbered carbon atoms in the molecule in the range 10 to 28. The carbon chain can be saturated or contain double bonds. The polyunsaturated fatty acids are important components of fish oils.

Feedstock Raw material used for chemical or biological processes, such as polymers to produce plastics.

Feedy fish Fish that have been feeding heavily such that the gut is full of food when caught. Consequently, the gut enzymes are very active and such fish are liable to belly burst unless they are gutted. The term is used especially with pelagic fish when they are feeding on plankton blooms.

Feingold diet Exclusion of foods containing synthetic colours, flavours, and preservatives and limitation of intake of fruits and vegetables such as oranges, apricots, peaches, tomatoes, and cucum-bers; intended to treat hyperactive children. There is little evidence either that these foods are a cause of hyperactivity or that the exclusion diet is beneficial.

Fenfluramine Anorectic drug used in the treatment of obesity.

Feng kuo Savoury filling of chopped roast meat or seafood with mushrooms and bamboo shoots, flavoured with oyster sauce and wrapped in dough made from flour, cornflour, and lotus root flour, then fried.

Fennel 1. Aromatic seeds and feathery green leaves of the perennial plant, Foeniculum vulgare, used to flavour a variety of dishes.

2. Foeniculum dulce. Annual plant, also called Florence fennel or finnochio; the swollen bases of the leaves are eaten as a vegetable, raw or cooked. Seeds used as flavouring.

Fenton reaction A reaction that generates hydroxyl radical (HO^-) through the combination of H_2O_2 with Fe^{2+}.

Fenugreek Trigonella feonumgraecum, a leguminous plant eaten as a vegetable; the seeds are used for flavouring. It is traditionally eaten by women in the Orient to help gain weight.

Ferment As a noun, the old name for enzyme. As a verb, to carry out the process of fermentation.

Ferment To change chemically through the action of living substances, such as yeast or bacteria.

Fermentation A form of anaerobic respiration occurring in certain microorganisms, e.g. yeasts. Alcoholic fermentation comprises a series of biochemical reactions by which pyruvate is converted to ethanol and carbon dioxide. It is the basis of the baking and brewing industries. In lactic-acid fermentation, which occurs in many microorganisms and in animal cells, the end product is lactic acid. Microorganisms display a range of fermentations, producing not only ethanol or lactic acid, but other products, such as propionic and butyric acids, acetate, and methane.

Fermenting, Fermented fish Refers usually to the enzymic liquefaction of fish, small crustaceans and squid to produce edible sauces or soft pastes particularly in Asian countries. In southeast Asia, these sauces are be made from salted fish in which bacterial action is inhibited by the salt but autolytic enzymic digestion takes place. Some products are mixed fermentations of fish and/or shellfish with vegetables or cereals by the use of lactic acid bacteria, moulds or yeasts.

Fermentograph An instrument for measuring the gas-producing power of a dough.

Fernet Branca Bitter digestif; herbs and spices steeped in white wine and brandy.

Ferric ammonium citrate The chemical form in which iron is sometimes added to foods. Occurs as brown-red scales and as green scales.

Ferritin Iron bound to the protein apoferritin, one of the main forms of body iron storage.

Feta Semi-hard white cheese from ewe's or goat's milk; preserved in salt water.

Fettucini Ribbon-shaped pasta.

Fiber Dietary fiber generally refers to parts of fruits, vegetables, grains, nuts and legumes that can't be digested by humans. Meats and dairy products do not contain fiber. Studies indicate that high-fiber diets can reduce the risks of heart disease and certain types of cancer. There are two basic types of fiber - insoluble and soluble. Soluble fiber in cereals, oatmeal, beans and other foods has been found to lower blood cholesterol. Insoluble fiber in cauliflower, cabbage and other vegetables and fruits helps move foods through the stomach and intestine, thereby decreasing the risk of cancers of the colon and rectum.

Fibrin The blood protein formed from fibrinogen which is responsible for the clotting of blood.

Fibrinogen One of the proteins of the blood plasma which is responsible for the clotting of blood. When prothrombin is activated to thrombin in response to injury, it converts fibrinogen to fibrin, which is deposited as strands that trap the red cells and form the clot.

Fig The fruit of Ficus carica; eaten fresh or dried. Figs have mild laxative properties, e.g. syrup of figs is a medicinal preparation.

Figgy pudding Pudding made with raisins; originally dried figs stewed in wine. Figgy duff or figgie hobbin is the Cornish name for a pastry containing raisins.

Figueredas Spanish; a spiced seafood dish similar to paella, but served on a bed of pasta rather than rice.

File powder Dried powdered young leaves of the sassafras tree; very aromatic, an essential ingredient of gumbo.

Filleting and Fillet Cutting by hand or by machine, roughly parallel with the backbone, a strip or slice of largely boneless flesh from along the length of a fish to produce from one side of the fish a fillet or, more precisely especially from roundfish, a single fillet; it is also called a side especially with reference to salmon. In tuna, the four strips of light muscle cut from the fish, two from each side, are called loins. Filleting machines are available for many species but hand filleting is still used widely, and may be preferred even when machines are available because the yield of fillet is better by hand. The amount of belly flap and lug left on the fillet depends on the way in which the fish was gutted or headed and gutted and the way in which the fillet is subsequently trimmed. The rib bones remaining at the anterior (neck) end of a fillet from round white fish are known as the pin bones. The number of bones remaining depends on subsequent trimming of the fillet by V-cutting or J-cutting to give a boneless fillet. The rib or chest bones, some of which are severed by filleting and remain in the edible part of a herring or kipper fillet, for example, are sometimes called the barrel bones. A block fillet is flesh cut from both sides of the fish so that the pieces from each side of the fish remain joined along the back. It may be specified as having all bones removed except the pin bones. This cut or fillet is also known as an angel fillet, butterfly fillet, cutlet, double fillet or, when smoked, a golden cutlet.

Filter To strain a liquid through a porous paper, fine cloth, etc., to remove small particles, as in making coffee or clarifying fruit juice to make jelly. Water may be filtered through charcoal to remove unpleasant flavours and colours; bacterial filters for water have pores fine enough to remove bacteria.

Filth test Name given to a test originated in the USA for determining the contamination of a food with rodent hairs and insect fragments as an index of the hygienic handling of the food.

Filtration The process of passing a liquid through a filter to remove any solid particles.

Fines herbes Mixture of chopped parsley, tarragon, chives, chervil, marjoram, and sometimes watercress.

Fingerware Edible seaweed, Laminaria digitata.

Fining agents Substances used to clarify liquids by precipitation, e.g. egg albumin, casein, bentonite, isinglass, gelatine, etc.

Finnan haddock Smoke-cured haddock.

Fino Very dry sherry.

Finocchio Variety of fennel with swollen leaf base; Foeniculum vulgare var. azoricum.

Fire point The temperature at which a frying oil will sustain combustion. It ranges between 340 and 360°C for different fats.

Firkin A quarter of a barrel of beer, 9 imperial gallons (40 L); also 56 lb (25.5 kg) of butter.

Firmi Rice dessert with almonds and pistachio nut

Firming agents Fresh fruits contain insoluble pectins as a firm gel around the fibrous tissues which keeps the fruit firm. Breakdown of cell structure allows conversion of pectin to pectic acid, with loss of firmness. The addition of calcium salts forms a calcium pectate el which protects the fruit against softening; these are known as firming agents. Alum is sometimes used to firm pickles.

Firmness The property of not yielding easily to pressure. It can be assessed by the fingers, for example, when assessing the quality of raw fish, or in the mouth when assessing cooked fish. A component of texture.

Fish cakes Chopped or minced fish, bound with egg and flour and seasoned with onion, pepper, and sometimes herbs, then deep fried.

Fish content The amount of the fish component of a composite dish or product containing fish. Usually expressed as a percentage. Fish content is usually referred back to the amount of raw fish used in the product, and is not the same as fill applied to canned products. Regulations might specify the minimum fish contents of some products such as fish cakes and fish spreads, and custom or Codes of Practice might require minimum fish contents in products such as coated and glazed products. It is not easy to measure accurately the fish content of finished products and quality control is best effected by control of the amounts of ingredients used in preparation of a product. The fish content of coated products can be measured by carefully scraping off the coating under prescribed conditions. Fish content of glazed products can be determined as the glazed weight. Fish content can be estimated from the nitrogen content of the blended product by multiplying the nitrogen content by a factor representing the average nitrogen content of the fish component of the mix. This procedure has some uncertainties associated with it. Components of the mix other than fish are likely to contain nitrogen and this non-fish nitrogen has to be allowed for by determining the amounts of these other components and correcting for their nitrogen contents. Another factor is that the nitrogen content of a particular type of fish varies intrinsically, or perhaps by some pre-processing operation before the fish is incorporated in the product, so that the fish content of a sample can be over-or underestimated when the average nitrogen content is used. This can have important consequences if regulations relating to minimum fish contents of a product are in force and end-products are being monitored by regulatory agencies.

Fish finger and Fish portion Rectilinear or cuboidal portion (blank), typically 18 g in weight with a length at least three times the breadth, usually cut automatically after tempering from a regular block of frozen fish flesh, often nominally 7.4 kg in size and usually from white fish. The block may be composed entirely of skinless, boneless fillets. The block may be jumble-packed or orientated and layered, and composed of skinless, boneless fillets plus the recovered fish flesh from the V-cuts, or composed entirely of recovered fish flesh or minced fish. The blank is often coated with a batter and crumb coating, flash fried, packed refrozen, distributed and sold frozen. The coated fish finger may contain about 50% fish by weight, but this will vary according to the product, and must of course conform with any relevant food labelling or other legislation. The fish portion is a larger size, perhaps up to 100 g, also cut from the larger, 7.4 kg, regular block, and may be cuboidal, trapezoidal, or some other shape. It may be coated in batter and crumb, batter only, covered in a suitable sauce, or form the basis of a more complex frozen recipe dish, or sold naked without any form of coating.

Fish tester The name given to an electronic instrument manufactured by Intelectron GmbH, Hamburg, for determining the freshness of fish. The measuring head consisted of a pair of electrodes attached to arms of a caliper device, and the electrodes were applied to the fish just behind the head. Properties of the current passing between the electrodes changed systematically as the fish aged and were measured as an index of freshness.

Fish, fatty or oily Anchovies, herring, mackerel, pilchard, salmon, sardine, trout, tuna, whitebait, containing about 15% fat and containing 10-40 ug vitamin D per 100 g, as distinct from white fish, which contain 1-2% fat and only a trace of vitamin D.

Fish, white Non-oily fish, i.e. 1-2% fat, e.g. cod, dogfish, haddock, halibut, plaice, saithe, skate, sole, and whiting. All are similar in nutrient composition; an exceptionally rich source of iodine; a rich source of protein, niacin, and selenium.

Fishroom An enclosed space, usually partly or wholly insulated and sometimes refrigerated, in a fishing vessel or factory ship for chilled storage of the catch.

Five-spice powder A mixture of star anise, anise pepper, fennel, cloves, and cinnamon, and sometimes also powdered dried orange peel.

Flageolet Small green variety of haricot bean.

Flambe Brandy or other spirit is poured over the food and set alight before serving. Also the process of singeing poultry before cooking.

Flan Open fruit tart, on a base of pastry or sponge.

Flapjack 1. Biscuit made from fat, sugar, rolled oats and syrup.

2. A thick pancake.

Flash point With reference to frying oils, the temperature at which the decomposition products can be ignited, but will not support combustion; range between 290 and 330°C.

Flat sours Bacteria such as Bacillus stearothermophilus render canned food sour by fermenting carbohydrates to lactic, formic, and acetic acids, without gas production. This means that the ends of the can are not swelled out but remain flat. Economically they are the most important of the thermophilic spoilage agents; some species can grow slowly at 25°C and thus spoil products after long storage periods

Flatfish Fish with a flattened shape, including dab, flounder, halibut, plaice, sole and turbot.

Flatogens Substances that cause gas production, flatulence, in the intestine, by providing fermentable substrate for intestinal bacteria. Those identified include small oligosaccharides such as raffinose, stachyose, and verbascose in a variety of beans.

Flatulence Production of gas in the intestine; hydrogen, carbon dioxide, and methane. May be caused by a variety of foods including beans, Brussels sprouts, cabbage, cauliflower, onion, radishes, melon, and avocado, which contain flatogens.

Flavedo The coloured outer peel layer of citrus fruits, also called the epicarp or zest. It contains the oil sacs, and hence the aromatic oils, and numerous plastids which are green and contain chlorophyll in the unripe fruit, turning yellow or orange in the ripe fruit, when they contain carotene and xanthophyll.

Flavin adenine dinucleotide A coenzyme in oxidation reactions, derived from vitamin B2, phosphate, ribose, and adenine.

Flavin mononucleotide A coenzyme in oxidation reactions, chemically the phosphate of vitamin B 2 (riboflavin).

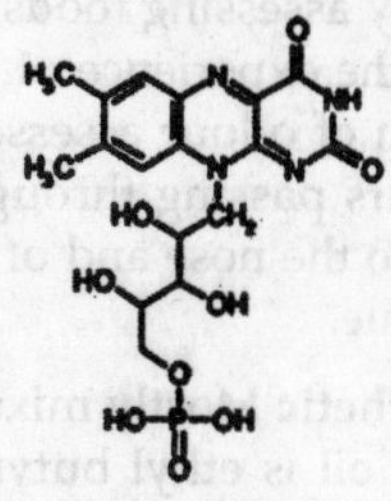

Fig. Flavin mononucleotide

Flavin The group of compounds containing the iso-alloxazine ring structure, as in riboflavin, and hence a general term for riboflavin derivatives.

Flavonoids Phytochemicals widely distributed in plants containing a characteristic aromatic trimeric heterocyclic nucleus. Several flavonoids such as catechin, epicatechin, proantocyanidin, and quercetin have antioxidant properties.

Flavoproteins Enzymes that contain the vitamin riboflavin, or a derivative such as flavin adenine dinucleotide or riboflavin phosphate, as the prosthetic group. Mainly involved in oxidation reactions in metabolism.

Flavour potentiator A substance that enhances the flavours of other substances without itself imparting any characteristic flavour of its own, e.g. monosodium glutamate and ribotide as well as sugar, salt, and vinegar in small quantities.

Flavour profile An experimental procedure used in the sensory evaluation of foods. The assessor describes the flavour of a product typically using descriptors from a prepared list of terms, though some systems - free choice flavour profiling - allow each assessor in a panel to use their own descriptors. The assessor also rates the perceived intensity of each flavour 'note' identified. The results are combined over members of a panel of assessors and the resulting assemblage of terms and intensities forms the

Flavour/odour The sensory experience perceived by assessing foods in the mouth, other than the experience of texture. It is a combination of odour assessed in the nose from vapours passing through the back of the mouth to the nose and of taste assessed on the tongue.

Flavours, synthetic Mostly mixtures of esters, e.g. banana oil is ethyl butyrate and amyl acetate; apple oil, ethyl butyrate, ethyl valerianate, ethyl salicylate, amyl butyrate, glycerol, chloroform, and alcohol; pineapple oil is ethyl and amyl butyrates, acetaldehyde, chloroform, glycerol, and alcohol.

Fleuron Small crescent-shaped piece of puff pastry used as a garnish.

Flip Drink made with beaten egg and milk, with added wine or spirit, and sweetened.

Flitch Side of bacon; half a pig, slit down the back, with the legs and shoulders removed.

Floats Fried biscuits made with yeast dough.

Florence oil Name given to a high grade of olive oil.

Florentine 1. Thin biscuits with nuts and dried fruit coated with chocolate.

2. Garnished with spinach.

Floridean starch A polysaccharide resembling glycogen, obtained from red algae.

Flounder Small flatfish, Platiclithys spp., also called fluke. flour Most commonly refers to the ground wheat berry, although also used for other cereals and applied to powdered dried matter such as fish flour, potato flour, etc. The ground wheat berry yields wholemeal flour; whiter flours are obtained by separation of the bran and the germ from the starchy endosperm.

Flour enrichment The addition of certain vitamins and minerals to flour, to contain not less than: in the UK, vitamin B 1, 0.24 mg; niacin, 1.6 mg; iron, 1.65 mg; calcium, 120 mg/100 g; in the USA, vitamin B 1, 0.44-0.56 mg; vitamin B 2, 0.2-0.33 mg; niacin, 3.6-4.4 mg; iron, 2.9-3.7 mg/100 g; calcium not specified.

Flour strength A property of the flour proteins enabling the dough to retain gas during fermentation to give a 'bold' loaf. 'Strong' flour is higher in protein content, has greater elasticity and resistance to extension, and greater ability to absorb water. A 'weak' flour gives a loaf that lacks volume.

Flour The powdered grain of wheat or other cereals, used in baking. The chief use of flour is in making bread. When the two proteins in wheat, glutenin and gliadin, are mixed with water they form gluten, which permits the dough to expand and retain the carbon dioxide resulting from fermentation of the yeast in bread dough. Different types of flour are made by varying the percentage of flour separated from the wheat. The principal commercial flours are whole wheat (100%), wholemeal and stone round (92%), wheat eal (80-90%), and white flours (70-72%). Many nutritionists consider it important to eat food made from whole wheat and wholemeal flour, rather than white flour. The former retains more of the bran and has more iron and calcium than white flour. They are also a good daily source of dietary fibre. Flour with a high gluten content is best when yeast is called for, as in dough for bread. On the other hand a softer flour with a lower gluten content is used for cakes, shortbread, etc.

Plain flour is all-purpose, with moderate gluten content. Self-raising flour is plain flour with the addition of raising agents.

Flow diagram Links the sequence of processing operations as a series of units.

Fluid bed dryer A bed of solid particles is supported on a cushion of hot air jets and the material may be conveyed this way, while being dried. The method achieves intimate mixing without mechanical damage; applied to cereals, tabletting granules, salt, coffee, and dried vegetables.

Fluke Small flatfish, Platichthys spp., also called flounder.

Fluoridation The addition of fluoride to drinking water. fluoride The ion of the element fluorine, one of the halogens. Although it occurs in small amounts in plants and animals, and has effects on the formation of dental enamel and bones, it is not considered to be a dietary essential and no deficiency signs are known.

Drinking water containing about 1 part per million of fluoride protects teeth from decay, and in some areas fluoride is added to drinking water to achieve this level. Naturally, the fluoride content of water ranges between 0.05 and 14 ppm.

Water containing more than about 12 ppm fluoride can lead to chalky white patches on the surface of the teeth, known as mottled enamel. At higher levels there is strong brown mottling of the teeth and inappropriate deposition of fluoride in bones known as fluorosis.

Fluoride Fluoride is a natural component of minerals in rocks and soils. Widespread use of fluoride in water supplies and oral health products is credited with the dramatic decline in dental caries among children and adults alike. All water contains fluoride, but it is sometimes necessary to add it to some public supplies to attain the optimal amount for dental health. Fluoride makes tooth enamel stronger and more resistant to decay. It also prevents the growth of harmful bacteria and interferes with converting fermentable carbohydrates to acids in the mouth.

Fluorosis Damage to teeth and bones caused by an excessive intake of fluoride.

FMN Flavin mononucleotide, one of the coenzymes derived from vitamin B 2.

Foam-mat drying A method of drying food. The liquid concentrate is whipped to a foam with the aid of a foaming agent, spread on a tray and dried in a stream of warm air. It reconstitutes very rapidly with water because of the fine structure of the foam.

Folding-in A method of combining a mixture of flour and other ingredients so that it retains its lightness; it is used for mixing meringues, souffies, and some cakes.

Fole gras The liver of goose or duck that has been specially fed and fattened; may be cooked whole or used as the basis of pate de foie gras.

Folic acid Folic acid, folate, folacin, all form a group of compounds functionally involved in amino acid metabolism and nucleic acid synthesis. Good dietary sources of folate include leafy, dark green vegetables, legumes, citrus fruits and juices, peanuts, whole grains and fortified breakfast cereals.

Recent studies show, if all women of childbearing age consumed sufficient folic acid, 50 to 70 percent of birth defects of the brain and spinal cord could be prevented, according to the U.S. Centers for Disease Control and Prevention (CDC.) Folic acid is critical from conception through the first four to six weeks of pregnancy when the neural tube is formed. This means adequate diet or supplement use should begin before pregnancy occurs.

Recent research findings also show low blood folate levels can be associated with elevated plasma homocysteine and increased risk of coronary heart disease.

Folinic acid The 5-formyl derivative of the vitamin folic acid; more stable to oxidation than folic acid itself, and commonly used in pharmaceutical preparations. The synthetic compound is known as leucovorin.

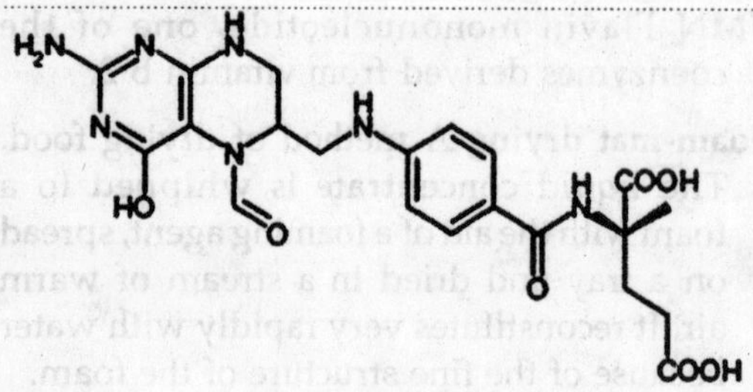

Fig. Folinic acid

Fondant Minute sugar crystals in a saturated sugar syrup; used as the creamy filling in chocolates and biscuits and for decorating cakes. Prepared by boiling sugar solution with the addition of glucose syrup or an inverting agent.

Fondue Cheese melted with wine and herbs, eaten by dipping small squares of bread into the hot mixture. Fondue bourguignonne is small cubes of marinated meat, cooked on a long fork in a vessel of hot oil at the table.

Fonduta Dish made from Fontina cheese and truffles.

Food Any solid or liquid material consumed by a living organism to supply energy, build and replace tissue, or participate in such reactions. Defined by the FAO/WHO Codex Alimentarius Commission as a substance, whether processed, semi-processed, or raw, which is intended for human consumption and includes drink, chewing gum, and any substance that has been used in the manufacture, preparation, or treatment of food but does not include cosmetics, tobacco, or substances used only as drugs.

Food additives An additive is defined as: any natural or synthetic material, other than the basic raw ingredients, used in the production of a food item to enhance the final product or any substance that may affect the characteristics of any food, including those used in the production, processing, treatment, packaging, transportation or storage of food.

Food additives are often referred to as E-numbers , as in the EU countries additives are numbered with a prefix E. The E thus refers to an approved additive.

Additives are not used to mask problems in the food, but are often used to prevent spoilage or other loss of quality.

There are many categories of food additives, such as : - acididity regulators and buffers

- anti-caking agents
- anti-oxidants
- bread enhancers
- emulsifiers
- food colours
- preservatives
- stabilisers
- thickening agents

All are tested for toxicity and safety. However, side effects can never completely be excluded.

Food allergy A food allergy causes a specific immunological reaction in the body that can be observed and diagnosed by various tests. An immunological reaction is a reaction against foreign molecules entering the body; normally, its purpose is to protect the body. In the case of food allergy, this reaction is extreme and is provoked by foods that have no adverse effects on most people. These adverse reactions are immediate and localised; common sites are the mouth, digestive tract, the airways and the skin.

Food borne Illness Also called food poisoning. Illness contracted by eating contaminated food. Effects range from mild fever, upset stomach, and diarrhea to dehydration and death. Examples of food-borne illnesses include hepatitis, botulism, and salmonellosis.

Food chain The chain between green plants through a sequence of organisms in which each eats the one below it in the chain, and is eaten in turn by the one above. Also used for the chain of events from the original source of a foodstuff through all the stages of handling until it reaches the table.

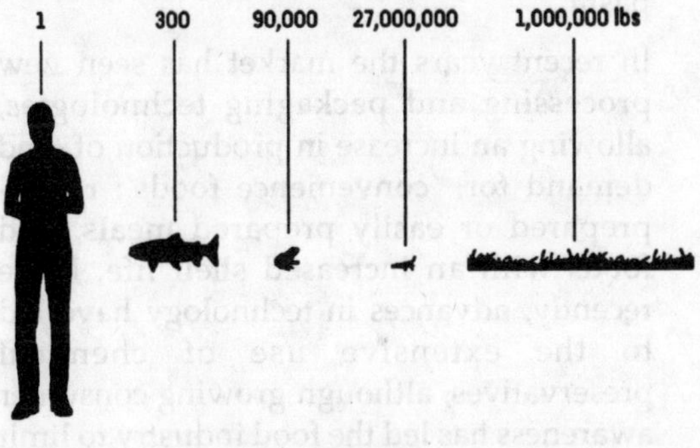

Fig. Food chain

Food composition tables Tables of the chemical composition, energy, and nutrient yield of foods, based on chemical analysis. Although the analyses are performed with great precision, they are, of necessity, only performed on a few samples of each type of food. There is, however, considerable variation, especially in the content of vitamins and minerals, between different samples of the same food. Therefore, calculation of energy and nutrient intakes based on use of food composition tables, even when intake has been weighed, can only be considered to be accurate to within about +10%, at best.

Food Hygiene, Hygienic The science and practice of ensuring food is safe to eat, that is, will not cause harm to the consumer of the food. Hygienic practices for fish processing and handling cover a wide range of activities, but the main elements are control of the origins of raw materials to ensure they do not contain noxious substances; control over the construction, equipping, running and maintenance of premises where fish is handled processed and stored to prevent contamination of the products with noxious agents; ensuring that persons coming in contact with the product adopt good hygienic practices; ensuring that fish, inputs to the process, and finished products are packaged and stored under conditions that preclude contamination.

Food irradiation A high-energy ionising radiation process used to preserve foods.

Food poisoning Illness as consequence of consuming food containing a noxious agent. Food poisoning is frequently associated with gastroenteritis but, in the wider context of food safety, illnesses other than those involving the digestive tract should be included. A wide variety of agents cause food poisoning, but they can be considered in two broad classes biological, and chemical. Biological agents are those than can grow and replicate in the body, and it is this growth that results in illness. Illnesses resulting from growth of organisms in the body are infections. Examples in fishery products are bacteria, viruses, and parasites. A general term for a chemical that can cause harm is poison, but this term applies to any injurious chemical whatever the route of exposure. Poisons that are generated by biological processes are referred to as toxins. The illness caused by a toxin is a toxicosis. Examples of poisons responsible for food poisoning that are not toxins are heavy metals such as cadmium and mercury. These can be present in the natural environment from both natural processes and from industrial source Others are highly chlorinated manufactured compounds like some pesticides, PCBs and dioxins. Examples of toxins are chemicals produced by bacterial growth in the food before consumption such as staphylococcal and botulinum toxins, and substances formed by plants or animals which are not human food, but are food of fish, including shellfish, and transmitted to humans who eat the contaminated products. Microbiological and viral agents in fish typically cause acute illnesses, that is, illnesses with moderately rapid onsets, i.e., hours, after consumption of the contaminated food. Parasites cause chronic diseases, illnesses with a slow onset that persist over years. Toxins poisoning typically results in acute illnesses with rapid onsets, whereas non-biological poisons can cause both acute and chronic illnesses.

Food preservative The treatment of food to prevent its deterioration and to maintain its nutritional value. Breakdown of food tissues is caused by enzymes, either contained within the food or produced by microorganisms—bacteria, yeasts, and fungi—growing in the food. These organisms can also produce substances that can cause food poisoning. Oxidation and dehydration also contribute to spoilage. Food preservation therefore aims to alter the condition of food to stop the activities of microorganisms and any chemical change. One of the oldest methods is drying or dehydration—used for meat, vegetables, cereals, milk products, etc.

Food processing Using food as a raw material and changing it in some way to make a food product.

Food retailing Businesses where the consumer can buy food products.

Food standard A set of criteria that a food must meet if it is to be suitable for human consumption, such as source, composition, appearance, freshness, permissible additives, and maximum bacterial content.

Food standards committee Permanent advisory body to the Ministry of Agriculture, Fisheries, and Food in the UK.

Food technology The application of scientific knowledge to the preparation, preservation, and storage of food. Food technology can slow or halt the natural degradation processes of certain foodstuffs and thus make foods available out of season. The palatability or nutritional qualities of many raw food materials can be enhanced by selective processing. For example, cereals may be crushed by milling to produce flour, a process which removes or breaks down the indigestible outer husk of the cereal seeds. The flour can be made more palatable by making it into bread or pasta.

In recent years the market has seen new processing and packaging technologies, allowing an increase in production of, and demand for, 'convenience foods': ready-prepared or easily prepared meals, and foods with an increased shelf life. More recently, advances in technology have led to the extensive use of chemical preservatives, although growing consumer awareness has led the food industry to limit the use of such food additives.

Modern food processing technology has also allowed the production of 'new' foods. possible to produce very good meat analogues from mycoprotein and vegetable proteins.

Foodborne disease Infectious or toxic disease caused by agents that enter the body through the consumption of food. The causative agents may be present in food as a result of infection of animals from which food is prepared or contamination at source or during manufacture, storage, and preparation.

There are three main categories: 1. diseases caused by micro-organisms that invade and multiply in the body;

2. Diseases caused by toxins produced by micro-organisms growing in the gastro-intestinal tract;

3. Diseases caused by the ingestion of food contaminated with poisonous chemicals or

containing natural toxins or the toxins produced by micro-organisms in the food.

Foodstuffs Defined in EU Directives as products intended for human consumption in an unprocessed, processed, or mixed state, with the exception of tobacco products, cosmetics, and pharmaceuticals.

Foo-foo Caribbean, West African; small dumplings made by pounding boiled green plantain.

Fool A puree of fruit with cream or custard.

Forcemeat A highly seasoned stuffing made from chopped or minced veal, pork, or sausage meat mixed with onion and a range of herbs.

Formic acid An organic acid, formula H-COOH. The simplest of the fatty acids. Used in fish technology for the preparation of silage.

Formula diet Composed of simple substances that do not require digestion, are readily absorbed, and leave a minimum residue in the intestine: glucose, amino acids or peptides, mono- and diglycerides rather than starch, proteins, and fats.

Fortification The deliberate addition of specific nutrients to foods in order to increase their content, sometimes to a higher level than normal, as a means of providing the population with an increased level of intake. Generally synonymous with enrichment, supplementation, and restoration; in the USA enrichment is used to mean the addition to foods of nutrients that they do not normally contain, while fortification is the restoration of nutrients lost in processing.

Fortified food A food with added nutrients, which were not originally present. May be interchanged with "enriched."

Fractional test meal A method of examining the secretion of gastric juices; the stomach contents are sampled at intervals via a stomach tube after a test meal of gruel. It is usual to test for total and free acidity, and in addition peptic activity may be measured.

Frail A rush basket for raisins, figs, etc.; also a quantity of raisins, usually about 75 lb (34 kg).

Frangipane (frangipani) Originally a jasmine perfume which gave its name to an almond cream flavoured with the perfume. The term is used for cake-filling made from eggs, milk, and flour with flavouring, and also for a pastry filled with an almond-flavoured mixture.

Frankfurter A seasoned, smoked beef and pork sausage (hot dog). A good source of protein and niacin; a source of vitamin B 1 and iron.

Frappe 1 Iced, frozen, or chilled. 2 Egg-white and sugar syrup whipped until so aerated that the density reaches 5 lb per gallon (100 g in 200 mL).

Free from For a food label or advertising to bear a claim that it is free from fat, saturates, cholesterol, sodium, or alcohol it must contain no more than a specified (low) amount. The precise levels at which such claims are permitted differ from one country to another. In the USA the food so described must contain only a trivial or physiologically insignificant amount of the specified nutrient.

Free radical A molecule with one or more unpaired electrons in its outer orbitals.

Freezant A refrigerant suitable for contact with food and used to freeze fish especially by spraying or by immersion.

Freeze concentration Concentration of a liquid by freezing out pure ice, leaving a more concentrated solution; it requires less input of energy, and causes less loss of flavour, than concentration by evaporation. Used especially in the concentration of fruit juices, vinegar, and beer.

Freeze drying Dehydrating fish by first freezing it and then subliming the ice that is formed under reduced pressure. In freeze drying highly efficient vacuum pumps attain very low pressures in a sealed chamber that contains the fish in contact with refrigerated plates to freeze the fish. Ice sublimes to water vapour at pressures below 0.64 kPa and the water vapour is removed by the vacuum pump.

Freezer Correctly refers to equipment for freezing fish either in a batch or continuously. Loose use has also established a meaning for the term as equipment for the cold or frozen storage of frozen fish, e.g. home freezer, walk-in freezer, upright freezer, chest freezer, and for derived terms such as freezer burn. A batch freezer is loaded completely at the start of freezing and emptied when freezing is completed; the product remains stationary. In a continuous freezer the fish is moved continuously on a conveyor belt (belt freezer), which may be an open mesh belt, a link belt or a flat continuous stainless steel sheet, or moved on trucks (truck freezer). But, confusingly, truck freezer also refers to a batch air blast freezer in which the fish remain stationary on trucks. There is in use a wide variety of methods of freezing and types of freezing equipment. In an air blast freezer, fish is frozen in a stream of high velocity cold air either in a batch or continuously, typically in a duct or tunnel in which a stream of cold air is guided over the product on shelves (batch) or on a conveyor; also called blast freezer, freezing tunnel, tunnel freezer. Mesh or slatted conveyor belts may mark the product with indentations and are difficult to keep clean, whereas a solid stainless steel belt does not mark the fish and is easy to keep clean. In good practice, air blast freezing can be achieved by using air at -35°C in contact with the product and moving over the product at an air speed or air velocity of 3 to 25 m/s. The air speeds at the upper end of the range are used on factory ships. The speed selected depends on the size and shape of the product, whether the process is batch or continuous, the stability of the product on the carrier, the final temperature of the product, and the economics. Prolonged higher air speeds can cause severe dehydration on the surface of the product, resulting in freezer burn. Generally, the aim is to quickly reduce the temperature of the warmest part of the product to -20°C at the thermal centre, so that the average temperature of the product leaving the freezer is at or below the desired temperature of -30°C. A continuous air blast freezer in which the cold air flows at right angles to the direction of movement of the fish is termed a crossflow freezer. In a series flow freezer, the air flows in the opposite direction to the product. A spiral freezer is a belt, air blast freezer in the form of a spiral to reduce floor space. The multi-pass freezer was also designed to reduce the floor space required. The product is transferred in the freezer from one belt to another, travelling backwards and forwards along the length of the freezer, usually in three passes, before discharge. In a fluidised bed freezer, freezing is done in a trough with a perforated base through which an upward flow of cold air entrains the produce, usually of small unit size such as shrimp, and causes it to behave like a liquid. In a contact freezer or plate freezer the fish is frozen by direct contact with a refrigerated surface, typically between two hollow metal plates cooled by a refrigerant, such that the distance between the plates can be varied up to 100 mm or more. One type in which fish, especially in flat packs such as laminated blocks, is frozen between two or more hollow, horizontal, parallel plates through which refrigerant passes is known as a horizontal plate freezer. In a vertical plate freezer, the refrigerated, parallel plates are vertical, and it is used mainly at sea or onshore for freezing large 25 or 50 kg blocks of whole, gutted, or headed and gutted fish. The immersion freezer or spray freezer is mostly used for IQF products of small size frozen within minutes. The fish product to be frozen is in direct contact with the refrigerated liquid (freezant). One type is brine freezing by immersing fish such as tuna in a tank of recirculating cold sodium chloride brine at about -15°C in a brine immersion freezer that may take about 3 days to freeze a large tuna completely. In a liquid nitrogen freezer, freezing is by spraying liquid nitrogen at -196°C onto, or into the air around, a fish product, or by a combination of these. The product is usually moved continuously through the freezer on a conveyor, and the system is designed to avoid stress cracking of the product as a result of cooling too rapidly. Similarly, in a carbon dioxide freezer there is continuous rapid freezing by spraying liquid carbon dioxide onto, or into the air around, a fish product, or by a combination of these, or by placing the product on a bed of solid carbon dioxide on a conveyor belt under a spray of liquid carbon dioxide. A temperature of about -75°C can be achieved. Freezing in both these systems is very rapid within minutes and drip losses on thawing are less than 1%.

Freezer burn Damage to frozen fish caused either by excessive drying during cold (frozen) storage as a result of undesirable migration of water from the frozen fish, or by excessive drying during air blast freezing at high air speeds for excessive lengths of time. It is usually indicated by matt white patches on the surface of the frozen fish and a dull, non-glossy surface on the thawed product. Also termed dehydration, dessication and drying.

Freezer capacity The weight of a specified product that a freezer can hold and also called the freezer load, whereas the freezer output is the weight of a specified product that can be frozen in unit time to the required storage temperature under specified operating conditions.

Freezer temperatures For long-term storage of frozen foods, domestic freezers run at -18°C; in the UK this is a 3-star rated deep freeze. A freezing compartment of a refrigerator is between -11°C, 2-star rated, for storage up to 4 weeks; and -4°C, 1-star rated, for storage up to a week. A 3-star rated deep freeze with a snowflake symbol is one that is suitable for freezing foods, as opposed to storing ready-frozen food; it has a higher cooling capacity than a simple deep freeze storage cabinet.

Freezing Extracting heat and reducing the temperature of fish in a freezer to a point at which almost all of the water in it has solidified. About 321 kJ of heat have to be removed from 1 kg of fillets from a lean fish such as cod to freeze them from 5°C to -30°C. The temperature of the fish falls rapidly to the initial freezing point at about -1°C when the phase change begins and ice crystals are first formed, because salts and other substances that are present naturally result in depression of the freezing point from 0°C. Thus, fish does not have a precise freezing point. The temperature then falls slowly between -1°C and -5°C, whilst a large amount of heat is extracted to freeze the bulk of the water in the flesh to ice; heat removed at this stage is termed latent heat, and this temperature range is known as the critical freezing zone, or the thermal arrest. After about 75% of the water in the fish is frozen and the critical zone is passed, the temperature again falls rapidly as the frozen fish is cooled further. The recommended temperature to which fish should be reduced during freezing is -30°C, at which temperature the fish should be stored. In practice, with a freezer operating at a temperature of -35°C to -40°C or even lower, the aim is to lower the temperature in the warmest part of the fish, the thermal centre usually near the centre of the thickest part of the fish, as quickly as possible through the thermal arrest range to -20°C, so that the average temperature of the product leaving the freezer is at or below the recommended temperature of -30°C within 4 hours or less. This is known typically as quick freezing, but a precise definition is difficult. Another way of describing it is the recommendation for fish that the temperature range of the thermal arrest should be passed in 2 hours or less, but this period should be much shorter (minutes) for products of small unit size such as shrimp. In contrast, slow freezing describes a freezing process that is undesirably long. It is sometimes referred to as sharp freezing, as in the bad practice of blowing cold air around produce in a cold store, and it is a term to be avoided. The total time required to reduce the equilibrium temperature of the fish to that at which it is to be stored is the freezing time. Other imprecise terms used to describe the freezing process include deep freezing used especially in the context of quick freezing, to describe freezing of fish to a temperature of -18°C or below; flash freezing, refers to the quick freezing of small fish products such as shrimp, typically in a few minutes, and also called snap freezing. Bulk freezingrefers to freezing loose unpackaged products or to freezing large quantities of fish for storage prior to further processing. Individual quick freezing (IQF) is the quick freezing of fish products, e.g. fillets, in such a manner that each unit of product remains separate when frozen, in contrast to a frozen block of product (e.g. laminated block). For methods of freezing or types of freezing equipment such as air blast, contact, fluidised bed, immersion and spray.

French toast North American breakfast dish; slices of bread dipped in beaten egg, fried, and served with cinnamon and sugar. Known in France as pain perdu.

Frenching Breaking up the fibres of meat by cutting, usually diagonally or in a criss-cross pattern.

Fresh Describes the state, for fish, of being recently caught or harvested, and hence showing no signs of spoilage. A meaning, which should be discouraged, is the condition of not having been frozen; this meaning does not preclude the product's being stale.

Freshness A term that is not used with consistent meaning in fish technology. Most frequently is applied to the condition of fish throughout the complete sequence of changes during storage above freezing temperature in going from completely fresh to utterly spoiled. This sequence has two main successive stages. During the initial period of storage some of the characteristics of the fresh product reduce in intensity or are lost; in the later stage bacterial degradation of tissues, i.e. spoilage, becomes evident. It might be considered more correct to apply 'freshness' only to the condition of fish in the first stage, and, say, 'staleness' to the condition in the later stage, but this makes for cumbersome terminology, and it would be more convenient to put pedantry aside and apply 'freshness' to the condition during the complete sequence. Freshness is most accurately measured by sensory methods, though various chemical, microbiological, and physical tests have been proposed, studied, and occasionally applied in quality control as surrogates.

Friandises A variety of small sweets, preserved fruits, etc., served as petits fours or desserts.

Fricadelle Minced meat balls.

Fricandeau Dish made from the long fillet of veal, braised or roasted.

Fricassee A combination of sauteing and stewing; food is fried briefly with a small amount of fat, then stewed.

Frijole bean Phaseolus acutifolius, also known as Mexican haricot bean, tepary, or pinto. Able to withstand drought.

Frijoles A Mexican dish based on boiled fava or lima (butter) beans which have been left to cool, then fried. Also known as refried beans.

Fritter A portion of sweet or savoury food, coated in batter and fried.

Fritto misto Small thin pieces of a variety of types of meat, fish, or vegetables, coated with egg and breadcrumbs or batter, and deep fried.

Frizzante Lightly sparkling wines, equivalent to French petillant.

Frogs' legs The back and legs of the edible frog, Rana esculenta.

Fromage a la creme Sour milk cheese. The curd is drained, mixed with cream, and pressed in heart-shaped moulds.

Fromage frais French equivalent of quark. A soft, unripened cheese, 80% water, made from skim milk or semi-skim milk, and may include added cream; 1-8% fat; a rich source of vitamin B 12, and a good source of vitamin B 2.

Frosting 1. American name for icing on cakes; in the UK icing made from sugar and egg white.

2. A way of decorating the rim of a glass in which a cold drink is to be served; the edge is coated with whipped egg white, dipped into caster sugar, and allowed to dry.

3. For a margarita cocktail the edge of the glass is frosted by dipping it into lemon juice, then salt.

Frothing Dredging the surface of roast joints and poultry with flour before basting, in order to give an attractive brown finish.

Frozen at sea Fish frozen soon after capture, either onboard a fishing vessel with freezing plant, such as a freezer trawler, or on a factory ship that is equipped to process its own catch onboard and/or the catch of other fishing vessels.

Fructosan A general name for polysaccharides of fructose, such as inulin. Not digested, and hence a part of dietary fibre or non-starch polysaccharides.

Fructose Fructose is a monosaccharide found naturally in fruits, as an added sugar in a crystalline form and as a component of high-fructose corn syrup (HFCS) and inulin.

Fructose syrups Glucose syrups made by hydrolysis of starch and half the glucose then being converted into fructose, similar to invert syrup produced from sucrose but cheaper. Also known as iso-syrups, high fructose syrups, and high fructose corn syrups.

Fruit Fruit is the usually edible reproductive body of a seed plant, especially one having a sweet pulp associated with the seed.

Fruitarian A person who eats only fruits, nuts, and seeds; an extreme form of vegetarianism.

Frumenty An old English pudding made from whole wheat stewed in water for 24 hours until the grains have burst and set in a thick jelly. then boiled with milk. Similar to flummery.

Frying Cooking foods with oil at temperatures well above the boiling point of water. Deep frying, in which a food is completely immersed in oil, reaches a temperature around 185°C. Nutrient losses are less than in roasting, about 10-20% thiamin, 10-15% riboflavin and nicotinic acid from meat; about 20% thiamin from fish.

Frying Cooking foods with oil at temperatures well above the boiling point of water. Deep frying, in which a food is completely immersed in oil, reaches a temperature around 185°C. Nutrient losses are less than in roasting, about 10-20% thiamin, 10-15% riboflavin and nicotinic acid from meat; about 20% thiamin from fish.

Fumet Concentrated stock from meat, fish, or vegetables prepared by boiling down to a syrupy consistency, used to give flavour and body to sauces.

Funchi Caribbean; corn meal pudding.

Functional foods The term "functional foods" refers to a food that provides a health benefit as well as nutrients. The term can refer to whole foods, to fortified, enriched or enhanced foods, and dietary supplements that have the potential to improve mental and physical well being and reduce the risk of diseases.

The ingredients responsible for this benefit can be naturally present or may have been added during processing. The levels of nutrients in foods can be increased beyond their natural levels to create an enriched product. Fortified products contain nutrients or ingredients that were not present in the original food.

Fungus Organisms including yeasts, molds, smuts and mushrooms.

Furcellaran Danish agar; an anionic, sulphated polysaccharide extracted from the red alga, Furcellaria fastigiata, structurally similar to carrageenan; used as a gelling agent.

Fusel oil Alcoholic fermentation produces about 95% alcohol and 5% fusel oil: a mixture of organic acids, higher alcohols, aldehydes, and esters, known collectively as congeners. It is present in low concentration in wines and beer, and in high concentration in pot-still spirit. On maturation of the liquor fusel oil changes and imparts the special flavour to the spirit Many of the symptoms of hangover can be attributed to fusel oil in alcoholic beverages.

Fusion Joining the membrane of two cells of different origin to create a cell that contains the parent cells' nuclear material. Used in monoclonal antibody technology to make hybridomas - fusing an immortal cell and an anti body-producing cell.

Fussol Monofluoroacetamide, a systemic insecticide for treating fruit.

Fu-yung Chinese dishes prepared with egg-white and cornflour mixed with minced chicken; commonly applied to a variety of dishes with egg scrambled into the mixture.

G

Galactose A six-carbon sugar differing from glucose only in position of the hydroxyl group on carbon-4. It is about onethird as sweet as sucrose. The main dietary source is as part of the disaccharide lactose in milk, and it is important in formation of the galactolipids of nerve tissue.

Gammon Hind legs of bacon pig, cured while still part of the carcass. Ham is the same part of the pig but is cured after removal from the carcass.

Gaping The separation of blocks of muscle in a raw fillet at the myocommata, the thin membrane that separates the blocks. In severe cases the fillet will fall apart. The propensity to gape differs markedly among species. Generally, roundfish gape more than flat fish, but there are important exceptions to this generalisation; ling and catfish, for example, rarely show gaping.

Garam masala A mixture of aromatic spices widely used in Indian cooking; contains powdered black pepper, cumin, cinnamon, cloves, mace, cardamom seeds, and sometimes also coriander and/or bay leaf.

Garibaldi biscuit Square or rectangular biscuit with a layer of currants inside.

Garlic mustard A common wild plant of hedgerows and woodland; the leaves have a garlic-like flavour and can be used in salads or cooked as a vegetable.

Garlic The bulb of Allium sativum with a pungent odour when crushed, widely used to flavour foods. There is some evidence that garlic has a beneficial effect in lowering blood cholesterol.

Garnish To decorate food.

Garrafeira Portuguese; aged, strong table wines.

Gas storage, controlled Storage of fruits, vegetables, and prepacked red meat in a controlled atmosphere in which a proportion of the oxygen is replaced by carbon dioxide, sometimes with the addition of other gases. In modified atmosphere (MA) storage the control is less precise.

Gasohol Fuel; blend of ethanol and unleaded gasoline, usually 10 percent ethanol and 90 percent gasoline.

Gastric acidity A number of digestive disorders are popularly called 'gastric acidity', but the principal nutritional problem in this area is the reduction in gastric acid secretion that occurs with advancing age; this limits the absorption of iron and vitamin B 12.

Gastric inhibitory peptide A hormone secreted by the mucosa of the duodenum and jejunum in response to absorbed fat and carbohydrate which stimulates the pancreas to secrete insulin. Also known as glucose-dependent insulinotropic polypeptide.

Gastric secretion Gastric juice contains the enzymes chymosin and lipase, the inactive precursor of pepsin, intrinsic factor, mucin, and hydrochloric acid. The acid is secreted by the parietal cells at a strength of 0.16 mol/L (05-0.6% acid); the same cells also secrete intrinsic factor, and failure of acid secretion is associated with a failure to absorb vitamin B 12, because of lack of intrinsic factor. Pepsinogen is secreted by the 'chief' cells of the gastric mucosa, and is activated to pepsin by either gastric acid or the action of existing pepsin; it is a proteolytic enzyme. The only function of chymosin is to coagulate milk; the lipase hydrolyses a small proportion of dietary fat.

Gastrin Polypeptide hormone secreted by the stomach in response to foods which stimulate gastric and pancreatic secretion.

Gastritis Inflammation of the mucosal lining of the stomach; may result from infection or excessive alcohol consumption.

Gastro-enteritis Inflammation of the mucosal lining of the stomach and/or small or large intestine, normally resulting from infection.

Gastroenterology The study and treatment of diseases of the gastrointestinal tract.

Gastro-intestinal tract The whole of the digestive tract, from the mouth to the anus.

Gastrostomy feeding Feeding a liquid diet directly into the stomach through a tube that has been surgically introduced through the abdominal wall.

Gateau Elaborate cake with sponge, biscuit, or pastry base and topped with fruit, jelly, and cream.

Gavage The process of feeding liquids by stomach tube. Also feeding an excessive amount.

Gazpacho Spanish; vegetable soup made by blending green pepper, cucumber, onions, and tomatoes, uncooked, with bread crumbs, olive oil, vinegar, and garlic; served ice-cold with a garnish of chopped onion, cucumber, red and green peppers, etc.

Gean Scottish name for the fruit of Prunus avium avium; also known as wild cherry, sweet cherry, and mazzard.

Gefillte fish The dish is of Russian or Polish origin where it is commonly referred to as Jewish fish. The whole fish is served and the filleted portion chopped and stuffed back between the skin and the backbone. More frequently today, the fish is simply chopped and made into balls, which are either fried or boiled. In the UK it has been legally referred to as 'fish cutlets in fish sauce' rather than fish cake.

Geist German name for liqueur made by macerating soft fruit in alcohol, then distilling.

Gel A sol or colloidal suspension that has set to a jelly.

Gelatine A soluble protein prepared from collagen or bones by boiling with water: used for sugar confectionery, in canned meats, for table jellies, and in pharmaceutical capsules. Gelatine from fish is isinglass. As a protein it is of poor nutritive value, since it lacks tryptophan.

Gelatinise Thicken.

Gelatinous Something that is gelatinous is thick and like jelly.

Gellan gum A polysaccharide produced by fermentation of the bacterium, Pseudomonas elodea, used in some foods.

Gene A sequence of DNA that represents a fundamental unit of heredity. Most genes code for RNA that is translated into protein, but some genes simply code for certain forms of RNA.

Gene splicing Inserting new genetic information into a chromosome using recombinant DNA techniques.

Generic descriptor The name used to cover the different chemical forms of a vitamin; compounds that have the same vitamin activity.

Generoso Portuguese, Spanish; fortified wines.

Genetic analysis Studying how traits and genes for traits are passed from generation to generation, and how genes and the environment interact to result in traits.

Genetic code Information coded within nucleotide sequences of RNA and DNA that specifies the amino acid sequence in protein synthesis and on which heredity is based.

Genetic counseling Providing current or prospective parents with information on the probabilities of inherited diseases occurring in their children, and on diagnosis and treatment of such diseases.

Genetic diseases Also known as inborn errors of metabolism. Diseases due to a single defective gene, with a characteristic pattern of inheritance in families. Many affect the ability to metabolize individual amino acids or carbohydrates and can be treated by dietary restriction.

Genetic engineering Using recombinant DNA techniques and related methods to move one or several genes from one organism to another, to rearrange one or several genes within a cell, or to alter gene-controlled processes. Transferring a DNA segment from one organism and inserting it into the DNA of another organism to modify, amplify, transform and express genetic information. The two organisms can be totally unrelated.

Genetic modification A change in the genes in a living organism, as occurs in nature, and which has been used for many years in selective breeding, or, more quickly and specifically, in the laboratory.

Genoa cake A rich, dark, fruit cake containing glace cherries and decorated with almonds or brazil nuts.

Genoese Whisked sponge mixture of eggs, sugar, flour, and soft creamed butter. Victoria sponge is a type of genoese.

Gentiobiose A disaccharide consisting of two molecules of glucose joined 1,6-ß.

Gentleman's relish Trade name for a paste of anchovies, butter, cereal, salt, and spices developed in the UK in the nineteenth century; also called patum peperium.

Genus Category of organisms ranking above a species and below a family.

Gerber test An analytical test for the fat.

Germ, wheat The embryo or sprouting portion of the wheat berry, comprising about 2.5% of the seed. Contains 64% of the vitamin B 1, 26% of the vitamin B 2, 21% of the vitamin B6, and most of the fat of the wheat grain. It is discarded, with the bran, when the grain is milled to white flour.

German pound cake Similar to genoa cake, but containing less fruit and no almonds.

Gewürztraminer A grape variety widely used for wine making, although not one of the classic varieties; the wines have a characteristic spicy flavour and aroma.

Gherkin Young green cucumber of a small variety, used mainly for pickling.

Giardiasis Intestinal inflammation and diarrhoea caused by infection with the protozoan parasite Giardia lamblia.

Giberellins Plant growth substances derived from giberellic acid, originally found in the fungus Gibberella fujikori growing on rice. About thirty giberellins are known; they cause stem extension and allow mutant dwarf forms of plants to revert to normal size, induce flower formation, and break bud dormancy. They are used to accelerate the germination of barley for malting.

Gibier wild animals that are hunted for food, game.

Giblets The edible part of the entrails of a bird; gizzard, liver, heart, and neck.

Gigot Leg of lamb or mutton. In Ireland gigot chops are neck chops used for stewing. gill Obsolete British measure of liquid, 5 or 10 fl oz (1/4 or 1/2 pint), varying regionally. Still used as the basis for measure of spirits in the UK, a standard measure is 1/5 or 1/6 of a 5 fl oz gill.

Gin Alcoholic drink made by distilling fermented starch or other carbohydrate, flavoured mainly with juniper berries together with coriander seeds, angelica, cinnamon, and orange and lemon peel. Distillate is diluted to 40% alcohol by volume, 220 kcal (925 kJ) per 100 mL. The name is derived from the French genièvre (juniper); originally known as geneva, schiedam, or hollands, since it is Dutch in origin. There are two types of English gin: Plymouth gin, with a fuller flavour, and London gin. Dutch and German gins are more strongly flavoured than English or American; steinhäger and schinkenhäger are distilled from a mash of wheat, barley, and juniper berries; wacholder is made from neutral spirit flavoured with juniper. Dutch gin may be jonge (young) or oude.

Ginger The rhizome of Zingiber officinale, used as a spice. Preserved ginger is made from young fleshy rhizomes boiled with sugar and either packed in syrup or crystallized.

Gingerbread Cake flavoured with ginger and treacle, often baked in the shape of an animal or person, and glazed.

Gingivitis Inflammation, swelling, and bleeding of the gums; may be due to scurvy, but most commonly the result of poor dental hygiene.

Gipping Partial evisceration to remove intestines but not pyloric caeca which contains enzymes responsible for the characteristic flavour of herring when it is subsequently salted.

Girdle Griddle.

Girolle French name for the chanterelle, an edible wild mushroom, Cantharellus cibarius.

GLA A long-chain polyunsaturated fatty acid from the omega-6 family. It is used for increasing the body's PGE1 anti-inflammatory system.

Glace 1. Iced or frozen.

2. Having a smooth glossy surface or glaze.

Glace cherry Cherry preserved in a heavy syrup of sugar and glucose, generally also with added colour. Used in confectionery and fruit cakes. Little or none of the original vitamin C content of the cherry will remain.

Glamorgan sausage A dish based on Caerphilly cheese, breadcrumbs, and egg, fried in sausage shape.

Glayva Scottish whisky-based liqueur, flavoured with herbs and sweetened with heather honey.

Glaze To brush the surface of sweet or savoury dishes to give a glossy surface or to assist in browning.

Glazing and Glaze, Glazed product Applying a protective coating of ice, an ice glaze, to unwrapped frozen fish, fillets, tailmeat of crustaceans, adductor muscle of scallops or other suitable frozen fish product, to reduce undesirable drying or dehydration of the fish during frozen or cold storage. Excessive drying during frozen storage results in freezer burn. The glaze is usually applied by dipping, spraying or brushing with water, which must be of drinking quality. The amount of glaze picked up by the product depends on the temperature of the fish, the temperature of the water, the size and shape of the product, its surface area, and the glazing time. The application of glaze is difficult to control and it is often applied in an uncontrolled way so that the amount added is not constant and the thickness is not uniform. Consequently, the amount of protection afforded by the glaze is unpredictable, and there can be problems with product specifications and the legal requirements for retail products concerning description of the glazed product, especially the declared glazed weight. See Drained weight.

Gliadin One of the proteins that make up wheat gluten. Allergy to, or intolerance of, gliadin is coeliac disease.

Globins Proteins that are rich in the amino acid histidine, relatively deficient in isoleucine, and contain average amounts of arginine and tryptophan. Often found as the protein part of conjugated proteins such as haemoglobin.

Globulins Class of proteins that are heat-coagulated and soluble in dilute solutions of salts; they differ from albumins in being relatively insoluble in water. They occur in blood, milk, and some plants, e.g. edestin from hemp seed and amandin from almonds.

Gloss The sheen on raw fillets and smoked fillets. The loss of gloss, especially in the gloss of smoked fillets, is an indication that the product has been held under poor frozen storage conditions.

Glossitis Inflammation of the tongue; may be one of the signs of riboflavin deficiency.

Gloucester cheese Originally there were single and double Gloucester versions with minor differences; by far the commonest now is double Gloucester, similar in chemical and nutrient composition to Cheddar cheese.

Glucagon A hormone secreted by the a-cells of the pancreas which causes an increase in blood sugar by increasing the breakdown of liver glycogen and stimulating the synthesis of glucose from amino acids.

Glucans Soluble but undigested complex made of glucose units; found particularly in oats, barley, and rye.

Glucaric acid Alternative name for saccharic acid, the dicarboxylic acid derived from glucose.

Glucide Name occasionally used for saccharin.

Glucitol Obsolete name for sorbitol, one of the sugar alcohols or glycitols.

Glucocorticoids The steroid hormones secreted by the adrenal cortex which regulate carbohydrate metabolism.

Glucomannan A polysaccharide consisting of glucose and mannose.

Gluconeogenesis The metabolic processes involved in the formation of glucose from non-carbohydrate precursors, such as glycerol, lactate, and a variety of amino acids.

Gluconic acid The acid formed by oxidation of the hydroxy group on carbon-1 of glucose to a carboxylic acid group. Also termed dextronic acid, maltonic acid, and glycogenic acid.

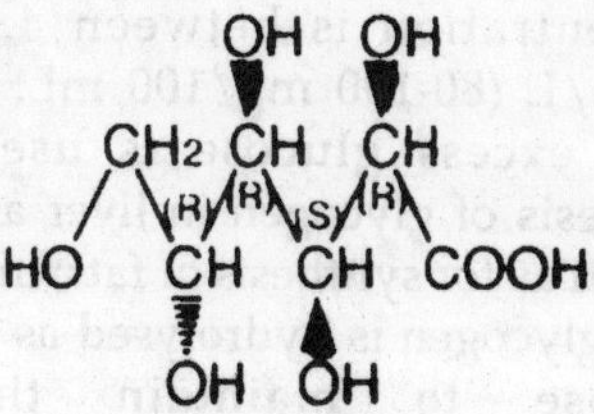

Fig. Gluconic acid

Glucono-o-lactone A derivative of gluconic acid; slowly liberates acid at a controlled rate; used in chemically leavened bread to liberate carbon dioxide from bicarbonate.

Glucosamine The amino derivative of glucose, a constituent of a variety of complex polysaccharides.

Glucosan A general term for polysaccharides of glucose, such as starch, cellulose, and glycogen.

Glucose A six-carbon monosaccharide sugar, with the chemical formula $C_6H_{12}O_6$, occurring free in plant and animal tissues and formed by the hydrolysis of starch and glycogen. Also known as dextrose, grape sugar, and blood sugar.

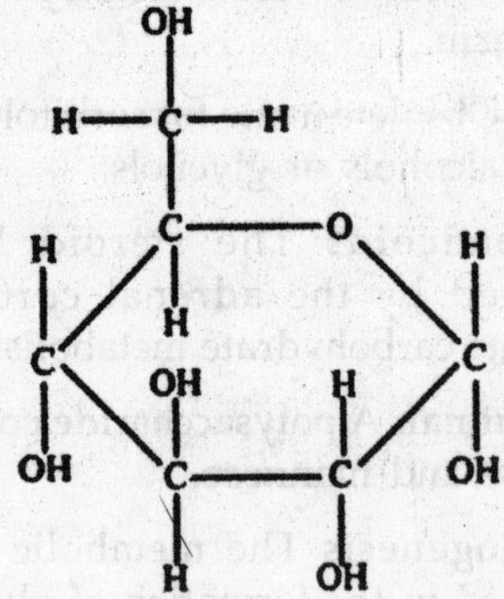

Fig. Structural formula for Glucose

The major dietary carbohydrates are starches, which are polymers of glucose and disaccharides: sucrose; lactose; maltose and isomaltose, which are dimers of glucose.

The circulating carbohydrate in the bloodstream is glucose; the normal concentration is between 4.5 and 5.5 mmol/L (80-100 mg/100 mL. In the fed state excess glucose is used for the synthesis of glycogen in liver and muscle, as well as for synthesis of fats; in the fasting state glycogen is hydrolysed as a source of glucose to maintain the blood concentration.

It is used in the manufacture of confectionery, since its mixture with fructose prevents sucrose from crystallizing

Glucose metabolism Process through which glucose is oxidized to carbon dioxide and water as a metabolic fuel. The oxidation of glucose to carbon dioxide and water yields 3.9 kcal (16.4 kJ)/g, or 686 kcal. The first series of reactions does not require oxygen and is referred to as glycolysis or glucose fermentation, yielding two molecules of the three-carbon compound pyruvic acid. Under anaerobic conditions this can be reduced to lactic acid.

Pyruvic acid is normally oxidized to acetyl CoA, which is then oxidized to carbon dioxide and water in a series of reactions known as the citric acid or Krebs cycle. Both glycolysis and the citric acid cycle are linked to the formation of ATP from ADP and phosphate, as a metabolically usable energy source.

Glucose oxidase Enzyme that oxidizes glucose to gluconic acid, with the formation of hydrogen peroxide. Used for specific quantitative determination of glucose, including urinary glucose excreted in diabetes, and also to remove traces of glucose from foodstuffs. Originally isolated from the mould Penicillium notatum and called notatin.

Glucose tolerance The ability of the body to deal with a relatively large dose of glucose is used as a test for diabetes mellitus. The fasting subject ingests 50 or 75 g of glucose and the concentration of blood sugar is measured at intervals. In normal subjects the fasting sugar concentration is between 4.5 and 5.5 mmol/L, and rises to about 7.5 mmol/L, returning to the starting level within 1-11/2 hours. In diabetics, the blood sugar concentration rises considerably higher and takes longer to return to the baseline value. The graph of the results forms a glucose-tolerance curve.

Glucose, confectioners' Glucose syrups are known as glucose in confectionery making.

Glucosides Complexes of substances with glucose. The general name for such complexes with other sugars is glycosides.

Glucosinolates Substances occurring widely in plants of the family Crucifera, genus Brassica; broken down by the enzyme myrosinase to yield, among other products, the mustard oils which are responsible for the pungent flavour . Several of the glucosinolates interfere with the metabolism of iodine by the thyroid gland, and hence are goitrogens; chemically they are thio-esters. There is evidence that the various glucosinolates in vegetables may have useful anti-cancer activity, since they increase the rate at which a variety of potentially toxic and carcinogenic compounds are conjugated and excreted.

Glucostatic mechanism A theory that appetite depends on the difference between arterial and venous concentrations of glucose; when the difference falls low enough, the hunger centres in the hypothalamus are stimulated.

Glucosuria; appearance of glucose in the urine, as in diabetes and after the administration of drugs that lower the renal threshold.

Glucuronic acid The acid derived from glucose by the oxidation of the hydroxyl group on carbon-6. Many substances, including hormones and potentially toxic ingested substances, are excreted as conjugates with glucuronic acid, known as glucuronides. It is present in various complex polysaccharides.

Glucuronides A variety of compounds are metabolized by conjugation with glucuronic acid to yield water-soluble derivatives for excretion from the body.

Glühwein Spiced hot wine punch, generally fortified by the addition of rum.

Glutamate Salts of glutamic acid.

Glutamic acid A non-essential amino acid; it is acidic since it has two carboxylic acid groups; its amide is glutamine.

Glutamine A non-essential amino acid, chemically the amide of glutamic acid. glutathione A tripeptide of glycine, glutamic acid, and cysteine which is involved in oxidation-reduction reactions, the conjugation of foreign substances for excretion and possibly the transport of amino acids into cells.

Glutathione A tripeptide composed by glutamic acid, cysteine, and glycine, widely distributed in cells. It participates in various redox reactions, such as the destruction of peroxides and free radicals. When reacting with a free radical it si oxidized to GSSG, and is reconverted back to the reduced GSH form by glutathione reductase. It is the substrate of gutathione peroxidase.

Glutathione peroxidase An enzyme of the oxidoreductase class that catalyzes the detoxifying reduction of hydrogen peroxide and organic peroxides using glutathione as substrate.

Glutelins Proteins insoluble in water and neutral salt solutions but soluble in dilute acids and alkalis, e.g. wheat glutenin.

Gluten A protein which is contained in wheat.

Gluten enables dough to rise when baking, by trapping bubbles of carbon dioxide produced by the yeast in the mixture. Some people are allergic to gluten, and have to have a gluten-free diet.

Gluten-free foods Formulated without any wheat or rye protein for people suffering from coeliac disease.

Glutose A six-carbon sugar with a keto group on carbon-3; it is not metabolized and non-fermentable.

Glycaemic index The ability of a carbohydrate to increase blood glucose, compared with the same amount of glucose: the increase in blood glucose over 2 hours after ingesting 50 g of available carbohydrate, expressed as a percentage of that after 50 g of glucose.

Glycerides Esters of glycerol with fatty acids. Since glycerol has three hydroxyl groups, it can be esterified with three molecules of fatty acid to form a triglyceride, the main type of fat in the diet and body, sometimes known as simple fats or neutral fats. If all three molecules of fatty acid are the same, the product is a simple triglyceride, e.g. tristearin, triolein; mixed glycerides may also be formed such as distearo-olein and stearo-oleopalmitin.

Glycerol A trihydric alcohol, chemically 1,2,3-propane triol, also known as glycerine. Simple or neutral fats are esters of glycerol with three molecules of fatty acid, i.e. triacylglycerols, sometimes known as triglycerides. Glycerol is a clear, colourless, odourless, viscous liquid, sweet to taste; it is made from fats by alkaline hydrolysis. Used as a solvent for flavours, as a humectant to keep foods moist, and in cake batters to improve texture and slow down staling.

Glyceryl lactostearate Also known as lactostearin. Formed by glycerolysis of hydrogenated soya bean oil followed by esterification with lactic acid, which results in a mixture of mono- and diglycerides and their lactic mono-esters. Used as an emulsifier in shortenings.

Glycine A non-essential amino acid, chemically the simplest of the amino acids, it is amino-acetic acid, CH_2N_2COOH. It has a sweet taste and is sometimes mixed with saccharin as a sweetening agent. Known at one time as collagen sugar.

Glycogen A complex carbohydrate found in animal tissues and acts as an energy reserve. It is present in the liver and muscle of fish and shellfish. It is readily converted in the muscle to glycogen which provides energy for muscle activity. The glycogen content of fish muscle is influenced by the nutritional status of the fish and is low when the fish is not feeding. On the death of the fish, the glycogen in the muscle is converted to lactic acid. This causes the pH of the muscle to drop from its typical value in the living animal of around 7.2. The drop in pH is particularly large in species such as tunas the muscles of which contain large amounts of glycogen.

Glycogen storage diseases A group of rare genetic diseases characterized by excessive accumulation of glycogen in liver and/or muscles and, in some forms, profound hypoglycaemia in the fasting state. Treatment is by feeding small frequent meals, rich in carbohydrate.

Glycogenesis The synthesis of glycogen from glucose in liver and muscle after a meal, stimulated by the hormone insulin.

Glycogenolysis The breakdown of glycogen to glucose for use as a metabolic fuel and to maintain the normal blood concentration of glucose in the fasting state. Stimulated by the hormone glucagon.

Glycolysis The first sequence of reactions in glucose metabolism, leading to the formation of two molecules of pyruvic acid from each glucose molecule.

Glycoproteins Proteins conjugated with carbohydrates such as uronic acids, polymerized glucosamine-mannose, etc., including mucins and mucoids; found in the vitreous humour of the eye, cornea, cartilage, and gastric mucosa.

Glycosides Compounds of a sugar attached to another molecule. When glucose is the sugar, they are called glucosides. A wide variety occur in plants and some, such as digitalis and turin rutin, are useful medicinally.

Glycosuria Appearance of glucose in the urine, as in diabetes and after the administration of drugs that lower the renal threshold.

Glycyrrhizin Triterpenoid glycoside extracted from liquorice root Glycyrrhiza glabra; 50-100 times as sweet as sucrose but with liquorice flavour. Used to flavour tobacco and pharmaceutical preparations, and as a foaming agent in some non-alcoholic beverages.

Gnocchi Square, round, or other shaped pasta used to garnish soups or served as a savoury dish with cheese sauce. May also be made from potato flour.

Gob stopper Large, spherical, hard sugar sweet, usually flavoured with mint and containing a caraway seed in the centre.

So called because it fills the mouth for a long time while it is slowly dissolving.

Gofio Flour milled from toasted wheat grains.

Goflo Bread made from powdered bracken roots and barley meal.

Goitre Enlargement of the thyroid gland, seen as a swelling in the neck, most commonly due to deficiency of iodine in the diet or to the presence of goitrogens in foods. In such cases there is commonly underproduction of the thyroid hormones, i.e. hypothyroid goitre. Euthyroid goitre is a condition in which the enlargement of the thyroid gland is sufficient to compensate for a limited dietary deficiency of iodine and permit normal production of thyroid hormones.

In infancy, iodine deficiency can also lead to severe mental retardation, goitrous cretinism. Supplementation with iodide often prevents the condition, hence the use of iodized salt. Rarely, goitre may be due to other causes, including excessive stimulation of the thyroid gland. In this case there is overproduction of the thyroid hormones, i.e. hyperthyroid goitre.

Goitrogens Substances found in foods which interfere with the synthesis of thyroid hormones or the uptake of iodide into the thyroid gland, and hence can cause goitre, especially when the dietary intake of iodide is marginal. Although goitrogens are associated with underactivity of the thyroid in animals, it is not established that they are a cause of goitre in human beings.

Golden syrup Light-coloured syrup made by evaporation of cane sugar juice.

Goldwasser Literally 'gold water'; an aniseed- and caraway- or cuminflavoured liqueur containing minute specks of gold leaf.

Gomez classification One of the earliest systems for classifying protein-energy malnutrition in children, based on percentage of expected weight for age: over 90% is normal, 76-90% is mild malnu-trition, 61-75% is moderate malnu-trition and less than 60% is severe malnutrition.

Good manufacturing practice Part of a food and drink control operation aimed at ensuring that products are consistently manufactured to a quality appropriate to their intended use.

Goose Domesticated water-fowl. A rich source of protein, iron, vitamins B 2, B6, B 1 2, and niacin.

Gooseberry Berry of the shrub, Ribes grossularia. The British National Fruit Collection contains 155 varieties.

Gorny dubnya Bitter liqueur flavoured with ginger, angelica, and cloves.

Gossypol Yellow toxic pigment found in some varieties of cottonseed. When included in chicken feed, it causes discoloration of the yolk, but has not been found to be toxic to human beings, and has been investigated as a possible male contraceptive agent.

Gouda Dutch semi-hard cheese with random holes formed by gases during the ripening period.

Gougere Savoury choux pastry containing cheese.

Goujon Small, deep-fried pieces of fish. The name is derived from gudgeon, a small fresh-water fish. The term is now also used for small pieces of chicken breast.

Goulash Hungarian; literally 'cowherd'; beef stewed with potatoes, tomatoes, onions, peppers, and paprika. Distinct from Austrian gulasch.

Gourds Vegetables of the family Cucurbitaceae, including calabash or bottle gourd, ash gourd, snake gourd, cucumber, vegetable marrow, pumpkin, squash, coocha or chayote, cantaloup melon, water melon. All contain more than 90% water and have little food value apart from vitamin C at 10 mg per 100 g. In addition, yellow pumpkin contains 900 gg carotene per 100 g. Melons are sometimes grown for their seeds, which contain 20-40% oil and 20% protein.

Gout Painful disease caused by accumulation of crystals of uric acid in the synovial fluid of joints; may be due to excessive synthesis and metabolism of purines, which are metabolized to uric acid, or to impaired excretion of uric acid. Traditionally associated with a rich diet, although there is little evidence for dietary factors in causing the condition. May be exacerbated by alcohol.

Grading The allocation of a defined category, or grade, of quality to a sample, or collection of similar samples, of a product. Grading can be, and usually is, done simultaneously with sorting though sorting can be done without grading. A grade can be defined as a single attribute of quality, for example size or freshness, or by some weighted function of several attributes.

Graham bread Wholewheat bread in which the bran is very finely ground. Graham cakes are made from wholemeal flour and milk. The name is that of a miller of wholemeal flour who advocated its use in the USA.

Graining Crystallization of refined sugar when boiled. Prevented by adding glucose or cream of tartar as sugar doctors.

Gram-negative, gram-positive A method of classifying bacteria depending on whether or not they retain crystal-violet dye after staining and decolorizing with alcohol.

Grams, indian Various small dried peas (legumes), e.g. green gram (Phaseolus aureus), black gram (Phascolus mungo), red gram (Cajanus indicus), Bengal gram or chick pea.

Granita Water-ice or sorbet; sparkling wines made by the tank method.

Grape Fresh fruit of a large number of varieties of Vitis vinifera. One of the oldest cultivated plants. Can be grouped as dessert grapes, wine grapes, and varieties that are used for drying to produce raisins, currants, and sultanas.Of the many varieties of grape that are grown for wine making, nine are considered 'classic varieties': cabernet sauvignon, chardonnay, chenin blanc, merlot, pinot noir, riesling, sauvignon blanc, semillon, syrah.

Grapefruit Fruit of Citrus paradisi; thought to have arisen as a sport of the pomelo or shaddock, a coarser citrus fruit, or as a hybrid between pomelo and sweet orange. It contains 35-40 mg vitamin C per 100 g. The pith contains naringin, which is very bitter. The name is said to have arisen because the fruit is borne on the tree in clusters.

Grattons Crispy remains of melted fatty tissues of poultry or pork. German equivalent is gribbens.

Gravadlax Scandinavian; pickled or marinated raw salmon.

Gravy browning Caramelized sugar and starch used to thicken and colour gravy and sauces; may be powder or liquid.

Gravy granules Seasoned and coloured granules of modified starch with a savoury flavour, used to make gravy; they form a gel on addition of boiling water.

Gravy Sauce made from the juices and extractives which run out from meat during cooking, normally thickened.

Gray The SI unit for ionizing radiation (= 100 rad). The Gray is equivalent to 1 J/kg.

Greasy haddock, Greasers Haddock, especially smoked haddock, that have been infected with the organism Ichthyophonus hoferi, are known as a greasers or greasy haddock. The flesh of infected fish has a soft and greasy feel, and a sweet sickly smell. The fillets have a blotchy appearance which show as white spots in the smoked products. Infections by the organism are not confined to haddock; herring for example are similarly affected. Ichthyophonus is usually described as a fungus, but its true taxonomic position is not clear.

Greengage Green variety of plum introduced into England in the early eighteenth century by Sir William Gage.

Greening A grey/green discoloration of canned tuna meat. The chemistry of the colour development is unclear, but trimethylamine and myoglobin are involved. A similar discoloration may occur in gas packed, sliced smoked salmon.

Grenache A grape variety widely used for wine making, although not one of the classic varieties.

Grenadin Small slice of fillet of veal, larded and braised. grenadine French; syrup made from pomegranate juice, used as a beverage and to flavour beer.

Gribbenschmaltz Dripping containing the crispy remnant of the fatty tissue of the animal.

Griddle Also girdle; iron plate used for baking scones, etc., on top of stove.

Grill To cook by radiant heat; some of the fat is lost. Barbecues cook by grilling.

Grillade A l'ardoise Andorran method of cooking on a red-hot roof slate heated over a wood fire.

Grilse Young salmon that has returned to fresh water after one year in the sea.

Grind To reduce hard foods such as nuts and coffee beans to small particles using a food mill, grinder, or sometimes a liquidizer.

Griskin Chine of pork, also used for a thin, poor, piece of loin.

Grissini Italian 'finger rolls' or stick bread, 15-45 cm (6-18 in) long, and normally crisp and dry.

Grist Cereal for grinding.

Gristle The connective tissue of the meat, consisting mainly of the insoluble proteins collagen and elastin. Usually inedible and accounts for the toughness of some cuts of meat. Prolonged slow cooking converts collagen to gelatine, but has no effect on elastin.

Groats Oats from which the husk has been entirely removed; when crushed, Embden groats result. Used to make gruel and porridge.

Grog British naval drink; sugared rum mixed with hot water. Named after Admiral Vernon whose nickname 'Old Grog' came from his grosgrain coat.

Ground meat American term for minced meat.

Grouse Game bird, Lagopus lagopus, eaten fresh or after being hung for 2-4 days to develop flavour. The whole bird weighs about 700 g; an extremely rich source of iron and vitamin B 2.

Gruel Thin porridge made from oatmeal, barley, or other cereal.

Guacamole Mexican; sauce made from very ripe avocado, mashed with garlic, lemon juice, and chilli.

Guanine One of the purines.

Guar gum Cyamopsis gum; from the cluster bean, Cyamopsis tetragonoloba. Member of Leguminosae, used in India as livestock feed. The gum is a water-soluble galactomannan; used in 'slimming' preparations, since it is not digested by digestive enzymes, and experimentally in the treatment of diabetes, since it slows the absorption of nutrients, and so prevents a rapid rise in blood sugar after a meal.

Guarana A substance from the seeds of a climbing shrub, Paullina cupana, native to the Amazon region. It contains caffeine and related compounds; used in the UK as an ingredient of drinks, chewing gum, a powder to be sprinkled on food, and capsules and tablets.

Guava Fruit of the Central and South American tropical shrub Psidium guajava, eaten raw or preserved as guava jelly.

Guinea fowl Game bird, Numida meleagris, not seasonal. Nutritionally similar to chicken.

Gulab jaman Indian dessert; deep-fried balls of ground almond and flour dough, served in syrup.

Gulasch Austrian; cubes of meat or poultry fried with onions, then simmered with paprika, tomato, and caraway seed. Distinct from Hungarian goulash or gulyas.

Gum arabic Exudate from the stems of several species of acacia; the best product comes from Acacia senegal. Used as thickening agent, as stabilizer, often in combination with other gums, in gum drops and soft jelly gums, and to prevent crystallization in sugar confectionery. Also used as the adhesive on postage stamps.

Gum drops Sugar confectionery based on sucrose and glucose with gum arabic or a mixture of gelatine and gum arabic.

Gum Substances that can disperse in water to form a viscous mucilaginous mass. Used in food processing to stabilize emulsions, as a thickening agent, and in sugar confectionery. The substances may be extracted from seeds, plant sap or exudates, and seaweeds, or they may be made from starch or ceilulose, or they may be synthetic, such as vinyl polymer. Most of these are not digested and have no food value, although they contribute to the intake of nonstarch polysaccharides.

Gum tragacanth Obtained from the trees of Astralagus spp., used as a stabilizer.

Gum, british Partly hydrolysed starch, dextrin.

Gumbo American; soup or stew made from okra, onions, celery, and pepper, flavoured with file powder, and containing chicken, meat, fish, or shellfish. Also a name for okra.

Gur Mixture of sugar crystals and syrup, brown and toffee-like, made by evaporation of juice of sugar cane; also called jaggery.

Gury Whole raw white cabbage soaked in brine with beetroot and red peppers.

Guthrie test Test for a number of genetic diseases based on measuring the concentrations of amino acids in a small sample of blood taken by pricking the heel of the child a few days after birth, by biological assay using mutated bacteria.

Gutting Removal by hand or by machine of the contents of the gut cavity or belly cavity, the guts, after slitting the fish open from throat to vent, to provide a gutted fish; also termed eviscerating. Fish without the guts removed are termed ungutted. The aim should be to thoroughly clean and wash out the belly cavity. To facilitate this, roundfish of sufficient size can be double naped and the belly opened up over its entire length; care should be taken to remove the entire contents of the gut cavity including the liver. The belly or gut cavity is the compartment containing stomach, intestine and other internal organs known collectively as the guts and the gonads of a fish. The guts or viscera may refer to the entire contents of the belly cavity, or may represent the stomach, intestine and associated organs but exclude the liver and gonads. The gonads are the reproductive organs, the roe or hard roe in female and the milt or soft roe in male fish. The long gut is the stomach and intestine which is pulled out while still attached to the head of the fish by nobbing to produce a nobbed fish; the head is partially severed and pulled without cutting open the belly so that the head and gut are removed in one operation, and the roe or milt is left in. In contrast, in the gibbing or gipping of fatty fish for pickle curing, the gills, long gut and stomach are removed in one operation, so that the roe or milt and some of the pyloric caeca remain. The latter are essential to obtain the desired flavour.

H

HACCP A quality management system that identifies critical points in production and sets up measures to ensure nothing to affect the safety of the product happens at these points.

Haff disease Acute paroxysmal myoglobinuria suffered by fishermen around the Koenigsberg Haff in eastern Germany; attributed either to a toxin in the seawater which entered the fish or to thiaminase in raw or incompletely cooked fish.

Hagberg test Measure of ? 1 -amylase activity of flour based on the change in viscosity of flour paste.

Haggis Traditional Scottish dish made from sheep's heart, liver, and lungs cooked and chopped with suet, onions, oatmeal, and seasoning, and stuffed into the stomach of a sheep. Said to have originated with the Romans when they were campaigning in Scotland; when they broke camp in an emergency, the food was wrapped in the sheep's stomach. A similar Norman-French dish was afronchemoyle. An exceptionally rich source of iron; a rich source of protein; a good source of vitamins B 1, B 2, niacin, calcium, and copper; a source of zinc.

Hake A white fish, Merluccius bilinearis.

Halal Food conforming to the Islamic (Muslim) dietary laws. Meat from permitted animals and birds. The animals are killed under religious supervision by cutting the throat to allow removal of all blood from the carcass, without prior stunning.

Halibut liver oil The oil from the liver of the halibut, one of the richest natural sources of vitamins A and D; contains 50 mg of vitamin A and 80 μg of vitamin D per gram.

Halophilic bacteria, Halophiles Bacteria that can grow in the present of salt, and perhaps requiring the presence of salt in order to grow.

Halverine Name sometimes given to low-fat spreads with less than the statutory amount of fat in a margarine.

Ham The whole hind leg of the pig, removed from the carcass and cured individually; sometimes the process is secret. Hams cured or smoked in different ways have different flavours, e.g. York, Bradenham, Suffolk, and Westphalian hams. A rich source of protein, niacin, and vitamin B 1.

Ham, green Ham that has been cured but not smoked.

Haman taschen Traditional Jewish cakes for the festival of Purim; pastry made from kuchen dough filled with poppy seed and honey, chopped stewed prunes, or cream cheese.

Haman's ears Biscuits; very thin pieces of egg dough, deep fried.

Hamburger A flat cake made from ground (minced) beef, seasoned with salt, pepper, and herbs, and bound with egg and flour. Commercial beefburgers are usually 80-100% meat, but must by law (in the UK) contain 52% lean meat, of which 80% must be beef. Cereal, cereal fibre, or bean fibre may be added as filler or 'meat extender'. A rich source of protein, vitamin B 12, niacin, copper, and iron; a good source of zinc; a source of vitamins B 2 and B 6.

Hand of pork The foreleg of pork; usually salted and boiled.

Hangover Headache and feeling of malaise resulting from excessive consumption of alcoholic beverages. The severity differs with different beverages and is not due to the toxic effects of alcohol alone, but to the presence of higher alcohols and esters, the substances that give different beverages their distinctive flavours.

Hansa can An all-aluminium can with easily opened ends.

Haram Food forbidden by Islamic law.

Hard sauce Butter and sugar, flavoured with brandy or rum, served with rich puddings.

Hardwood Wood from deciduous trees, especially oak, that when burned produces smoke that imparts a more desirable flavour to fish than softwood. Oak is generally regarded as producing an excellent flavour but mahogany, hickory, cherry, apple and beech, amongst others, are also used. Wood treated with preservatives must not be used for smoking.

Hare Game animal, similar to rabbit but larger; caught wild but not farmed commercially. Lepus europaeus is the common hare, but some twenty Lepus species occur in Europe.

Haricot French for beans; also stewed chopped meat, as haricot mutton.

Hartnup disease A genetic disease affecting the absorption and tissue uptake of the amino acid tryptophan, with considerable loss in the urine. Characterized by development of pellagra, and treated by administration of supplements of niacin.

Harvard standard Tables of height and weight for age used as reference Values for the assessment of growth and nutritional status in children, based on data collected in the USA. Now largely replaced by the N.H.standards.

Harvey wallbanger A cocktail based on half a measure of Galliano floating on four measures of orange juice and half a measure of vodka.

Hash browns American; cooked potatoes, chopped, formed into small cakes, and fried.

Hash Dish of cooked meat reheated in highly flavoured sauce. In the USA canned corned beef is known as corned beef hash.

Haslet Old English country dish made from pig's offal cooked in small pieces with seasoning and flour. Also known as pig's fry.

Hasty pudding English, sixteenth century; made from flour, milk, butter, and spices, which, since they were usually readily available, could be quickly made into a pudding for unexpected visitors. Made in the USA with maize (corn) flour instead of wheat flour.

Hausa groundnut Legume grown in West Africa, Kerslingiella geocarpa; 20% protein, 60% fat.

Hay diet A system of eating based on the concept that carbohydrates and proteins should not be eaten at the same meal, for which there is no scientific basis. It ignores the fact that almost all carbohydrate-rich foods also contain significant amounts of protein. In any case, in the absence of adequate carbohydrate, protein is oxidized as a metabolic fuel and therefore not available for tissue building. Also called combining diet or food combining.

Haybox cooking The food is cooked for only a short time, then placed in a well-lagged container, the haybox, where it remains hot for many hours, so cooking continues without further use of fuel. Also known as the fireless cooker.

Haze Term in general use in brewing to indicate cloudiness of the beer. Chill haze appears at 0°C and disappears at 20°C; permanent haze remains at 20°C but there is no fundamental difference. It is caused by gums derived from the barley, leucoanthocyanins from the malt and hops, and glucose, pentoses, and amino acids.

Hazel nut Fruit of the tree Corylus avellana; cultivated varieties include Barcelona nut, cob nut, and filbert. A rich source of copper and vitamin E; a source of protein, niacin, vitamins B 1 and B 2, calcium, zinc, and iron.

HDL High-density lipoproteins, one of the classes of plasma lipids.

HDPE High-density polyethylene.

Headcheese Mock brawn.

Heading and gutting Removal by hand or by machine of both head and guts to produce headed and gutted (H&G) fish, especially in white fish, in preparation for freezing at sea, or for the processing of klippfisk (klipfish), and for the processing of tuna. For klipfish the nape bone or collar bone is retained on the fish. But, for other purposes the cut is made behind the collar bone which remains with the head, and the cut may be angled to remove varying portions of the belly flaps. These are the side walls of the belly cavity. The flap is usually still attached to a fillet when the fillet is removed during filleting of the fish; also called flap, lug flap, wing.

Health foods Substances whose consumption is advocated by various reform movements, including vegetable foods, whole grain cereals, food processed without chemical additives, food grown on organic compost, 'magic' foods, and pills and potions. Numerous health claims are made but rarely is there any evidence to support these claims.

Heartburn A burning sensation in the chest usually caused by reflux of acid digestive juices from the stomach into the oesophagus. A common form of indigestion, treated by antacids.

Heat of combustion Energy released by complete combustion, as for example, in the bomb calorimeter. Values can be used to predict energy physiologically available from foods only if an allowance is made for material not completely oxidized in the body. For example, the end products of protein oxidation in the body are carbon dioxide, water, and urea; the latter contains non-available energy.

Heat sealing A method of sealing plastic containers by heating two oining layers or portions of the container until they melt together thereby forming a good seal.

Heat sterilisation See **Canning**

Heat-treated These terms apply to a variety of thermal treatments and processes. The heat-treated fish or shellfish may be ready-to-eat either hot or cold, it may require reheating or further cooking before eating, or it may be subject to further cooking as an ingredient in a composite product such as fish-in-sauce with vegetables, or fish pie. The degree of heat treatment received affects the microbiological status and stability of the product. Blanching or scalding are not intended to cook the fish but to aid removal of the skin or shell. Cooking of krill at 95°C for 1 to 3 minutes coagulates the protein, inhibits enzymic autolysis and melanosis, and subsequent cooling permits storage of the cooked krill for up to 5 days at 0°C before noticeable spoilage takes place. These mild heat treatments and others such as pasteurisation and sous vide processing will kill most vegetative bacteria, but not the heat resistant strains and spores. Growth of any surviving organisms is minimised by chilled or frozen storage, acidity, salt or some other controlling factor. Molluscan shellfish harvested from sewage contaminated waters may contain bacterial pathogens as well as viruses that can cause gastroenteritis and hepatitis if contaminated shellfish are consumed. Under regulations adopted in Europe, for example, one option is that these shellfish be subjected to an approved heat process in approved premises to ensure that microbiologically contaminated molluscs are rendered safe to eat. The approved heat processes are sterilisation in hermetically sealed containers, i.e. canning; animals in the shell and not frozen are immersed in boiling water such that the internal temperature of the flesh is raised to at least 90°C, and this minimum temperature is maintained for at least 90 seconds; animals in the shell and not frozen are cooked for 3 to 5 minutes in an enclosed space in which the temperature is 120 to 160°C and the pressure is 2 to 4 kg/cm 2, followed by shelling and freezing of the flesh to a core temperature of -20°C. Long term microbiological stability for non-dried products without the use of refrigeration is obtained by prolonged heating in hermetically sealed containers, such as cans or retortable pouches. The severity of the heat process is greater for low acid foods, which must receive a minimal thermal process, equivalent to a botulinum cook, sometimes called a full heat sterilisationprocess. Products manufactured in, or destined for, countries with hot climates may require a much more severe thermal process. See Canning.

Heavy cure Refers to the degree of smoking, salting or drying, or a combination of these processes, for preservation of the product.

Hedonic methods Experimental procedures used in sensory evaluation of foods in which assessors express their liking/disliking for samples in rating or in discrimination tests. In the former, a typical scale would be one ranging from 'like extremely' to 'dislike extremely'; in the latter, assessors would select the most liked, or disliked, from a pair of samples or would rank samples by the degree of liking for them.

Heel-prick test Test for genetic diseases based on measuring the concentrations of amino acids in a small sample of blood taken by pricking the heel of the child a few days after birth.

Helicobacter pyloris Bacterium commonly infecting the gastric mucosa in patients with ulcers. Believed to be the underlying cause of ulcers, and also implicated in the development of gastric cancer.

Hemicelluloses Complex carbohydrates included as dietary fibre, composed of polyuronic acids combined with xylose, glucose, mannose, and arabinose. Found together with cellulose and lignin in plant cell walls; most gums and mucilages are hemicelluloses.

Hemoglobin American spelling of haemoglobin; similarly, hematin = haema-tin, heme = haem, hemosiderin = haemosiderin.

Hepatitis Inflammatory liver disease, characterized by jaundice, abdominal pain, and anorexia. May be due to bacterial or viral infection, alcohol abuse, or various toxins. Treatment is usually conservative, with a very low fat diet and complete abstinence from alcohol.

Even after recovery, people may continue to be carriers of the virus, especially for hepatitis B and C, which are transmitted through blood and other body fluids. Liver cancer and cirrhosis are more common among people who have suffered from hepatitis B or C.

Herb butter A mixture of butter and mayonnaise blended with parsley, tarragon, dill, watercress, thyme, green pepper, garlic, and other herbs, used as a savoury spread on biscuits or bread.

Herb tea Or tisane, an infusion made from any kind of herb, fruit, or flower. Camomile, lime blossom, and fennel seeds are commonly used. Medicinal or health claims are sometimes made, largely on traditional rather than scientific grounds.

Herbs Soft-stemmed, aromatic plants used fresh or dried to flavour and garnish dishes, and sometimes for medicinal effects. Not clearly distinguished from spices, except that herbs are usually the leaves or the whole of the plant while spices are only part of the plant, commonly the seeds, or sometimes the roots or rhizomes.

Hermesetas Trade name for saccharin tablets. herring Oily fish, Clupea harengus; young herrings are sild. Sprat is Clupea sprattus; young are brislings. Pilchard is Clupea pilchardus; young are sardines. Kippers, bloaters, and red herrings are salted and smoked herrings; bucklings are hot-smoked herrings. Gaffelbitar are preserved herring. An exceptionally rich source of vitamins D, B 12, and selenium; a rich source of protein, niacin, and vitamin B 6.

Hermetic The sealing of a package that forms a closure that is absolutely impervious to micro-organisms, gas, water molecules and dust.

Herring worm Common name given to parasitic nematodes of the genus Anisakis found in the flesh of herring, and other species of fish.

Hesperidin A flavonoid found in the pith of unripe citrus fruits; chemically a complex of glucose and rhamnose with the flavonone hesperin. At one time called vitamin P, since it affects the fragility of the capillary walls, although there is no evidence that it is a dietary essential.

Hess test A test for capillary fragility in scurvy. A slight pressure is applied to the arm for 5 minutes and a shower of petechiae appear on the skin below the area of application.

Hexa-metaphosphate A polymerised form of orthophosphate consisting of six condensed phosphate groups. Available under its trade name of Calgon. It is a sequestrant for calcium, and is included in detergents.

Hexamethylenetetramine Also called heaxamine. An antiseptic, it is generally not permitted as an additive in foods, though there may be exceptions for a very limited list of foods. But, it has been used as a preservative in marinated fishery products.

Hexamic acid Trade name for cyclohexyl sulphamic acid, the free acid of cyclamate, a synthetic sweetener about 27 times as sweet as sugar.

Hexoses Six-carbon sugars such as glucose or fructose.

Hexuronic acid The acid derived from a hexose sugar by oxidation of the hydroxyl group on carbon-6. The hexuronic acid derived from glucose is glucuronic acid.

Hiatus hernia Protrusion of a part of the stomach upwards through the diaphragm. The condition occurs in about 40% of the population, most people suffering no ill-effects; in a small number of people there is reflux of stomach contents into the oesophagus, causing heartburn.

Hickory nut North American walnut, Carya spp.; the best known is the pecan nut.

High-density lipoproteins One of the classes of plasma lipids.

High-performance liquid chromatography Also known as highpressure liquid chromatography, and generally abbreviated to hplc. An extremely sensitive analytical technique, typically able to separate and measure nanograms or smaller amounts of compounds in samples of 10-100 μl.

High-temperature short-time treatment Sterilization by heat from times ranging from a few seconds to minutes; usually applied to flow sterilization, in which the process time is less than about 1 minute; based on the fact that at higher temperatures bacteria are destroyed more rapidly than damage can occur to nutrients and texture.

Himbeergeist German spirit distilled from raspberries with added alcohol.

Hindle wakes Very old English method of cooking chicken, stuffed with fruit and spices, including prunes. Possibly a corruption of hen de la wake, a feast of fourteenth-century Flemish introduction.

Hi-soy Trade name for full-fat soya flour.

Histamine The amine formed by decarboxylation of the amino acid histidine in the body, also found in small amounts in cheeses, beer, chocolate, sauerkraut, and wines. Excessive release of histamine from mast cells is responsible for many of the symptoms of allergic reactions. It also stimulates secretion of gastric acid, and administration of histamine is used as a test for achlorhydria.

Histamine, Histamine poisoning, Histamine fish poisoning Histamine is a biogenic amine. It is formed in fish muscle by decarboxylation of the amino acid histidine by bacteria. It is physiologically active and is released in allergic reactions such as hay fever when it is responsible for some of the symptoms of allergic responses. Histamine can reach high concentrations in muscles of fish after death that contain high concentrations of histidine, typically, though not exclusively, members of the mackerel, tuna and herring families. The bacteria responsible for the deamination are not active below about 10°C and chilling of the fish prevents production of histamine. Histamine is associated with scombrotoxin poisoning, though the aetiology of the illness is not completely clear. Food control regulations often specify the maximum concentration of histamine that may be permitted in fish and fish products for human consumption.

Histidinaemia Genetic disease due to a defect in the metabolism of the amino acid histidine. If untreated it leads to mental retardation and nervous system abnormalities. Treatment is by feeding a diet very low in histidine.

Histidine An amino acid found bound in proteins and in the free form in fish tissues. Can be decarboxylated by some bacteria to histamine.

Histones Proteins rich in arginine and lysine, soluble in water but not dilute ammonia. They occur mainly in the cell nucleus and are concerned with the regulation of DNA.

Hiyu Japanese name for Chinese spinach, also known as bhaji and callaloo.

HMG CoA reductase inhibitors Drugs which inhibit the enzyme hydroxymethylglutaryl CoA reductase, the controlling enzyme of cholesterol synthesis, used in the treatment of hypercholesterolaemia.

Hock 1. Generic term for white wines from the Rhine region of Germany, known in the USA as Rhine wines; bottled in brown glass, to distinguish from Moselle wines (in green glass).

2. The knuckle of pork; also used in the USA for foreleg pork shank.

Hogget One-year-old sheep.

Hogshead A traditional UK measure of volume or size of barrel: for beer or cider contains 54 gallons (243 L); for wine contains 52½ gallons (236 L).

Hoisin Sauce spicy vegetable-based brownish-red sauce.

Holishkes Cabbage leaves stuffed with rice, minced meat, and sultanas.

Hollandaise sauce Rich sauce made from egg yolks, butter, and lemon.

Holocellulose Mixture of cellulose and hemicellulose in wood, the fibrous residue that remains after the extractives, the lignin, and the ash-forming elements have been removed.

Holoenzyme An enzyme protein together with its coenzyme or prosthetic group.

Holosides Complexes of sugars that yield only sugars on hydrolysis. As distinct from heterosides which yield other substances as well as sugars on hydrolysis, e.g. tannins, anthocyanins, nucleosides.

Hominy Prepared maize kernels. Lye hominy has the pericarp and germ removed by soaking in caustic soda. Pearled hominy is degermed hulled maize. Corn grits are ground hominy.

Homocysteine A sulphydryl-containing aminoacid derived form the metabolic demethylation of dietary methionine.

Homogenise The process in which the size of the fat globules is reduced to small uniform particles, which are then distributed evenly throughout the liquid. For example, the cream in homogenised milk is distributed throughout the liquid rather than rising to the top to form a layer.

Homogenization Emulsions usually consist of a suspension of globules of varying size. Homogenization reduces these globules to a smaller and more uniform size. In homogenized milk the smaller globules adsorb more of the protein, which acts as a stabilizer, and the cream does not rise to the top.

Honey Syrupy liquid made by bees from the nectar of flowers. The flavour and colour depend upon the flowers from which the nectar was obtained and the composition also varies with the source. Average composition: water 18% (12-26%), invert sugar, i.e. glucose and fructose, 74% (69-75%), sucrose 1.9% (0-4%), ash 0.18% (0.1-0.8%), organic acids 0.1-0.4%. If the ratio of fructose to glucose is high, there is a tendency for the honey to crystallize.

Honeycombing A textural defect in tuna characterised by pitted, spongy-looking meat, generally localised near the head of the fish which appears on cooking the meat, for example, in the pre-cooking stage before packing during the canning of tuna. It is a result of time/temperature abuse, appearing in flesh that has started to decompose. Off-odours usually accompany the honeycombing.

Honeydew honey During periods of prolonged drought, bees may supplement their nectar supplies with honeydew, the sweet fluid excreted on leaves by leaf-sucking insects. The resultant honey is dark, with an unpleasant taste.

Hops Perennial climbing plant, Humulus lupulus; the dried female flowers contain bitter resins and essential oils and are added to beer both to preserve it and to enhance the flavour. The tender shoots are eaten as a vegetable in some countries.

Horchata de chufa Spanish; aqueous extract of tiger nut used as a drink.

Hordein A protein in barley; one of the prolamines.

Hordenin Alkaloid found in germinated barley, sorghum, and millet which can cause hypertension and respiratory inhibition.

Hormone Chemical that acts as a messenger relaying instructions to start certain physiological activities. Hormones are synthesized in one type of cell, and released to direct the function of other cell types.

Hormones, sex Male hormones, or androgens, include testosterone, dihydrotestosterone and androsterone; female hormones, or oestrogens, include oestradiol, oestrone, and progesterone. Chemically, all are steroids, derived from cholesterol.

The synthetic female hormones stilboestrol and hexoestrol have similar biological activities to the oestrogens, but are quite different chemically. Apart from clinical use the oestrogens have been used for chemical caponization of cockerels and to enhance the growth rate of cattle

Horse Equus caballus; an exceptionally rich source of iron; a rich source of protein and niacin, and a source of vitamins B 1 and B 2.

Horseradish The root of Armoracia lapathifolia. Its pungency is caused by volatile oils. Used as a condiment, usually as a creamed sauce, or grated and mixed with beetroot, it is also an important ingredient in wasabe powder.

Horse's neck A long drink based on ginger ale with ice, bitters, and a spirit.

Hortvet freezing test Test for the adulteration of milk with water by measuring the freezing point; milk normally freezes between -0.53 to -0.55°C; if diluted with water it will freeze above -0.53°C.

Hot breads American term for waffles and pancakes.

Hot cross bun Spiced yeast bun traditionally eaten in the UK at Easter; the top is decorated with a cross of dough.

Hot dog Frankfurter sausage in a long bread roll.

Hot sauce A tomato sauce with pungent flavour due to cayenne.

Hot springs conference International Conference held in 1943 at which the Food and Agriculture Organization (FAO) of the United Nations originated.

Hotch-potch Thick soup or stew of meat and vegetables.

Hotpot Baked stew or casserole of meat or fish, topped with sliced potatoes.

Hot-water crust Pastry made by melting lard into boiling water and pouring the mixture onto the flour, then kneading it into a dough.

Hovis Trade name for a mixture of brown flour and wheatgerm; from Latin hominis vis, 'strength of man'; originally, in the 1880s, called Smith's Old Patent Germ Bread. It is not baked centrally, but franchised to local bakers. Four large slices (140 g) are a rich source of vitamin B 1, iron, and niacin; a good source of protein, calcium, and zinc; supply 4.5 g of dietary fibre and 300 kcal (1260 kJ).

Howard mould count Standardized microscopical technique for measuring mould contamination.

Howtowdie Scottish; boiled chicken with poached egg and spinach.

Hsiao mai Fish-filled wraplings or wuntun.

HTST A process by which food is sterilised at very high temperatures but only for a very short period.

Huckleberry Wild North American berry, the fruit of Gaylussacia baccata and other species. Similar to blueberry but has larger seeds; used in tarts, pies, and preserves.

Huff paste Northern British name for pastry made from suet, flour, and water, used to enclose meat, fish, or poultry while baking.

Human growth hormone Human somatotropin, a protein hormone made in the pituitary gland. Lab-grown (HGH) bacteria given a copy of the gene for human growth hormone can then produce large amounts of HGH, which can be purified and used to treat certain kinds of human dwarfismpa pathological condition of growth arrested by various causes.

Humbug Hard boiled sweet, normally peppermint-flavoured, cushionshaped.

Humectants Substances such as glycerol, sorbitol, invert sugars, honey which prevent loss of moisture from foods, especially flour confectionery, which would make them unappetizing; they also prevent sugar crystallizing and the growth of ice crystals in frozen foods. They are used in other products too, such as tobacco, inks, glues, etc.

Humidity The moistness of air. Weight of water per unit weight of air is absolute or specific humidity. Saturation humidity is the absolute humidity of air that is saturated with water vapour at a given temperature. Relative humidity is the degree of saturation: the ratio of water vapour pressure in the atmosphere to water vapour pressure that would be exerted by pure water at the same temperature.

Hummus Middle Eastern; hors d'œuvre, a puree of chickpeas and tahini with garlic, oil, and lemon juice.

Hurricane A New Orleans cocktail based on rum, passion fruit juice, and lime juice, in equal measures.

Hush puppies Southern USA; fried, seasoned, batter cakes, served especially with fish.

Husk The outer woody cellulose covering of seeds and grains. In wheat it is loosely attached and removed during threshing; in rice it is firmly attached. High in fibre content and of limited use as animal feed.

Hydrogen peroxide A free radical mainly generated by mitochondria or by dismutation of superoxide anion by superoxide dismutase. Through the Fenton reaction H_2O_2 in the presence of Fe^{2+} generates hydroxyl radical. H_2O_2 is catalyzed by catalase producing oxygen and water.

Hydrogenated oils Liquid oils hardened by hydrogenation.

Hydrogenation Conversion of liquid oils to semi-hard fats by the addition of hydrogen to the unsaturated double bonds; used for margarines and shortenings intended for bakery products.

Hydroxyl radical A powerful cytotoxic oxidant produced by the Fenton reaction between hydrogen peroxide and Fe^{2+}.

Hydroxylysine Amino acid found only in connective tissue proteins; incorporated into the protein as lysine and then hydroxylated in a vitamin-C-dependent reaction.

Hydroxyproline Amino acid mainly in connective tissue proteins; incorporated into the protein as proline and then hydroxylated in a vitamin-C-dependent reaction. Peptides of hydroxyproline are excreted in the urine and the output is increased when collagen turnover is high, as in rapid growth or resorption of tissue.

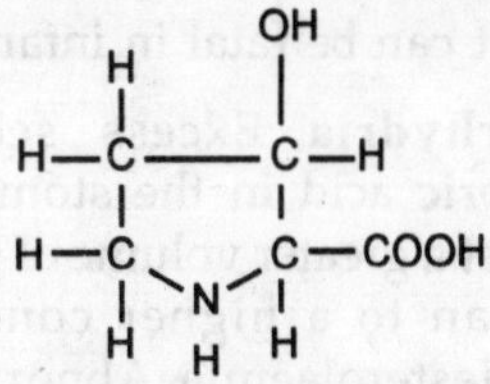

Fig. Hydroxyproline

Hydroxyproline index Urinary excretion of hydroxyproline is reduced in children suffering protein-energy malnutrition. The index is the ratio of urinary hydroxyproline to creatinine per kg of body weight, and is low in malnourished children.

Hydroxytryptamine Also called serotonin. A neurotransmitter amine synthesized from the amino acid tryptophan, also formed in blood platelets; it acts as a vasoconstrictor. Found in plantains and some other foods, but metabolized in the intestinal mucosa.

Hygroscopic Readily absorbing water, as when table salt becomes damp. Materials such as calcium chloride and silica gel absorb water so readily that they are used as drying agents.

Hyper Prefix meaning above the normal range, or abnormally high.

Hyperalimentation Provision of unusually large amounts of energy, either intravenously or by nasogastric or gastrostomy tube.

Hyperammonaemia High blood ammonia concentrations, especially after protein intake, leading to coma, convulsions, and possibly death. May be due to a variety of genetic diseases affecting amino acid metabolism. Treatment is normally by severe restriction of protein intake.

Hypercalcaemia, idiopathic Elevated plasma concentrations of calcium believed to be due to hypersensitivity of some children to vitamin D toxicity. There is excessive absorption of calcium, with loss of appetite, vomiting, constipation, flabby muscles, and deposition of calcium in the soft tissues and kidneys. It can be fatal in infants.

Hyperchlorhydria Excess secretion of hydrochloric acid in the stomach due to secretion of a greater volume of gastric juice rather than to a higher concentration. hypercholesterolaemia Abnormally high concentrations of cholesterol in the blood. Generally considered to be a sign of high risk for atherosclerosis and ischaemic heart disease. Treatment is by restriction of fat and cholesterol intake and a high intake of non-starch polysaccharides, which increase the excretion of cholesterol and its metabolites in the faeces. In severe cases, ion-exchange resins may be fed, to increase further the excretion of bile salts, and drugs may be given to inhibit the synthesis of cholesterol in the body.

Familial hypercholesterolaemia is a genetic disease in which affected individuals have extremely high blood concentrations of cholesterol, frequently dying from ischaemic heart disease in early adulthood; treatment is as for other forms of hypercholesterolaemia, but more rigorous.

Hypocalcaemia Low blood calcium, leading to vomiting and uncontrollable twitching of muscles if severe; may be due to underactivity of the parathyroid gland, kidney failure, or vitamin D deficiency.

Hyposite Little used word, from the Greek, for low-energy food.

Hypothermia Low body temperature. Occurs among elderly people far more readily than in younger adults, often with fatal results. Also used in connection with deliberate reduction of body temperature to 28°C to permit heart and brain surgery.

Hypothyroidism A condition characterized by lowered metabolic rate and general loss of vigor caused by deficient activity of the thyroid gland.

Hypotonic A solution more dilute than the body fluids.

Hypovitaminosis Vitamin deficiency.

Hypoxanthine A compound formed by chemical reactions occurring in fish muscle after death of the fish. Its concentration in a sample of fish muscle is a measure of the freshness of the fish.

Hyssop Pungent aromatic herb, Hyssopus officinialis, used in salads, soups, and in making liqueurs.

I

Ice cream A frozen confection made from fat, milk solids, and sugar. Some European countries permit the use of non-milk fat and term the product ice cream, while if milk fat is used, it is termed dairy ice cream.

According to UK regulations, contains not less than 5% fat and 7% other milk solids; according to US regulations, 10% milk fat and 20% other milk solids. Stabilizers such as carboxy methylcellulose, gums, and alginates are included, and emulsifiers such as polysorbate and monoglycerides. Mono- and diglycerides bind the looser globules of water and are added in 'non-drip' ice cream.

Icing Putting ice among fish, as in boxing, to chill it to 0 to -0.5°C. The ice may be derived from block ice, i.e. ice made by freezing water in a mould to form a block typically weighing about 50 kg, which is stored whole and normally has to be crushed before use to give small irregularly shaped pieces known as crushed ice. The ice may be small ice produced in small pieces ready for use, as in the case of flake ice, which is probably the most suitable and widely used for chilling fish. This is produced by freezing a thin layer of water on a refrigerated cylinder, and scraping it off in fragments or flakes; also known as scale ice or slice ice. It may be small ice cut into small pieces immediately after manufacture, as in the case of tube ice made by freezing water inside vertical tubes, cut into short lengths immediately on release, and sometimes crushed before use. Small ice refers to the many kinds of ice made in small pieces in addition to those described above such as snow or powder ice, plate ice, ribbon ice and many others. Weight for weight, the cooling capacity or refrigeration effect of ice is always the same for all practical purposes, irrespective of the origin of the ice. A slushy mixture of ice and water should never be compared with an equal weight of ice alone. One kg of slushy ice has a lower cooling capacity than 1 kg of fresh crisp ice, to an extent that depends on the degree of wetness of the slush. One kg of ice at -5°C has only about 3% more cooling power than 1 kg of ice at 0°C. The quantity of ice required initially to cool a unit weight of fish from a given temperature (T°C) to 0°C is T divided by 80 times the specific heat of the fish. Additional ice is required to absorb the heat input from the surroundings in order to keep the mass of fish at the chill temperature. For example, in practice for cold water fisheries, 3 parts fish to 1 part ice by weight should be sufficient in properly iced boxes in a properly operated fishroom to cool the fish and leave sufficient ice after a 7 to 8 day trip to indicate to the buyer that the fish have been properly chilled. Warm water fisheries may require a ratio of 2:1 or even 1:1. Binary ice, known also as flow ice or slurry ice, is made from brine and is composed of very small ice crystals suspended in brine or seawater. The proportion of ice crystals and the brine strength can be varied to control the temperature of the mixture, which can lead to superchilling or partial freezing of the fish at temperatures less than -1°C. The mixture can be pumped if the ice fraction is 60% or less. Drained binary ice crystals are sometimes referred to confusingly as snow ice.

Idli South-East Asian; small cakes made from a mixture of cooked rice and black gram, fermented with the aid of mould.

IEP The Individualized Education Plan is the written report for each school year which delineates the goals and special services necessary to meet the child's needs. Parents, teachers and administrators take part in the preparation of the child's IEP for the academic year. This begins when your child turns 3 years old, as that is the time your child's education comes under the auspices of the Board of Education and no longer the Birth to Three program.

IFSP Individualized Family Service Plan. This is the special service plan agreed upon between the Birth to Three-Service Coordinator and the family. This report is done annually until your child reaches the age of 3 years old.

Ileostomy Surgical formation of an opening of the ileum on the abdominal wall, performed to treat severe ulcerative colitis.

Ileum Last portion of the small intestine, between the jejunum and the colon.

Immigration The movement of people from one country, to live in another country.

Immune system Series of defence mechanisms of the body. There are two major parts: humoral, mediated through antibodies secreted into the circulation; and cell-mediated. Lymphocytes produce antibodies against, and bind to, the antigens of foreign cells, leading to death of the invading organisms; other white blood cells are phagocytic and engulf the invading organisms.

Immunoglobulins Specific antibodies produced in the blood in response to foreign proteins or other antigens. Five classes, IgA, IgE, IgG, IgM, and IgI; present in circulating blood as a result of previous exposure to the antigens, and also present in breast milk to confer some immunity on the baby.

In vitro Literally 'in glass'; used to indicate an observation made experimentally in the test-tube, as distinct from the natural living conditions, in vivo.

In vivo In the living state, as distinct from in vitro.

Inanition Exhaustion and wasting due to complete lack or nonassimilation of food; a state of starvation.

Incaparina A number of protein-rich dietary supplements developed by the Institute of Nutrition of Central America and Panama (INCAP), based on cottonseed flour, or soya and vegetables, with various nutrient supplements. All versions contain 27.5% protein.

Inclusion The practice of including handicapped children in regular school and preschool and other community environments.

Indian rice grass Perennial, growing wild in the USA, Oryzopsis hymenoides; tolerant to drought. The seeds, which resemble millet, are small, round, and dark in colour, covered with white hairs. Traditionally used by native Americans for flour, now used almost exclusively for forage.

Indicator organism In the context of food hygiene, a microorganism, or group of microorganisms, which are not themselves considered to be harmful in the food, but whose presence might suggest that the food could have been contaminated by harmful microorganisms existing in the same environment as the indicator organism. The presence of the indicator organisms is tested for rather than the pathogenic organisms because it is simpler and quicker to test for them. The presence of indicator organisms, or presence in high numbers, in a product could prompt further testing for the pathogenic organisms, or could be a reason for rejecting the product. Examples of indicator organisms used in quality control of fishery products are coliforms, faecal coliforms, and Escherichia coli.

Indigestion Discomfort and distension of the stomach after a meal, also known as dyspepsia, including heartburn. Persistent indigestion may be a symptom of a digestive disorder such as hiatus hernia or peptic ulcer.

Indigo carmine Blue food colour (E-132), derivative of indigotin, which comes from tropical leguminous plants Indigofera spp.

Induction period The lag period during which a fat or oil shows stability to oxidation because of its content of antioxidants, natural or added, which are oxidized preferentially. After this there is a sudden and large consumption of oxygen and the fat becomes rancid.

Industrialisation Use of mechanisation and automation technology.

Infant formula Modified milk products or milk substitutes for feeding infants in place of breast feeding. The nutritional composition is adjusted to approximate to that of breast milk, and is controlled by legal standards.

Special formula preparations are available for premature babies and those of very low birth weight.

For infants intolerant of cow's milk, there are special formula preparations based on soya bean or other products. For those with genetic diseases a range of special formulae are available.

Infuse To extract the flavour from herbs, spices, etc., by steeping them in a liquid, usually by pouring boiling liquid over them, covering them and leaving them to stand without farther cooking or heating, as in making tea.

Ingredient Any substance used in the manufacture or preparation of a foodstuff and still present in the finished product, even if in an altered form. Contaminants and adulterants are not considered to be ingredients.

Inorganic Materials of mineral, as distinct from animal or vegetable, origin. Apart from carbonates and cyanides, inorganic chemicals are those that contain no carbon.

Inosotol A carbohydrate derivative which is an essential nutrient for micro-organisms and many animals and sometimes classified as a vitamin, although there is no evidence that it is a dietary essential for human beings. Deficiency causes alopecia in mice and 'spectacle eye' in rats. Obsolete names are inosite and meat sugar.

Chemically hexahydroxyclohexane (CHOH) 6 ; there are nine isomers, but only one, meso- or myo-inositol, is of physiological importance. It is a constituent of many phospholipids involved in membrane structure and as part of the signalling mechanism for hormones which act at the cell surface.

The insecticide gammexane is hexachlorocyclohexane, and appears to function by competing with inositol. The hexaphosphate ester of inositol is phytic acid.

In-plant chlorination The practice of adding chlorine into the water supply to a fish processing plant. The primary purpose is to ensure that products, premises and equipment are not contaminated with harmful waterborne microorganisms. The chlorine can be injected as chlorine gas or as solutions of hypochlorite. The gas or solutions are metered into the water supply according to the water flow. Sufficient chlorine should be added to achieve a low level of free chlorine. Some Codes of Practice, for example, those of the Codex Alimentarius Commission, allow concentrations of up to 10 mg/l of active chlorine, but many national regulations set maximum limits for water in contact with foods at much lower concentrations, often less than 1 mg/l, although there is no evidence of risk of harm at the higher level.

Inspection The examination of a product to ascertain its compliance with the required specification of quality. Inspection can be 100%, that is inspection of all products passing an inspection point in a processing line or all items in a batch of product, or be based on sampling. Sampling inspection is typically applied to batches of products entering a process or stage in a process, to checking finished products before warehousing or despatch, or to checking finished products being accepted for marketing. The process of inspecting a batch and the action to be taken as a result of sampling and testing are described in a sampling plan.

Instant foods Dried foods that reconstitute rapidly when water is added, e.g. tea, coffee, milk, soups, precooked cereal products, potatoes, etc. The dried powders may be agglomerated to control particle size and improve solubility. 'Instant puddings' are formulated with pregelatinized starch and disperse rapidly in cold milk.

Insulin Protein hormone that regulates blood sugar, made in cells of the pancreas. In the laboratory, microbes given a copy of the gene for human insulin can make insulin to treat diabetes mellitus, a shortage of insulin.

Integrating motor pneumotachograph Apparatus for measuring energy expenditure indirectly from oxygen consumption. It meters the expired air and removes a proportion for analysis.

Interesterification Fats are mixtures of triacylglycerols consisting of various fatty acids esterified with glycerol. Interesterification is the result of heat treatment that causes an exchange of some of the fatty acids between the glycerol molecules and alters the properties of the fat. Lard, for example, is not a good creaming agent for baked products until it has been so treated.

Intermediate hydrogen carrier Most oxidation reactions in metabolism involve the loss of hydrogen, which is transferred onto a coenzyme, commonly derived from either niacin or vitamin B 2. The coenzyme is an intermediate hydrogen acceptor, and is reoxidized by transfer of the hydrogen to another substrate, or to oxygen, forming water.

Intermediate moisture foods These are semi-moist with about 25% (15-50%) moisture but with some of the water bound (and so unavailable to micro-organisms) by the addition of glycerol, sorbitol, salt, or certain organic acids, so preventing the growth of microorganisms.

Intestinal juice Also called succus entericus. Digestive juice secreted by the intestinal glands lining the small intestine. It contains a variety of enzymes, including enteropeptidase, the enzyme that converts trypsinogen to active trypsin, aminopeptidase, nucleases, and nucleotidases.

Intestine The gastro-intestinal tract; more specifically the part after the stomach, i.e. the small intestine where the greater part of digestion and absorption take place, and the large intestine where water is absorbed.

Intrinsic factor A protein secreted in the gastric juice which is required for the absorption of vitamin B 12 ; impaired secretion results in pernicious anaemia.

Intrinsic quality In the context of fishery products, the quality of the fish at the time of capture or harvesting. It will include attributes such as species, size, condition, composition, damage, and presence of parasites, harmful microorganisms, taints, toxins and contaminants.

Inulin Soluble but undigested polymer of fructose found particularly in Jerusalemartichoke, and, to a lesser extent, other root vegetables. Included with non-starch polysaccharides. Also called dahlin and alant starch.

Inversion Applied to sucrose, means its hydrolysis to glucose and fructose.

Invert sugar The mixture of glucose and fructose produced by hydrolysis of sucrose, 1.3 times sweeter than sucrose. So called because the optical activity is reversed in the process. It is important in the manufacture of sugar confectionery, and especially boiled sweets, since the presence of 10-15% invert sugar prevents the crystallization of sucrose.

Invertase Enzyme that splits sucrose into glucose and fructose; also called sucrase and saccharase.

Iodoform odour A slightly irritating odour similar to that of iodoform. It is detected occasionally in vertebrate fish and in crustaceans, more often in the latter. It is typically confined to particular fishing grounds and seasons.

Ion An atom or group of atoms that has lost or gained one or more electrons, and thus has an electric charge. Positively charged ions are known as cations, because they migrate towards the cathode (negative pole) in solution, while negatively charged ions migrate towards the positive pole and hence are known as anions.

Ion-exchange resin An organic compound that will adsorb ions under some conditions and release them under other conditions. The bestknown example is in water-softening, where calcium ions are removed from the water by binding to the resin, displacing sodium ions. The resin is then regenerated by washing with a concentrated solution of salt, when the sodium ions displace the calcium ions. Ion-exchange resins are used for purification of chemicals, metal recovery, a variety of analytical techniques, and treatment of hypercholesterolaemia.

Ionization The process whereby the positive and negative ions of a salt or other compound separate when dissolved in water. The degree of ionization of an acid or alkali determines its strength.

Ionizing radiation Electromagnetic radiation that ionizes the air or water through which it passes, e.g. X-rays and y-rays. Used for the sterilization of food, etc., by irradiation.

Irish cream Liqueurs prepared from Irish whiskey blended with other spirits and cream, sweetened, and sometimes flavoured with coffee or chocolate.

Irish moss A red seaweed, Chondrus crispus; source of the polysaccharide carrageenan.

Iron A mineral used by the body to make hemoglobin, the part of the blood that moves oxygen from the lungs to the body tissue.

Iron An essential mineral. The average adult has 4-5 g of iron, of which 60-70% is present in the blood as haem in the circulating haemoglobin, and the remainder present in myoglobin in muscles, a variety of enzymes, and tissue stores. Iron is stored in the liver as ferritin, in other tissues as haemosiderin, and as the blood transport protein transferrin.

Iron balance Losses in faeces 0.3-0.5 mg per day, in sweat and skin cells 0.5 mg, traces in hair and urine, total loss 0.5-1.5 mg per day. Blood loss leads to a considerable loss of iron. The average diet contains 10-15 mg, of which 0.5-1.5 mg is absorbed. The haem iron of meat and fish is considerably better absorbed than the inorganic iron of vegetable foods. Reference intakes are 8.7 mg for adult men and 14.8 mg for women; women who have heavy menstrual blood losses may not be able to obtain enough from food, and supplements are necessary. Absorption of iron is aided by vitamin C taken at the same time as iron-containing foods, and reduced by phosphate and phytic acid. Iron content of foods per 100 g: liver 6-14 mg, cereals up to 9 mg, nuts 1-5 mg, eggs 2-3 mg, meat 2-4 mg. Iron is added to flour so that it contains not less than 1.65 mg per 100 g. Fortified cereals provide 35% of the iron of British diets. Prolonged deficiency gives rise to anaemia.

Iron storage Ferritin is the iron storage protein in the intestinal mucosa, liver, spleen, and bone marrow. It is a ferric hydroxide-phosphate-protein complex containing 23% iron. Haemosiderin is a long-term reserve of iron in tissues; colloidal iron hydroxide combined with protein and phosphate, probably formed by agglomeration of ferritin, the short-term storage form. Abnormally high levels of haemosiderin occur in siderosis.

Iron transport Iron is transported in blood plasma in combination with proteins, as transferrin or siderophilin.

Iron, reduced Metallic iron in finely divided form, produced by reduction of iron oxide. The form in which iron is sometimes added to foods, such as bread. Also known by its Latin name ferrum redactum.

Irradiation Ionizing radiation kills micro-organisms and insects, and also inhibits sprouting of potatoes.

Irritable bowel syndrome Also known as spastic colon or mucous colitis. Abnormally increased motility of the large and small intestines, leading to pain and alternating diarrhoea and constipation; often precipitated by emotional stress.

Ischaemic heart disease Or coronary heart disease. Group of syndromes arising from failure of the coronary arteries to supply sufficient blood to heart muscles; associated with atherosclerosis of coronary arteries.

Isinglass Gelatine prepared from the swim bladder of fish. Used commercially to clear wine and beer, and sometimes in jellies and ice cream. Japanese isinglass is agar.

Islets of langerhans The endocrine parts of the pancreas; glucagon is secreted by the alpha cells and insulin by the beta cells of the islets.

Isoenzymes Enzymes that have the same catalytic activity, but different structures, properties, and/or tissue distribution.

Isoleucine An essential amino acid, rarely limiting in food. It is one of the branched-chain amino acids, together with leucine and valine.

Isomalt A bulk sweetener, about half as sweet as sucrose, consisting of a mixture of two disaccharides, glucose-glucitol and glucose-mannitol. It is about 50% metabolized, yielding 9 kJ (2.4 kcal)/gram. It is thought to be less laxative than sorbitol or mannitol, and does not encourage tooth decay, so is used in tooth-friendly sweets.

Isomaltose A disaccharide composed of two glucose units, differing from maltose only in the way in which the two glucose units are linked; unlike maltose it is not fermentable. Also known as brachyose.

Isomerose Trade name of high-fructose corn syrup: 70-72% solids, 42% fructose, 55% glucose, 3% polysaccharides.

Isomers Molecules containing the same atoms but differently arranged, so that the chemical and biochemical properties differ.

1. In positional isomers the functional groups are on different carbon atoms; e.g. leucine and isoleucine, citric and isocitric acids.

2. D- and L-isomerism refers to the spatial arrangement of four different chemical groups on the same carbon atom (stereo-

isomerism or optical isomerism). R- and Sisomerism is the same, but determined by a set of systematic chemical rules.

3. Cis- and trans-isomerism refers to the arrangement of groups adjacent to a carbon-carbon double bond; in the cis-isomer the groups are on the same side of the double bond, while on the transisomer they are on opposite sides.

Isotonic Solutions with the same osmotic pressure; often refers to a solution with the same osmotic pressure as body fluids. Hypertonic and hypotonic refer to solutions that are more and less concentrated.

Isotopes Forms of elements with the same chemical properties, differing in atomic mass because of differing numbers of neutrons in the atomic nucleus. Thus, hydrogen has three isotopes, of atomic masses 1, 2, and 3, generally written as 1H, 2H , and 3H. 1H is the most abundant isotope of hydrogen; 2H is stable, while 3H is radioactive.

The incorporation of isotopes into compounds permits the metabolic fates of those compounds in the body to be followed easily.

Isotopes, radioactive Some isotopes are unstable, and decay to stable elements, emitting radiation in the process. This may be ? 1 -radiation, ßradiation, v-radiation, or X-rays, depending on the particular isotope. Radioactive isotopes can readily be detected by the radiation emitted. The time taken for half the radioactive isotope to decay is the half-life of the isotope, and can vary from a fraction of a second, through several days to years.

Isotopes, stable Some isotopes are stable, and can be detected only by their different atomic mass. Since they emit no radiation, they are considered completely safe for use in labelled compounds given to' human beings. Examples of stable isotopes commonly used in nutrition research include 2H, 13C, 15N, and 18O.

IU International Units. A standard unit of measurement of biological activity that is used for fat soluble vitamins A, D, & E (as well as for some hormones, enzymes, and vaccines). It is an internationally-recognized standard established by the International Conference for Unification of Formulae. One I.U. represents a different amount for different substances. The weight equivalents for fat soluble vitamins are as follows: Beta Carotene.

J

J-cutting An alternative to V-cutting. A simpler, but more extravagant way of cutting out most or all of the pin bones. A cut (the J-cut) is made through the skinned fillet from the neck end of the fillet, dorsal to and along the line of pin bones towards the last pin bone, and then curved sharply down to the ventral edge of the belly flap. The pieces cut out are the J-cuts to leave a J-cut fillet. See Filleting, Gutting, and Heading and gutting.

Jejunum Part of the small intestine, between the duodenum and the ileum.

Jelabi Indian sweet; deep-fried spirals of dough served with syrup.

Jello North American name for table jelly.

Jelly condition A defect in the texture of fish muscle. The muscle has a translucent appearance, and the texture to the feel is soft and gelatinous. The cooked muscle has a soft texture. The jelly condition is associated with flesh of low protein content, <12%, and high water content. The condition is more common in some species than others, fish from deep waters are especially prone to the condition, and nutritional status can affect the incidence; starving fish have a high incidence, well fed fish a low incidence.

Jerked beef South American dried meat similar to biltong.

Jesuit's bark Cinchona bark, source of quinine.

Jigger Measuring cup used by bar staff for spirits, etc., when the bottles do not have a dispenser fitted.

Jojoba oil Liquid wax of long-chain fatty acids esterified with long-chain alcohols from seeds of the shrub Simmondsia chinensis. Of interest in cosmetics as a replacement for sperm whale oil but also has food applications, e.g. coating agent for dried fruits.

Jonathan Calcined, ground oat chaff used as adulterant for maize and other cereals.

Jonge Young Dutch gin.

Joule The SI unit of energy; used to express energy content of foods and energy expenditure of man and animals. Gradually adopted as a replacement for the calorie from about 1970; 4.2 kilojoules (kJ) is equivalent to 1 kilocalorie (kcal).

Jovennuevo Spanish; young wines.

Jowar Indian name for sorghum, also known as great millet, kaffir corn, guinea corn.

Judas goat Sheep cannot readily be driven to slaughter but will follow a goat. A Judas goat is used to lead the sheep to the killing pens.

Jugged Food cooked in earthenware pot or jug.

Jug-jug Caribbean; minced meat and pigeon peas cooked with millet, served as an accompaniment to ham or roast chicken.

Jujube 1. Sweet made from gum and sugar.

2. Shrub (Ziziphus mauritania or Z. jujuba), important fruit crop in India; the fruit is reddish-brown, up to 2 cm in diameter, with a single stone; a good source of vitamin C.

Julep Whisky or brandy flavoured with mint and served over ice.

Jullenne 1. Vegetables cut into thin, match-like strips.

2. A clear vegetable soup.

Jumble Small rich biscuit flavoured with lemon or almond, formed into a letter S, or into tiny, rock-like heaps.

Juniper The ripened berries of the bush Juniperis communis, used as a flavouring in gin, and, together with other herbs, in stuffing or sauces, especially for game.

Junket Dessert made from milk by treating with rennet to curdle the protein.

Jus, au Dish served with the natural juices or gravy.

K

Kabinett German and Austrian wine classification, used for highquality wines. In earlier years monastic wine producers used to set aside their best wine in a special cupboard or small room.

Kabuní Albanian; rice sauteed in butter and stock with raisins, served with chicken or meat.

Kaccavia Greek fish soup with tomatoes, vegetables, and garlic; the origin of bouillabaisse. The name is derived from the kakavia, the earthenware pot in which fishermen cooked their fish in the middle of the boat.

Kale Scottish name for any type of cabbage; in England it means specifically open-headed varieties of cabbage with curly leaves, also known as curly kale or borecole. Distinct from sea kale or Swiss chard.

Kalia Chicken broth flavoured with juice from pickled cucumbers and garnished with diced chicken, celery root, parsley root, and carrots.

Kalonji Wild black onion seeds (Nigella indica), used as a spice.

Kalteszal Soup made from egg yolk, sugar, cinnamon, and beer, served cold.

Kamaboko Fish paste made from surimi, sometimes with added starch.

Kaoliang A drink made from the Chinese sorghum, Sorghum nervosum; also used for the grain itself.

Karaya gum Obtained from East Indian trees of the genus Sterculia. Used as a stabilizer, e.g. in frozen water ices; also used in combination with other stabilizers; sometimes used as a laxative. Also called sterculia gum.

Karo syrup Trade name for a dextromaltose preparation made from maize starch, used as carbohydrate modifier in milk preparations for infant feeding. Consists of a mixture of dextrin, maltose, glucose, and sucrose.

Karyotype The chromosomal complement of an individual, including the number of chromosomes and any abnormalities. The chromosomes are stained with special dyes that produce distinct banding (striped) patterns. The term is also used to refer to a photograph of an individual's chromosomes arranged according to size.

Kataifi Pastry in thin strands; the dough is squeezed through a perforated disc onto a hot metal plate, on which it is dried in long strands. Also the name for rolls made from these pastry strands, filled with chopped nuts and sugar, baked and served drenched with syrup.

Katemfe The Sierra Leone name for an intensely sweet African fruit, Thaumatococcus daniellii; also known as the miraculous fruit of Sudan (not the same as miracle berry). The active principle is a protein named thaumatin.

Kebab Turkish for roast meat. Shishkebab is small pieces of mutton rubbed with salt, pepper, etc., and roasted on a skewer (shish in Turkish) sometimes interspaced with vegetables. Shashlik is a Georgian version. Döner kebab is a Turkish speciality consisting of marinated mutton or lamb packed into a cylindrical mass and grilled on a vertical rotating spit.

Kedgeree dish of rice and pulses. Modified to Victorian breakfast dish of flaked fish with egg and rice.

Keftethes Minced meat rissoles.

Keliweli Ghanaian; spiced deep-fried banana or plantain.

Kelp Large brown seaweeds of the genus Laminaria. Occasionally used as food or food ingredient but mostly the ash is used as a source of alkali and iodine. Sometimes claimed as a health food with unspecified properties.

Kephalins Or cephalins; phospholipids containing ethanolamine, hence phosphatidylethanolamines. Found espe-cially in brain and nerve tissue.

Kepler extract of malt Trade name for one of the earliest of the malt extracts, intended as a dietary supplement and to aid the digestion, since it was rich in diastase compared with ordinary malt extracts.

Keratin The insoluble protein of hair, horn, hoofs, feathers, and nails. Not hydrolysed by digestive enzymes, and therefore nutritionally useless. Used as fertilizer, since it is slowly broken down by soil bacteria. Steamed feather meal is used to some extent as a supplement for ruminants.

Keshy yena Caribbean; baked Edam or Gouda cheese with a variety of fillings. The name derives from the Spanish queso relleno, stuffed cheese.

Kesp Trade name for a textured vegetable protein product made by the spinning process.

Ketchup From the Chinese koechap or kitsiap, originally meaning brine of pickled fish. Now used for spicy sauce or condiment made with juice of fruit or vegetables, vinegar, and spices. Tomato ketchup is a common sauce.

Ketogenic diet A diet poor in carbohydrate (20-30 g) and rich in fat; causes accumulation of ketone bodies in tissue; formerly used in the treatment of epilepsy.

Ketonaemia High concentrations of ketone bodies in the blood.

Ketone bodies Acetoacetate, [ß-hydroxybutyrate and acetone; acetoacetate and acetone are chemically ketones; although ßhydroxybutyrate is not, it is included in the term ketone bodies because of its metabolic relationship with acetoacetate.

In the fasting state, fatty acids are mobilized from adipose tissue as a metabolic fuel. Most tissues have only a limited capacity for fatty acid oxidation; however, the liver can oxidize more than is required for its own metabolic needs. Acetoacetate and ß-hydroxybutyrate are formed from fatty acids in the liver, and are transported in the bloodstream for use as metabolic fuels by other tissues. Acetoacetate is chemically unstable and breaks down to acetone, which is metabolically useless, and is excreted in the urine and on the breath.

Ketones Chemical compounds containing a carbonyl group (C=O), with two alkyl groups attached to the same carbon; the simplest ketone is acetone.

Ketonic rancidity Certain moulds of the genera Penicillium and Aspergillus species attack fats containing short-chain fatty acids and produce ketones with a characteristic odour and taste, so-called ketonic rancidity. Fats such as butter, coconut, and palm kernel are most susceptible.

Ketonil Trade name for protein-rich food low in phenylalanine for feeding patients with phenylketonuria.

Ketonuria Excretion of ketone bodies in the urine.

Ketosis High concentrations of ketone bodies in the blood.

Kibble To grind or chop coarsely.

Kichals Middle-European (especially Jewish); spherical biscuits made from beaten egg, caster sugar, and self-raising flour.

Kid Young goat usually under 3 months old; similar to lamb, but with a stronger flavour.

Kidney Usually from lamb, ox, or pig; a rich source of protein, niacin, iron, zinc, copper, selenium, vitamins A, B 1, B 2, B 12, and folate; a good source of vitamin B 6 and, unusually for a meat product, vitamin C; a source of iodine.

Kiln May be used for smoking and/or drying. See Smoking kiln, salting.

Kilo As a prefix for units of measurement, one thousand times (i.e. 10^3); symbol k.

Kimchi Korean; dish based on fermented cabbage with garlic, red peppers, and pimientos often with the addition of fish and other foods. Nutritional value depends on the ingredients but all preparations are good sources of vitamin C.

King, a la Dish served in a rich cream sauce (often flavoured with sherry) and including mushrooms and green peppers.

Kipper and Kippering Herring split along the back, brined, sometimes dyed, and cold smoked to produce a product known as a kipper. The brown colouring used is referred to as kipper dye and is usually brown FK. A kipper fillet may be a cold smoked, single fillet of herring (which may then be canned), or a fillet cut from a so-called boneless kipper which used to be produced from a kipper from which the head, belly wall and backbone was removed. A boneless kipper is now usually regarded as a cold smoked, block fillet of herring. Confusingly, kippering may refer to production of kippers from herring or, in the UK, to the process of splitting a fish in the same manner as a herring and cold smoking, as well as generic reference to cold smoking, or to the process of hot smoking in the USA.

Kirsch Brandy or eau-de-vie distilled from fermented cherries, including the crushed stones, giving a pungent flavour; not sweet.

Kitron Greek liqueur prepared by distillation of brandy with lemon leaves; sweetened.

Kiwano The horned melon, a New Zealand fruit with a spiky yellow skin and green flesh.

Kiwi Fruit of Actinidia sinensis, originally a native of China and also known as Chinese gooseberry.

Kizel Cold dessert made from red fruit juice thickened with cornflour or arrowroot. Polish kiziel is similar.

Kjeldahl method The most commonly used procedure for determining the nitrogen content of a sample. The sample is heated with concentrated sulphuric acid and a catalyst, commonly mercury, and the nitrogen is converted to ammonia. The digest is made alkaline, and the ammonia distilled off and measured. Named after the originator of the procedure.

Kleftiko Clay oven. Also a joint of lamb baked in a clay oven. Originally a method of cooking meat in the embers of a fire in a pit in the ground.

Kleik Gruel made from barley and bouillon.

Klim Trade name for dried milk.

Klipfish Salted and dried cod, mainly produced in Norway. The fish is boned, stored in salt for a month, washed, and dried slowly. It is known as bacalao in South America and Spain, bacalhau in Portugal.

Klippfisk Norwegian name for salted and dried cod (Gadus morhua) made briefly as follows. Typically, it is bled, headed with a so-called round cut and gutted, split and cleaned to leave a piece of the backbone at the tail to give strength. The fish are stacked in layers with solid salt, eventually taking up salt to about 18% by weight, while the saturated brine formed is allowed to drain away. The fish are removed from the stack, washed, drained and hung up for drying for a few days, then restacked and hung up for drying a number of times to reduce the moisture content further, or dried in driers.

Klondyking Traditionally refers to the practice of preserving fish for a short time by cooling them in a mixture of ice and salt. Also used to describe the transshipping of fish at sea from a catcher vessel to a carrier vessel on which the fish were formerly, and sometimes still are, preserved as above, but are now more likely to be frozen, or otherwise processed.

Kluski Light dumplings or strips of noodles, used to garnish soups.

Knead To press firmly and repeatedly with the hands and fingers, to shape dough or flour mixture by hand.

Knickerbocker glory Dessert made from layers of jelly of different colours, ice cream, and fruit, topped with whipped cream and served in a tall glass (a sundae glass).

Knödeln Dumplings made from flour or matzo meal; served like pasta or in soup. Nockerln are smaller; butternockerln are made with butter; lebernockerln contain finely chopped fried liver.

Knuckle of veal Lower part of the leg of the calf furthest from the fillet or shoulder.

Köche Light steamed or baked pudding.

Kofta Spiced meat balls. Kofta may also be moulded over sweetsour plums, minced dried apricots, and herbs or eggs (these latter are known as nargizi or narcissus).

Kohlrabi Swollen stem of Brassica oleracea gongylodes; there are green and purple varieties.

Kokoh In the Zen macrobiotic diet this is a mixture of ground seeds and cereals fed to young infants; it is deficient in a number of nutrients and can result in growth retardation unless supplemented.

Kokoretsi Highly seasoned Greek sausage made from lamb offal with the intestine wound around the coarsely chopped meat, skewered and grilled slowly.

Konjac Gum derived from tubers of Amorphophallus konjac; eaten in Japan as a firm jelly.

Korma meat or vegetables braised with water, stock, yoghurt, or cream.

Korn German spirit prepared by distillation of fermented maize, sometimes flavoured with juniper. Generally has a relatively low alcohol content; doppelkorn is 38% alcohol.

Kosher The selection and preparation of foods in accordance with traditional Jewish ritual and dietary laws. Foods that are not kosher are traife. The only kosher flesh foods are from animals that chew the cud and have cloven hoofs, such as cattle, sheep, goats, and deer; the hindquarters must not be eaten. The only fish permitted are those with fins and scales; birds of prey and scavengers are not kosher. Moreover, the animals must be slaughtered according to ritual before the meat can be considered kosher.

Koulouria Sweet bread baked in a large ring, flavoured with sesame seeds.

Kourabiethes Shortbread traditionally baked for Christmas and New Year.

Kraftbruehe Austrian; soup made from minced beef, carrot, onion, leek, and

Hamburg parsley root, cooked in consomme.

Krapiva Nettle soup.

Kräuterbutter Butter mixed with chopped herbs.

Kreatopita Greek pie made with three kinds of meat (usually lamb, goat or kid, and chicken), together with cheese, onions, tomatoes, and raisins, with a pastry top.

Krebs' cycle Or citric acid cycle, a central pathway for the metabolism of fats, carbohydrates, and proteins. Named for Sir Hans Krebs, who first described the pathway.

Kreplach Pasta envelopes filled with minced meat (similar to ravioli) cooked in, and served with, soup.

Krill Term that refers to many species of planktonic crustaceans but is mostly used for the shrimp Euphausia superba. This is the main food of whales, and some penguins and other seabirds; occurs in shoals in the Antarctic, containing up to 12 kg/m3. Collected in limited quantities for use as human food.

Kromenski Minced poultry, game, or meat, bound to a stiff paste with sauce, wrapped in bacon, coated in batter, and fried.

Kuchen dough Pastry dough, containing egg, margarine, and yeast.

Kudoa spp A genus of the zoological class Myxosporea found as parasites in the muscle tissue of many species of marine and freshwater vertebrate fish. Many species of Kudoa secrete a proteolytic enzyme to form, at the least, unsightly white cysts in the flesh and, in extreme cases, to result in softening and liquefaction of the flesh. The cysts become very obvious if the fillet is smoked. Some species of Kudoa are specific to particular species of fish, but Kudoa thyrsites is known to infect some 20 or more species, including economically important species such as salmons - wild and farmed - and Pacific hake. (In early papers on the milky condition of hake and other similar defects, the organism was named as Chloromyxum thyrsites and described as a protozoan parasite. There has been revision of the taxonomy of Kudoa and similar parasites and they are now classed as metazoans and placed in their own phylum, Myxozoa). (Moran et al, 1999).

Kugelhopf Cake made from yeast dough in the shape of an inverted flower pot, with a hole through the centre which is usually filled with raisins and currants.

Kulfi Ice cream containing rice flour or sesame seed and ground almonds.

Kümmel Liqueur flavoured with caraway seeds, fennel, and orris root; 40% alcohol.

Kumquat A citrus fruit of the genus Fortunella; widely distributed in S. China and now cultivated elsewhere; resembles other citrus fruits, but very small, of ovoid shape, with acid pulp and sweet, edible skin.

K-value An chemical index used as a measure of the freshness of fish. Adenosine triphosphate (ATP) present in fish muscle is converted on the death of the animal though a sequence of intermediates to hypoxanthine (Hy). The intermediates are, in sequence, adenosine diphosphate (ADP), adenosine monophosphate (AMP), inosine monophosphate (IMP), and inosine (Ino). K-value is defined as the ratio, (often multiplied by 100), of the sum of Ino and Hy contents to the sum of concentration of ATP and all degradation products, expressed on a molar basis. ATP is converted to IMP within a day or two after death of the animal, the fish being stored at 0ºC. The IMP is lost more slowly leaving just Ino and Hy, and the K-value approaches 1.0, (or 100). K-value is usually the best predictor of freshness measured either by time of storage or by sensory methods for many species of fish, including shellfish. The relationship between K-value and storage time or sensory freshness is not the same over all species of fish and must be determined for a particular species of interest. It is not a simple test to carry out and requires some analytical skills and expensive equipment, and has not found favour for quality control. It is probably used most widely in Japan.

Kvass Beer, normally homebrewed from rye, buckwheat, or wheat, with malt and yeast, often flavoured with mint; 2-4% alcohol by volume, 1-5% carbohydrate, 60-150 kcal (250-630 kJ) per 300 mL.

L

LA An essential fatty acid. It can not be produced by the body and must be obtained through diet. It is of the omega-6 family.

Labelled substances To follow the metabolic fate of foodstuffs, etc., in the body, they can be labelled by introduction of an unusual chemical substituent or with a stable or radioactive isotope, so that the products can readily be identified.

Lacquer With reference to canned foods, a layer of gum and gum resin coated on to tinplate and hardened with heat. The layer of lacquer protects the tin lining from attack by acid fruit juices.

Lactalbumin One of the proteins of milk (casein 3%, lactalbumin 0.5%, lactoglobulin 0.25%). Not precipitated from acid solution as is casein; hence, during cheese making the whey contains lactalbumin and lactoglobulin. They are precipitated by heat and a whey cheese can be made in this way.

Lactase Enzyme in certain yeasts and mammalian intestinal tracts that catalyzes converting lactose to glucose and galactose. Lactase tablets remove lactose from milk for people who are lactose intolerant.

Lactation The process of synthesizing and secreting milk from the breasts.

Lactoferrin Iron-protein complex in human milk, only partly saturated with iron; has a role inhibiting the growth of E. coli and other potentially pathogenic organisms.

Lactoflavin Obsolete name for riboflavin, so named because it was isolated from milk.

Lactogen A drug or other substance that increases the production and secretion of milk.

Lactose Milk sugar; also white crystal sugar made from whey used in baby food, baked goods, candies and pharmaceuticals.

Lactose The carbohydrate of milk, sometimes called milk sugar. A disaccharide of glucose and galactose. Used pharmaceutically as a tablet filler and as a medium for growth of micro-organisms. The fermentation of lactose to lactic acid by bacteria is responsible for the souring of milk. Ordinary lactose is ? 1 -lactose, which is 16% as sweet as sucrose; if crystallized above 93 ’C, it is converted to the ß-form which is more soluble and sweeter.

Lacto-serum Grandiloquent word for whey.

Lactulose A disaccharide of galactose and fructose which does not occur naturally but is formed in heated or stored milk by isomerization of lactose. About half as sweet as sucrose. Not hydrolysed by human digestive enzymes but fermented by intestinal bacteria to form lactic and pyruvic acids. Thought to promote the growth of Lactobacillus bifidus and so added to some infant formulae; in large amounts it is laxative. Because of the bacterial fermentation, it is also used to acidify the colon and hence prevent the absorption of ammonia in patients with liver failure.

Laetrile Name given to an extract of apricot kernels, arnygdalin, a glucoside of benzaldehyde and cyanide. Claimed as a cancer cure, although there is no supporting evidence, and sometimes called vitamin B 17, although there is no evidence that it is a dietary essential or has any metabolic function.

Lakerda Pickled raw fish.

Lamb Meat from sheep younger than 12-14 months. A rich source of protein, niacin, iron, zinc, copper, and vitamin B 12.

Lambrusco Wine with low alcohol content, lightly sparkling or frizzante, made from the lambrusco variety of grape.

Lamb's lettuce Or corn salad, a hardy annual plant, Valerianella locusta or V. olitoria used in salads in winter and early spring. A rich source of vitamins A (as carotene) and C.

Lamb's wool Old English drink made by pouring hot ale over pulped roasted apples and adding sugar and spices.

Laminate To combine two or more layers of material to form packaging. The layers are held together by an adhesive or heat bonding.

Laminated block A frozen, rectilinear, cuboidal block of closely controlled dimensions and weight, typically a nominal 7.4 kg, usually made from white fish as an intermediate product for the production of fish fingers (fish sticks) and portions. It is composed entirely of skinless, boneless fillets (fillet block), that may be jumble-packed or orientated and layered, or composed of skinless, boneless fillets plus the recovered fish flesh from the V-cuts, for example, or composed entirely of recovered fish flesh (minced fish), termed a mince block or minced fish block. A proprietary polyphosphate mixture is often added to the fillets and/or mince before packing. The blocks may be made at sea on factory ships or onshore. Originally, fish blocks were composed of layered, skinless, boneless fillets and termed laminated blocks, and produced for the purpose of manufacturing fish sticks (fish fingers). As demand expanded for fingers and portions, fish flesh recovered from pin bone trimmings by means of bone separators was mixed with the fillets in the block. Subsequently, minced fish blocks were made for economy and catering products. The blocks are formed typically by packing the fish in a one-piece, protective, waxed card carton folded inside a metal mould or frame on a tray, and compacted to eliminate voids or air spaces within the mass of fish . The cartons packed in the frames are frozen in a horizontal plate freezer, ejected from the frame after freezing in a block press, and packed usually in fours in a semi-rigid, cardboard master carton (outer), for storage and distribution at -30°C. The blocks are a common item of international trade in frozen fish products.

Lamprey Cartilaginous fish resembling eels; sea lamprey is Petromyzon marinus, river lamprey or lampern is Lampetra fluviatilis.

Langue de chat Thin, flat, crisp biscuit, shaped like a cat's tongue, sometimes with the edges dipped in chocolate.

Lanolin The fat from wool. Consists of a mixture of cholesterol oleate, palmitate and stearate, and not useful as food; used in various cosmetics.

Lapskaus Norwegian; boiled salted meat.

Larch gum A polysaccharide of galactose and arabinose (ratio of 1: 6), found in the aqueous extract of the Western larch tree (Larix occidentalis); a potential substitute for gum arabic, since it is readily dispersed in water.

Lard compounds Blends of animal fats, such as oleostearin or premier jus, with vegetable oils, to produce products similar to lard in consistency and texture.

Lard Originally rendered fat from pig carcasses (sheep and cattle are also used). The best quality is from the fat surrounding the kidneys; neutral lard is the highest quality, prepared by agitating the minced fat with water at a temperature below 50°C; kidney fat provides No. 1 quality; back fat provides No. 2 quality.

Lard substitutes Vegetable shortenings made from mixtures of partially hardened vegetable fats with the consistency of lard.

Larding Method of adding fat to lean meat so that it does not dry during long slow cooking. Narrow strips of bacon fat (lardoons) are threaded into the surface of the meat with a special 'larding needle' before cooking. Barding is the process of tying a thin sheet of bacon fat or a rasher of fat bacon over the meat.

Lasagne Wide ribbons of pasta; lasagne verdi is flavoured with spinach. Narrow ribbons are lasagnette.

Lassi Beverage made from yoghurt or buttermilk, mixed with water.

Latent heat The heat lost or gained during a change of phase, for example, from liquid to solid during freezing or from solid to liquid during thawing.

Laterality The tendency found in most school-age children to be dominant in either their right or left side (especially for writing). A young child's developing inner sense of the body's two sides. Leads to early gross motor skills involving the cooperative work of both sides (rolling, crawling, walking), fine motor skills, and preparation for complex eye-hand coordination.

Lathyrism Nerve damage associated with high intakes of Lathyrus sativus (Kesari dhal, chickling pea, chickling vetch). The crop is often grown in dry districts in Asia and North Africa together with wheat. Normally little is eaten, but when there is a drought and the wheat crop is poor, the dhal predominates and is eaten as a major food.

Latkes Pancakes made from grated potatoes.

Lauric acid A medium-chain saturated fatty acid (12 carbons) in butter, coconut oil, and palm oil.

Laver Edible seaweed. Laver bread is made from the seaweed Porphyra spp., by boiling in salted water and mincing to a gelatinous mass. It is made into a cake with oatmeal or fried.

Laxative A substance that helps the expulsion of food residues from the body. If 'strongly' laxative it is termed purgative or cathartic. Dietary fibre and cellulose function because they retain water and add bulk to the contents of the intestine; Epsom salts (magnesium sulphate) also retain water; castor oil and drugs such as aloes, senna, cascara, and phenolphthalein irritate the intestinal mucosa.

Lead A mineral of no nutritional interest, since it is not known to have any function in the body. It is toxic and its effects are cumulative. May be present in food from traces naturally present in the soil; as contamination of vegetables grown near main roads, which absorb volatile lead compounds from car exhaust fumes; from shellfish that have absorbed it from seawater; from lead glazes on cooking vessels; and in drinking water where lead pipes are used. Traces are excreted in the urine.

Leaf Lard is made from the residue of kidney and back fat after the preparation of neutral lard by heating w.th water above 100°C in an autoclave. Prime Steam Lard is fat from any part of the carcass, rendered in the autoclave.

Lean body mass Measure of body composition excluding adipose tissue, i.e. cells, extracellular fluid, and skeleton.

Lean cuisine Trade name for a range of frozen meals prepared to a specified energy content.

Lean fish Used in a number of contexts. It may refer to white fish, or to spent fish, i.e. fish that has recently spawned, usually in poor condition, or to recovering spent fish, or it may refer to fatty fish with a seasonally low fat content. More strictly, it may be defined as fish that contain 2% or less fat in the flesh in the live state.

Leathers, fruit. Fruit purees dried in air in thin layers, 4-5 mm thick, then built up into thicker preparations.

Leaven To add a substance to (bread and other things made with flour) to make it get bigger when it is cooked. Yeast is added to yeast-leavened bread to make it rise and increase in volume.

Lebkuchen Gingerbread, often baked in carved moulds, traditionally eaten at Christmas.

Lecithin Chemically lecithin is phosphatidyl choline; a phospholipid containing choline. Commercial lecithin, prepared from soya bean, peanut, and maize, is a mixture of phospholipids in which phosphatidyl choline prędominates. Used in food processing as an emulsifier, e.g. in salad dressing, processed cheese, and chocolate, and as an anti-spattering agent in frying oils. Is plentiful in the diet and not a dietary essential.

Leckeli Biscuit made with honey, almonds, candied peel, cloves, nutmeg, and ginger.

Lectins Proteins,from legumes and other sources which bind to the carbohydrates found at cell surfaces. They therefore cause red blood cells to agglutinate in vitro, hence the old names haemagglutinins and phytoagglutinins.

Raw or undercooked beans of some varieties of Phaseolus vulgaris cause vomiting and diarrhoea within 2 hours of consumption due to the high level of lectins, but they are rapidly inactivated by boiling.

Legislation Law passed by parliament.

Legumes include the groundnut Arachis hypogaea, and soya bean, Glycine max, grown for their oil and protein, the yam bean Pachyrrhizus erosus, and African yam bean Sphenostylis stenocarpa, grown for their edible tubers as well as seeds.

Legumes Members of the family Leguminosae eaten by man and domestic animals. Consumed as dry mature seeds (grain legumes or pulses) or as immature green seeds in the pod. On boiling, the dried seeds double in weight, so a 100-g cooked portion is approximately 50 g as a dried product.

Legumin Globulin protein in legumes.

Lekach Sponge cake made with ginger and honey.

Lemon curd Cooked mixture of sugar, butter, eggs, and lemons. Legally (UK regulations) must contain 4% fat, 0.33% citric acid, 1% dried egg or equivalent, 0.125% oil of lemon or 0.25% oil of orange, and not less than 65% soluble solids.

Lemon grass Lemon-scented grasses (Cymbopogon spp.), native to South East Asia, widely used in Thai, Indonesian, and Malay cooking.

Lemon oil The peel oil of the lemon, 0.15-0.3% of the weight of the fruit.

Lemon Sour yellow fruit of Citrus limon. A rich source of vitamin C.

Lemonade Originally a beverage made from lemon juice with water and sugar; now also a wide variety of carbonated beverages.

Lentils Legumes; dried seeds of many varieties of Lens esculenta, they may be green, yellow, or orange-red. When ground, they are frequently used to thicken soup. A rich source of copper and selenium; a good source of iron; a source of protein, vitamin B 6, folate, and zinc,

Lettuce Leaves of the plant Lactuca sativa; many varieties are grown commercially.

Leucine An essential amino acid; rarely limiting in foods. Chemically, amino-isocaproic acid.

Leucocytes White blood cells, normally 5000-9000/mm3; includes polymorphonuclear neutrophils, lymphocytes, monocytes, polymorphonuclear eosinophils, and polymorphonuclear basophils. A 'white cell count' determines the total; a 'differential cell count' estimates the numbers of each type.

Fever, haemorrhage, and violent exercise cause an increase; starvation and debilitating conditions a decrease.

Leucocytosis Increase in the number of leucocytes in the blood.

Leucopen ia Decrease in the number of leucocytes in the blood.

Leucosin One of the water-soluble proteins of wheat flour. leucovorin The synthetic 5-formyl derivative of folic acid; more stable to oxidation than folic acid itself, and commonly used in pharmaceutical prepara-tions. Also known as folinic acid.

Levans Polymers of fructose (the principal one is inulin) that occur in tubers and some grasses.

Leveret A young hare.

Levitin One of the proteins of egg yolk; about one-fifth of the total, the remainder being vitellin. Rich in sulphur, it accounts for half of the sulphur in the yolk.

Licoroso Portuguese; fortified sweet wines. lieben German and Austrian classification for semi-sweet wines.

Liebfraumilch German sweet wine from the Rhine valley; originally from the vineyard beside the Liebfrauenkirche in Worms.

Light As applied to foods usually indicates: 1. a lower content of fat compared with the standard product (e.g. breadspreads, sausages);

2. Sodium chloride substitutes lower in sodium;

3. Low-alcohol beer or wine. US legislation restricts the term 'light' to modified foods that contain one-third less energy or half the fat of a reference unmodified food, or to indicate that the sodium content of a low-fat, low-calorie food has been reduced by half.

Light box See Candling

Light cure Refers to the mild degree of smoking or salting

Lignin Indigestible part of the cell wall of plants. It is included in measurement of dietary fibre, but not of non-starch polysaccharide.

Lignocellulose Alternative name for lignin.

Lillet French light vermouth made from red or white Bordeaux wines in which fruit peel and herbs are steeped; aged in oak casks.

Lima beans US term for flat, kidney-shaped bean, smaller ones are butter limas, called butter beans in the UK, and larger ones are potato limas.

Lime The fruit of Citrus aurantifolia, cultivated almost solely in the tropics, since it is not as hardy as other citrus fruits. Used to prevent scurvy in the British Navy (replacing, at the time, lemon juice) and so giving rise to the nickname of 'Limeys' for British sailors and for British people in general. Contains about 10-20 mg vitamin C per 100 g fruit or fresh juice.

Limit dextrin When a branched polysaccharide such as glycogen or amylopectin is hydrolysed enzymically, glucose units are removed one at a time until the branch point is reached. The hydrolysis then stops, leaving what is termed a limit dextrin; further hydrolysis requires a different enzyme.

Limmits Trade name for a 'slimming' preparation composed of wholemeal biscuits with a methyl cellulose mixture as filling, containing some vitamins and minerals; intended as a meal replacement.

Limonin The bitter principle in the albedo of the Valencia orange. Isolimonin is the bitter principle of the navel orange. Both are present as a non-bitter precursor which is liberated into the juice during extraction and is slowly hydrolysed, making the juice bitter.

Limpet A shellfish, Patella vulgata.

Linamarin Cyanogenic glucoside found in cassava (manioc) which may be a cause of neuropathies in areas where cassava is a major food; the cyanide is removed in traditional processing by grating and exposing to air.

Linie aquavit Norwegian aquavit which is aged in oak casks that are shipped to Australia and back before bottling, thus crossing the equator (linie) twice.

Linoleic acid An essential polyunsaturated fatty acid, predominant in most edible vegetable oils.

Linolenic acid A non-essential polyunsaturated fatty acid which has some pharmacological actions. Found in oils from the seeds of evening primrose, borage, and blackcurrant.

Linolenic acid An essential polyunsaturated fatty acid.

Linseed oil Vegetable oil from the seeds of flax, Linum usitatissimum.

Linzer torte Austrian; flan made with eggs, ground almonds, and mixed peel, with a covering of jam and a lattice of pastry on top.

Liothyronine Alternative name for triiodothyronine (T3), the most potent of the thyroid hormones.

Lipaemia Increase in blood lipids, as occurs normally after a meal.

Lipase Enzyme that hydrolyses fats to glycerol and fatty acids. Most lipases have low specificity and will attack any triglyceride or longchain ester. Present in the pancreatic juice, liver, and adipose tissue, and in many seeds and grains. Sometimes responsible for the development of rancidity in stored foods.

Lipase, hormone-sensitive The lipase of adipose tissue; it is stimulated by the action of hormones such as glucagon and adrenaline, which are secreted when there is a need for release of free fatty acids from adipose tissue for use as a metabolic fuel.

Lipectomy Surgical removal of subcutaneous fat.

Lipid composition The components of a mixture of lipids or of the constituents of a particular lipids. The fat extracted from fish contains several components, such as triglyceride lipid, phospholipids, sterols, and others, which can be separated to give an analysis of the components of the lipid. More usually, but loosely, the term lipid composition refers to analysis of the lipid for the constituents fatty acids.

Lipid peroxidation Radical oxidation of lipids. An autocatalytic process whereby polyunsaturated fatty acids and phospholipids undergo degradation by a chain reaction to form lipid hydroperoxides or lipoperoxides.

Lipidema Condition in which fat deposits accumulate in the lower extremities, from hips to ankles, with tenderness of the affected parts.

Lipids A general term for fats and oils (chemically triglycerides), waxes, phospholipids, steroids, and terpenes. Their common property is insolubility in water and solubility in hydrocarbons, chloroform, and alcohols.

Lipochromes Plant pigments soluble in fats and organic solvents, such as chlorophyll, carotenoids.

Lipodystrophy Abnormality in the metabolism or deposition of fats; abnormal pattern of subcutaneous fat deposits.

Lipofuscin A group of pigments that accumulate in several body tissues, particularly the myocardium, during life and are consequently associated with the ageing process.

Lipoic acid Chemically, dithio-octanoic acid, a coenzyme (together with vitamin B 1) in the oxidative decarboxylation reactions of glucose metabolism and the citric acid cycle. Although it is an essential growth factor for various micro-organisms, there is no evidence that it is a human dietary requirement.

Lipolysis The hydrolysis (splitting) of fats to yield glycerol and fatty acids.

Lipolytic rancidity Spoilage of foods as a result of hydrolysis of fats to free fatty acids on storage (by the action of lipase, either bacterial lipase or the enzyme naturally present in the food). Since the enzyme is inactivated by heat, this type of rancidity occurs only in uncooked foods.

Lipoperoxidation The process by which toxic forms of oxygen damage fats in the body, most notably the fatty acids in cell membranes.

Lipoproteins Complexes of protein and lipids.

Liposuction Procedure for removal of subcutaneous adipose tissue in obese people using a suction pump device.

Lipovitellenin A lipoprotein complex in egg comprising about one-sixth of the solids of the yolk.

Liquefied herring Herring reduced to liquid state by enzyme action at slightly acid pH; used as protein concentrate for animal feed.

Liqueurs Distilled, flavoured, and sweetened alcoholic liquors. For example, curacao (30% (weight/volume, w/v) alcohol, 30% sugar); cherry brandy (19% alcohol, 33% sugar); advocaat (13% alcohol, 30% sugar, 0.75% nitrogen).

Liquid diet Diet consisting of foods that can be served as liquids or strained purees, prescribed in acute inflammation of the gastrointestinal tract and for patients unable to consume normal foods, especially after surgery.

Liquorice Used in confectionery and to flavour medicines; liquorice root and extract are obtained from the plant Glycyrrhiza glabra; stick liquorice is the crude evaporated extract of the root. The plant has been grown in the Pontefract district of Yorkshire since the sixteenth century; hence the name Pontefract cakes for the sugar confection of liquorice.

Liquorose High strength, often fortified, wines.

Listeria A genus of bacteria commonly found in soil of which the commonest is Listeria monocytogenes. They can cause food poisoning (listeriosis). Listeria species are especially found in unwashed vegetables and some soft cheeses; they resist cold and the presence of salt and can multiply in a refrigerator. Symptoms of listeriosis are flu-like, with high fever and dizziness. Pregnant women, babies, and the elderly are especially at risk.

Listeria monocytogenes A species of bacteria pathogenic to humans. Listeriosis is the disease caused by infection by the organism. Its significance as a food pathogen has been realised only in the last two or three decades. It has a widespread occurrence in nature, being found in soils, vegetation, foods, including fishery products, internal environments of fish processing plants, and even commensally in humans. The route of infection is through the gut following consumption of contaminated food. Those most at risk are the foetus, pregnant women, neonates, the elderly, and immuno-compromised persons. Healthy persons are resistant to infection with no more than mild symptoms of gastrointestinal disturbance. At present, the number of organisms required to cause illness is not known. Fishery products have been implicated in incidents of listeriosis, but very rarely. The organism can grow at chill temperatures. Vigorous and rigorous sanitary and disinfection procedures are required to eliminate the organism from a fish processing plant. Other species of Listeria are often used as indicator organisms for Listeria monocytogenes, although they rarely cause food-borne illness. (Ben Embarek, 1994).

Lithium Metal not known to have any physiological function, although it occurs in food and water; lithium salts are used in the treatment of manic-depressive illness.

Liver Usually from calf, pig, ox, lamb, chicken, duck, or goose; an exceptionally rich source of iron and vitamins A, D, B 2, B 6, and B 12 ; a rich source of protein, zinc, copper, selenium, niacin, and vitamin B 1 ; also, unusually for meat, a good source of vitamin C. The vitamin A content of liver is high enough for it to pose a possible hazard to unborn children, and pregnant women have been advised not to eat liver. Fish liver is a particularly rich source of vitamins A and D, and fish liver oils (especially cod and halibut) are used as sources of these vitamins as nutritional supplements.

Liver sausage Sausage based on liver. An exceptionally rich source of iron, vitamin A, protein, vitamin B 2.

Livetin A water-soluble protein fraction of egg yolk.

Lobio Bean salad.

Lobscouse Sailor's stew of meat and vegetables, classically thickened with ship's biscuit.

Lobster A crustacean, Homerus vulgaris. A rich source of protein, niacin, vitamin B 12, and copper.

Locasol Trade name for a low-calcium milk substitute.

Lockshen Thin noodles.

Locksoy Fine-drawn rice macaroni.

Locoweed Astralagus and Oxytropus spp., common in arid areas of the western USA. Toxic to cattle, causing locoism: neurological damage, abortion, and birth defects. Apparently caused by an alkaloid, swainsonine, which is also found in mouldy hay.

Lofenalac Trade name for food low in phenylalanine for treatment of phenyl-ketonuria.

Loganberry Cross between European raspberry and Californian blackberry, Rubus ursinus var. loganobaccus, named after L. H. Logan.

Loin The back portion of animals.

Lollipop Boiled sweet or toffee on a small wooden stick. Ice lollipop is water-ice on a stick.

Lonalac Trade name for a milk preparation free from sodium.

London broil American name for steak, broiled or grilled and sliced thinly against the grain.

Lonzo Corsican; boned fillet of pork, steeped in wine and herbs, then air dried, thinly sliced, and eaten raw.

Loonzein Rice from which the husk has been removed; also known as brown rice, hulled rice, and cargo rice.

Loquat The small pear-shaped fruit of Eriobotyra japonica, a member of the apple family, also known as Japanese medlar.

Lotus The sacred lotus of India and China, Nelumbium nuciferum, a water plant whose rhizomes and seeds are eaten.

Lovage Herb of the carrot family, Ligusticum scoticum, with a strong musky scent of celery. The stems can be candied like angelica or used as a vegetable, and the leaves and stems are used in soup. The seeds can also be used as a seasoning, with a flavour like dill or fennel seed.

Love apple Old name for tomato.

Low birth weight Infants born weighing significantly less than normal are considered to be premature; their chances of survival and normal development are considerably improved if they are fed special formula preparations to meet their needs, rather than being breast-fed or fed normal infant formula. The normal range of weight at birth is between 2.5 and 4.5 kg.

Low fat A food containing 3 grams or less of fat per serving. Defined by the Food and Drug Administration in food labeling law.

Low-density lipoproteins Are assembled in the liver and exported to other tissues, where they are taken up and used as a source of cholesterol (and other lipids). High-density lipoproteins (HDL) are returned from tissues to the liver.

Lozenges Shapes stamped out of a mixture of icing sugar, glucose syrup, and gum arabic or gelatine with flavourings, then hardened at 32-43°C.

Lucerne The plant, Medicago sativa; essentially a forage crop, but eaten by man to a small extent. Also known as alfalfa.

Lumpfish Large sea-fish, Cyclopterus lumpus, the pink eggs of which are salted, pressed, and coloured, as Danish or German caviare.

Lunch The midday meal (but in northern England the main meal is traditionally eaten at midday and is known as dinner).

Luncheon meat Precooked, canned meat, usually pork. A rich source of protein, niacin; a good source of iron; a source of vitamins B 1 and B 2.

Lupins Legumes of Lupinus spp. The ordinary garden lupin contains toxic alkaloids (quinolizidines) and tastes bitter; varieties selected for animal feed and grain crop are low in alkaloids and known as sweet lupins; they are rich in protein and fat.

Luting Strip of pastry placed round a dish to seal on the lid or pastry cover, used when preparing potted game, etc., in covered dishes.

Lycasin Trade name for hydrogenated glucose syrup, a bulk sweetener.

Lychee The fruit of Litchi chinensis, native to China; the size of a small plum, with a hard case and translucent, white, jelly-like sweet flesh surrounding the seed. A rich source of vitamin C.

Lycopene Red carotenoid pigment found in tomato, pink grapefruit, and palm oil. It has no vitamin A activity. The synthetic material is sometimes used as a food colour (E-160(d)).

Lye-peeling A method of removing skins from vegetables by immersion in hot caustic soda solution (lye) followed by 'tumbling' in a wash to remove the skin and chemicals.

Lymph The fluid between blood and the tissues; the medium in which oxygen and nutrients are conveyed from the blood to the tissues, and waste products back to the blood. Similar to blood plasma in composition.

Dietary fat is absorbed into the lacteals as chylomicrons which are formed in the intestinal mucosa, and enters the bloodstream through the thoracic duct. After a fatty meal the lymph is rich in emulsified fat and is called chyle.

Lymphatics Vessels through which the lymph flows, draining from the tissues and entering the bloodstream at the thoracic duct.

Lysine An essential amino acid of special nutritional importance, since it is the limiting amino acid in many cereals. Can be synthesized on a commercial scale, and when added to bread, rice, or cereal-based animal feeds, it improves the nutritive value of the protein.

M

MA Mental age.

Maatjes Cured herring made only from young female herrings. macadamia nut Fruit of Macadamia ternifolia.

Macaroon Cake made from ground almonds or coconut, sugar, and eggwhite, baked on rice paper.

Macedoine Mixture of fruits or vegetables, diced, or cut into small evenshaped pieces.

Macerases A group of enzymes (usually extracted from the mould Aspergillus) used to break down pectin in fruits to facilitate maximum extraction of the juice.

Mackerel An oily fish, Scomber scombrus. A rich source of protein, vitamins D, B 2, B 6, B 12, niacin, copper, iodine, and selenium.

Maconochie A canned meat stew much used in the First World War; made by Maconochie Brothers.

Macrobiotic diet A system of eating associated with Zen Buddhism; consists of several stages finally reaching Diet 7 which is restricted to cereals. Cases of severe malnutrition have been reported on this 'diet'. It involves the Chinese concept of yin (female) and yang (male) whereby foods, and even different vitamins (indeed, everything in life) are predominantly one or the other and must be balanced.

Macrocytes Large immature precursors of red blood cells found in the circulation in pernicious anaemia and in folic acid deficiency, due to impairment of the normal maturation of red cells; hence macrocytic anaemia.

Madeira cake Rich cake containing no fruit and flavoured with lemon; traditionally decorated with strips of candied citron peel.

Madeira wines Fortified wines from the Island of Madeira: sercial (dry); verdelho (semi-dry); bual (semi-sweet); malmsey (sweet).

Madeleine Small fancy sponge cake baked in a dariole mould (or sometimes a scallop-shaped mould). English version is victoria sponge mixture topped with jam and coconut.

Madras curry Pungent curry sauce with a large amount of chilli, but less than vindaloo.

Madrid stew Olla podrida or rotten pot; Spanish (Castilian) stew containing chickpeas, chicken, beef on the bone, smoked bacon, tocino (salted pork fat), chorizo, morcilla, onion, cabbage, carrots, turnips, and garlic.

Madrilene, a la Dish flavoured with tomato.

Magma Mixture of sugar syrup and sugar crystals produced during sugar refining.

Magnesium An essential mineral; present in all human tissues, especially bone. Involved in the metabolism of ATP. Present in chlorophyll and so in all green plant foods, and therefore generally plentiful in the diet. Deficiency in human beings leads to disturbances of muscle and nervous system; in cattle, to grass tetany. Magnesiumdeficient plants are yellow (or chlorosed).

Magnum Large wine bottle, 1.5 L.

Mahleb Spice prepared from black cherry kernels, Syrian in origin, widely used in Greek baked goods.

Maids of honour Small tartlets filled with almond-flavoured custard; said to have originated in the court of Henry VIII, where they were made by Ann Boleyn when she was lady-in-waiting to Catherine of Aragon.

Maigre, au Dish without meat and therefore traditionally suitable for Lent or a fast day.

Maillard reaction Two processes in foods can produce a brown colour. One is the enzymatic oxidation of phenolic substances such as occurs at the cut surface of an apple. The other is a reaction between lysine in proteins and sugars, and is variously known as the Maillard reaction, the browning reaction and non-enzymic browning. It takes place on heating or prolonged storage and is one of the deteriorative processes that take place in stored foods. It is accompanied by a loss in nutritive value, since the part of the protein that reacts with the sugar is not digested.

Maitre d'hotel 1. Simply prepared dishes garnished with butter creamed with parsley and lemon juice (maitre d'hotel butter); literally 'in the style of the chief steward'.

2. Especially in the USA, the head waiter.

Maize flour Highly refined and very finely ground maize meal from which all bran and germ have been removed.

Maize Grain of Zea mays, also called Indian corn and (in the USA) simply corn. Staple food in many countries, made into tortillas in Latin America, into polenta in Italy, and flaked as cornflakes commonly eaten as a breakfast cereal; various preparations in the southern states of the USA are known as hominy, samp, and cerealine.

Two varieties of major commercial importance are flint corn, which is very hard, and dent corn; there is also sweet corn, and a variety that expands on heating. The starch prepared from Z. mays dentata is termed corn flour; the ground maize is termed maize meal. There is a white variety; the usual yellow colour is partly due to cryptoxanthin.

Because of its low content of the amino acid tryptophan, diets based largely on maize are associated with the development of pellagra.

Maize rice Finely cut maize with bran and germ partly removed, also called mealie rice.

Maize starches, waxy Starch obtained from hybrids of maize consisting wholly or largely (99%) of amylopectin, compared with ordinary maize starch with 26% amylose and 74% amylopectin. The paste is semi-translucent, cohesive, and does not form a gel.

Maize, flaked Partly gelatinized maize used for animal feed. The grain is cracked to small pieces, moistened, cooked, and flaked between rollers.

Makhani Buttered rice prepared by steaming with butter, and aromatic spices.

Malabsorption syndrome Defect of absorption of one or more nutrients; signs include diarrhoea, steatorrhoea, abdominal distension, weight loss, and specific signs of nutrient deficiency.

Malai Cream prepared by boiling milk, leaving it to cool, and then skimming off the clotted cream.

Malic acid Organic acid occurring in many fruits, particularly in apples, tomatoes and plums. Molecular formula COOH–CHOH–CH_2–COOH. Used as a food additive to increase acidity.

Malnutrition Disturbance of form or function arising from deficiency or excess of one or more nutrients.

Malo-lactic fermentation The conversion of the malic acid in grape juice (and other fruit juices) into lactic acid, especially in red wines and cider, causing them to mellow and become less acid.

Malondialdehyde A chemical formed in the oxidation of lipids. The thiobarbituric acid test measures its concentration in a sample of fish or extracted lipid and provides a measure of the degree of oxidation.

Malt Mixture of starch breakdown products containing mainly maltose, prepared from barley or wheat.

The grain is allowed to sprout, when the enzyme diastase develops and hydrolyses the starch to maltose. The mixture is then extracted with hot water, and this malt extract contains a solution of starch breakdown products together with diastase. Malt extract may remain as the concentrated solution or evaporated to dryness.

Malt flour Germinated barley or wheat, in dried form. As well as dextrins, glucose, proteins, and salts derived from the cereal, it is rich in diastase and is added to wheat flour of low diastatic activity for bread making; used as an ingredient of 'malt' loaf.

Maltase Enzyme that hydrolyses maltose to yield two molecules of glucose; present in the brush border of the intestinal mucosal cells.

Malthus T. R. Malthus, author of an essay in 1798 postulating that any temporary or local improvement in living. conditions will increase population faster than the food supply, and that disasters such as war and pestilence, which check population growth, are inescapable features of human society.

Maltitol A sugar alcohol produced by hydrogenation of maltose. Slowly hydrolysed in the digestive tract to glucose and sorbitol and fairly completely utilized, providing 4 kcal (16 kJ)/gram; sweeter than maltose, and 90% as sweet as sucrose (E-965).

Maltol Also called laxarinic acid, palatone, veltol; chemically 3-hydroxy-2 methylpyrone. Found in the bark of young larch trees, pine needles, chicory, and roasted malt; synthesized for use as a fragrant, caramellike flavour for addition to foods; imparts a 'freshly baked' flavour to bread and cakes.

Maltose Malt sugar, or maltobiose, a disaccharide consisting of two glucose units linked 1-4. Hydrolysed by maltase. Does not occur in foods (unless specifically added as malt) but formed during the acid or enzymic digestion of starch. It is one-third as sweet as sucrose.

Malvasia A grape variety widely used for wine making, although not one of the classic varieties.

Manche Cutlet bone. Manchette is the paper frill used to decorate the bone.

Mandarin Loose-skinned citrus fruit, Citrus rericulata or C. nobilio. Varieties include satsumas and tangerines with various hybrids including tangelo, tangor, temple, clementine.

Mandelbrot Middle-European; sponge cake with almonds. Also known as komishbrot.

Mandeln Middle-European; small pieces of dough shaped like almonds, baked and added to soup before serving.

Mandolin Vegetable slicer.

Manganese An essential trace mineral which functions as the prosthetic group in a number of enzymes. Dietary deficiency has not been reported in man; in experimental animals manganese deficiency leads to impaired synthesis of mucopol-ysaccharides. Requirements are not known; intakes greater than 1.4 rog/day are considered safe and adequate.

Mangelwurzel, mangoldwurzel A root vegetable used as cattle feed, Beta vulgaris rapa; a cross between red and white beetroot.

Mange-tout Immature pods and embryo seeds of Pisum sativum var. macrocarpum (or macrocarpon) eaten whole; also known as sugar pea or snap pea. The name is sometimes used for immature French beans, Phaseolus vulgaris.

Mango A fruit, originally of Indo-Burmese origin and now grown widely throughout the tropics, Mangifera indica. The fruit is ovoid, 3-5 in diameter, with orange-coloured sweet aromatic flesh surrounding a central stone.

Mangosteen A fruit of Indian origin, Garcinea mangostana, the size of an orange with thick purple rind and sweet white pulp in segments.

Manna bread A cake-like product made from crushed, sprouted wheat without yeast; said to be a recipe of the Essenes who lived by the Dead Sea around the beginning of the Christian era.

Manna Dried exudate from the manna-ash tamarisk tree (Fraxinus omus). Abundant in Sicily and used as a mild laxative for children; it consists of 40-60% mannitol, 10-16% mannotetrose, 6-16% mannotriose, plus glucose, mucilage, and fraxin. This is thought to be the food eaten by the Children of Israel in the wilderness.

Mannitol Mannite or manna sugar, a six-carbon sugar alcohol found in beets, pumpkin, mushrooms, onions; 50-60% as sweet as sucrose. Extracted commercially from seaweed or made from the sugar mannose.

Mannosans Polysaccharides containing mannose.

Mannose A 6-carbon sugar found in small amounts in legumes, manna, and some gums.

MAP Is the term used for methods that will help to maintain the quality of a food product by changing the atmosphere inside its retail package. For example, reduce the availability of oxygen or manipulate the levels of carbon dioxide. It produces a gas mix to maximise shelf life.

Maple syrup Sap of the north American sugar maple tree, Acer saccharum. Evaporated either to syrup (63% sucrose, 1.5% invert sugar) or to dry sugar for use in confectionery.

Maple syrup urine disease A rare genetic disease affecting the metabolism of the amino acids leucine, isoleucine, and valine leading to accumulation of high concentrations of these three amino acids and the immediate products of their metabolism in the plasma and urine. The urine has a characteristic smell like that of maple syrup.

Maraschino 1. Sweetened spirit prepared by distillation of fermented maraschino cherries (both the-juice and the crushed kernels).

2. Cherries preserved in real or imitation maraschino liqueur, used to decorate cocktails, ice cream, or fruit salad.

Marasmic kwashiorkor The most severe form of protein energy malnutrition in children, with weight for height less than 60% of that expected, and with oedema and other symptoms of kwashiorkor.

Marc Spirit distilled from the fermented residue of grape skins, stalks, and seeds after the grapes have been pressed for wine making. The same as grappa, bagaciera, and aguardiente. Often a harsh raw spirit, drunk young, although some varieties are matured and smooth.

Marengo Chicken or veal sauteed in oil, and cooked in a sauce of white wine, tomatoes, mushroom, and garlic; said to have been invented in 1800 by Napoleon's chef after the battle of Marengo in Italy.

Margarine Emulsion of about 80% vegetable, animal, and/or marine fats and 20% water, originally made as a substitute for butter. Usually contains emulsifiers, anti-spattering agents, colours, vitamins A and D (sometimes E), and preservatives. Ordinary margarines contain roughly equal parts of saturated, monounsaturated, and polyunsaturated fatty acids; special soft varieties are rich in polyunsaturates. The energy yield is the same as that of butter.

Margarita Mexican cocktail based on tequila with lime or lemon juice; traditionally served in glasses frosted with salt around the rim.

Margherita A variety of pizza.

Marinating and marinades The preparation of marinades, typically involves the preservation of fish by treatment with solutions of dilute acid (usually acetic acid or vinegar) and salt, a process that gives the fish a characteristic flavour and texture. The salt contributes to the firmness of the flesh. In addition, the product is often seasoned with sugar, herbs and spices in the second stage of the process. Basically, there are two categories of marinade, cold marinades and cooked marinades, giving rise to a variety of products with different characteristics. Preservation of cold marinades depends entirely on the presence of acid and salt, but cooked marinades also involve some heat treatment and may be pasteurised. Shellfish are boiled to destroy the enzymes that would otherwise digest them during marinating. Unless the marinade is subsequently canned, or the equivalent, marinades are regarded as semi-preserves, and must be kept chilled, providing a shelf life of only about one to two months. The presence of acid is critical to preservation. The acid taste of the marinade can be reduced by substituting with citric or tartaric acid, or fermented vinegar, for some or all of the acetic acid but, the pH of the product must not exceed 4.5. At this pH, food poisoning bacteria and most spoilage organisms do not grow, but some bacteria and enzymes are still active and a small degree of ripening occurs which contributes to the texture and flavour.

Marine biotoxins Toxins in shellfish and marine fish, either produced naturally or accumulated by the fish from their diet (includes ciguatera and paralytic 'shellfish poisoning).

Marjoram Dried leaves of a number of aromatic plants of different species, used as seasoning for poultry, meats, and cheese dishes. The most widely accepted marjoram herbs are the perennial bush Origanum majorana and the annual sweet marjoram Majorana hortensis. Spanish wild marjoram is Thymus mastichina.

Marker gene Gene that is easy to find or observe. Attaching a marker gene to another gene that is hard to find, such as an antisense gene, is like putting a reflective collar on a dog so you can see the collar at night - even though you cannot see the dog, you know where it is.

Marmalade Defined by EU Directive as jam made from citrus peel; what was known as ginger marmalade is now known as ginger preserve. The name comes from the Portuguese marmalada, the quince, which was used to make preserves. Used in French and German for jam or preserves in general.

Marmite 1. The original form of pressure cooker used by Papin in 1681; it was an iron pot with a sealing lid.

2. Cookery term for a stock, or the pot in which stock is prepared.

Marmite Trade name for yeast extract flavoured with vegetable extract.

Marquise Chocolate mousse.

Marron glace Chestnuts preserved in syrup; semi-crystallized.

Marrow 1. Bone marrow; contents of bone cavities which are the site of formation of red blood cells.

2. Varieties of the gourd Cucurbita pepo. Grouped with courgettes, squashes, and pumpkins; 95% water.

Marsala Sicilian; fortified white wine.

Marshmallow Soft sweetmeat made from an aerated mixture of gelatine or egg albumin with sugar or starch syrup. Nougat is harder, containing less water, and usually incorporating dried fruit and nuts. Originally using the root of the marshmallow plant (Althaea officinalis), which provides a mucilaginous substance as well as starch and sugar.

Martini Trade name for a brand of vermouth; now also the name for a cocktail based on gin and vermouth. A sweet martini is made with Italian (sweet) vermouth; a dry martini with French (dry) vermouth; the proportion of gin may vary from 50 to 90%. A vodka martini (vodkatini) contains vodka rather than gin.

Marumillon 50 Trade name for mixture of the sweet glycosides extracted from stevia leaves.

Maryland, a la Dish with a butter and cream sauce, often containing wine. Chicken Maryland is deep-fried and has a garnish of sweet corn fritters, fried bananas, and fruit fritters.

Mascarpone Soft cream cheese from the Lombardy region, usually eaten as a dessert flavoured with fruit; also used in tiramisu.

Mash tun Vessel used for mashing.

Mashing In the brewing of beer, the process in which the malted barley is heated with water, to extract the soluble sugars and to continue enzymic reactions started during malting.

Mask To cover or coat a cooked meat dish with savoury jelly, glaze, or sauce; also to coat the inside of a mould with jelly.

Maslin, mashlum 1. Old term, still used in Scotland, for mixed crop of beans and oats used as cattle food.

2. In Yorkshire and the north of England, a mixed crop of 2-3 parts of wheat and 1 part of rye, used for bread making.

Mason jar Screw-topped glass jar for home bottling; patented 1858.

Massecuite The mixture of sugar crystals and syrup (mother liquor) obtained during the crystallization stage of sugar refining.

Mastic Resin from the evergreen shrub Pistacia lenticus and related species, with a flavour similar to liquorice, used in Greek and Balkan cookery.

Mastication Chewing, grinding, and tearing foods with the teeth while it becomes mixed with saliva.

Mastika Greek sweet, milky-white liqueur flavoured with anise and mastic.

Mastitis Inflammation of milk-producing glands in a cow's udder, caused by bacteria. Mastitis can be treated with antibiotics.

Mate Also yerba mate, or Paraguay or Brazilian tea. Infusion of the dried leaves of Ilex paraguayensis. Contains caffeine and tannin.

Matelote A rich, well-seasoned fish stew made with wine.

Matoke Steamed green banana or plantain.

Matsutake Edible wild fungus, Tricholoma matsutake, widely collected in Japan and exported canned or dried.

Matzo, motza Unleavened bread or Passover bread made as thin, flat, round or square water biscuits, and, according to the injunction in Exodus, eaten by Jews during the eight days of Passover in place of leavened bread.

Maw Fourth stomach of the ruminant.

MaxEPA Trade name for a standardized mixture of long-chain marine fatty acids, eicosapentaenoic and docosohexaenoic acids.

Maysin Coagulable globulin protein of maize.

Mazindol Anorectic drug used in the treatment of obesity.

Mazurki baked sweetmeat made from chopped raisins, almonds, walnuts, figs, prunes, candied fruit, flour, and eggs.

Mead A traditional wine made by fermentation of honey, sometimes flavoured with herbs and spices. One of the oldest alcoholic drinks.

Measuring cups Cups for the accurate measurement of dry ingredients.

Measuring jug Jug with scale on the side for the accurate measurement of liquids.

Measuring spoons Different sized sppons for the accurate measurement of small amounts.

Meat bar Dehydrated cooked meat and fat; a modern form of pemmican; 50% protein and 40% fat; provides 560 kcal (2350 kJ)/ 100 g.

Meat conditioning After an animal has been slaughtered, muscle glycogen breaks down and is metabolized to lactic acid, which tends to improve the texture and keeping qualities of the meat. Meat that has been left until these changes have occurred is 'conditioned'.

Meat extender Vegetable proteins (commonly textured soya protein) added to meat dishes to replace part of the meat.

Meat extract The water-soluble part of meat that is mainly responsible for its flavour. Commercially it is made during the manufacture of corned beef; chopped meat is immersed in boiling water, when the water-soluble extractives are partially leached out and concentrated. Rich in the B vitamins, meat bases, and potassium. Pavlov showed that meat extract is the most powerful oral stimulant of gastric acid secretion.

Meat factor Factor used to calculate the fat-free meat content of sausages and similar meat products from a nitrogen estimation.

Meat Generally refers to the muscle tissue of animal or bird, other parts being termed offal. Meat of all types, excluding bone, are rich sources of protein and niacin; most are rich sources of vitamin B 2 and iron; sources or good sources of vitamin B 1. Venison, horse meat, goose, and game birds are exceptionally rich in iron; pork is exceptionally rich in vitamin B 1. The fat content and proportions of fatty acids differ considerably between individual carcasses, species, and cuts of meat.

Meat speciation Identification of species of animal from which the meat originated.

Meat sugar Obsolete name for inositol.

Meat, curing Pickling with the aid of sodium chloride (salt), sodium nitrate, and some sodium nitrite, which permits the growth of only salt-tolerant bacteria and inhibits the growth of Clostridium botulinum. The nitrite is the effective preserving agent and the nitrate is converted into nitrite during the process. The red colour of cured meat is due to the formation of nitrosomyoglobin from the myoglobin of muscle.

Meat, reformed An artefact having the appearance of a cut slice or portion of meat, formed by 'tumbling' chopped meat, with or without the addition of finely comminuted meat, the soluble proteins of which bind the small pieces together.

Medicinal paraffin Liquid paraffin, a mineral oil of no nutritive value since it is not affected by digestive enzymes and passes through the intestine unchanged. Used as a mild laxative because of its lubricant properties.

Medlar The fruit of Mespilus germanica. Can be eaten fresh from the tree in Mediterranean areas but in colder climates, as in Britain, does not become palatable until it is half rotten.

Megavitamin therapy Treatment of diseases with very high doses of vitamins, many times the reference intakes; little or no evidence for its efficacy; vitamins A, D, and B6 are known to be toxic at high levels of intake.

Meio seco Portuguese; medium dry wines.

Melangeur Mixing vessel consisting of rollers riding on a rotating horizontal bed. Used to mix substances of pasty consistency (hence melangeuring).

Melangolo Italian name for the bitter orange.

Melba Peach poached in vanilla syrup, set in vanilla ice cream with a puree of raspberries. Created by Escoffier, 1892, in honour of the Australian Soprano Dame Nelly Melba.

Melba sauce Sweet sauce made from raspberries.

Melba toast Originally a split slice of toasted bread, retoasted; now a thin slice of bread dried in the oven to a golden-brown colour.

Melena Black tarry faeces, containing digested blood, the result of bleeding in the upper gastro-intestinal tract.

Melezitose Trisaccharide composed of two glucose and one fructose; hydrolysed to glucose plus the disaccharide turanose.

Melibiose x Disaccharide of glucose and galactose.

Melilot Sweet clover, a wild plant (Melilotus officinalis) which commonly grows in fields and on waste ground, especially on sandy soil; used as forage. The dried leaves have a sweet hay-like flavour.

Melitzanosalata Puree of aubergine with onion, garlic, and olive oil, served cold as a dip.

Mellorine US term for ice cream made from non-butter fat.

Melon Gourds, sweet fruit of Cucurmis melo. A rich source of vitamin C:

Melopita Honey and cheese cake.

Melting point The temperature at which a compound melts to a liquid. Often characteristic of a particular chemical and used as a means of identification. Particularly valuable as an index of purity, since impurities lower the melting point.

Membrane, semi-permeable One that allows the passage of small molecules but not large ones: e.g. pig's bladder is permeable to water but not salt; collodion is permeable to salt but not protein molecules.

Menadione, menaphthone, menaphthol-diacetate Synthetic compounds with vitamin K activity; vitamin K3, sometimes known as menaquinone-0.

Menaquinones Bacterial metabolites with vitamin K activity; vitamin K2

Menarche The initiation of menstruation in adolescent girls, normally occurring between the ages of 11 and 15. The age at menarche has become younger in Western countries, possibly associated with a better general standard of nutrition, and is later in less-developed countries.

Menhaden Oily fish, Brevoortia patronus, B. tyrannus, from Gulf of Mexico and Atlantic seaboard of the USA, a rich source of fish oils.

Mental retardation A condition where an individual has a lower-than-normal IQ, and thus is developmentally delayed.

Menu Commonly used to mean the list of foods and dishes served by a restaurant, but correctly a set meal (with options) or dish of the day, as opposed to a la carte.

Meringue Confection made from whisked egg-white and sugar, baked slowly in a cool oven; flavouring may be added, and the meringues may be filled with cream, ice cream, or fruit.

Merlot One of the 9 'classic' grape varieties used for wine making; important in some of the great fragrant, rich, red wines; widely grown throughout the world.

Mesomorph Description given to a well-covered individual with well-developed muscles.

Mesophiles, Mesophilic bacteria Members of a class of bacteria in a classification system based on the temperature range at which an organism can grow. Mesophiles have an optimum growth temperature, the temperature at which they multiply most rapidly, in the region of 30-40°C. Food poisoning organisms are adapted to grow in the body of warm-blooded animals and are mesophiles, though some can grow at chill temperatures.

Metabolic equivalent Unit of measurement of heat production by the body; 1 MET= 50 kcal/hour/m^2 body surface area.

Metabolic rate Rate of utilization of energy.

Metabolic water Produced in the body by the oxidation of foods. 100 g of fat produces 107.1 g; 100 g of starch 55.1 g; and 100 g of protein 41.3 g of water.

Metabolic weight Energy expenditure and basal metabolic rate depend on the amount of metabolically active tissue in the body, not the total body weight; body weight0.75 is generally used to calculate the weight of active tissue.

Metabolism The processes of interconversion of chemical compounds in the body. Anabolism is the process of forming larger and more complex compounds, commonly linked to the utilization of metabolic energy. Catabolism is the process of breaking down larger molecules to smaller ones, commonly oxidation reactions linked to release of energy. There is approximately a 30% variation in the underlying metabolic rate (basal metabolic rate) between different individuals, determined in part by the activity of the thyroid gland.

Metalloproteins Proteins containing a metal. For example, haemoglobin, cytochromes, peroxidase, ferritin, and siderophilin all contain iron; many enzymes contain copper, manganese, or zinc as a prosthetic group.

Methaemoglobin Oxidized form of haemoglobin which cannot transport oxygen to the tissues. Present in small quantities in normal blood, increased after certain drugs and after smoking; found rarely as a congenital abnormality.

It can be formed in the blood of babies after consumption of the small amounts of nitrate found naturally in vegetables grown in certain areas and in some drinking water, since the lack of acidity in the stomach permits reduction of nitrate to nitrite.

Methaglen A traditional British wine made from honey to which herbs are added before fermentation. Originally for medicinal purposes.

Methane Odorless, colorless, flammable gas; component of natural gas and important source of hydrogen and organic compounds. Efficient anaerobic bacteria decompose sludge, manure or landfill wastes to produce methane. Some sewage plants collect methane to fuel generators, digesters and air compressors.

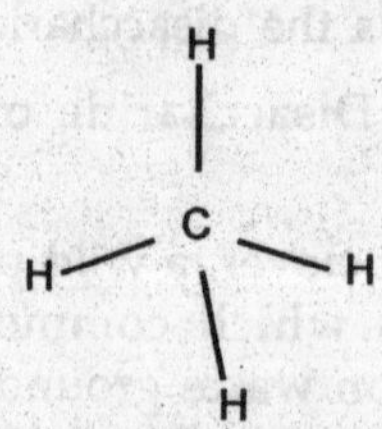

Fig. Methane

Methionine An essential amino acid; one of the three containing sulphur; cystine and cysteine are the other two. Cystine and cysteine are not essential, but can only be made from methionine, and therefore the requirement for methionine is lower if there is an adequate intake of cyst(e)ine. For this reason the sulphur amino acids are always considered together.

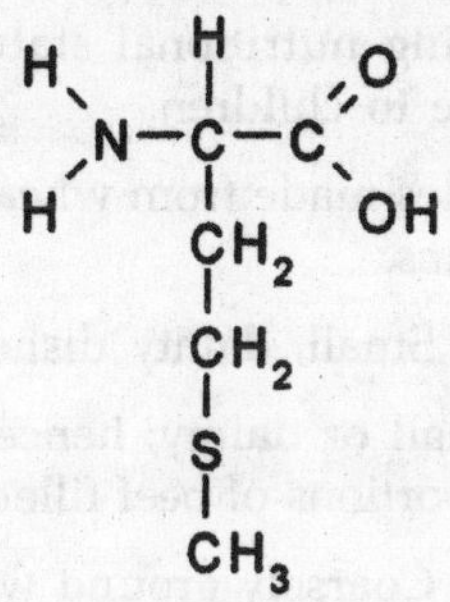

Fig. Methionine

Methocel Trade name for methyl cellulose.

Methode champenoise Sparkling wines made by a second fermentation in the bottle, as for champagne, but not in the Champagne region of north-eastern France.

Methofas Trade name for methyl hydroxypropyl cellulose.

Methyl alcohol The first member of the alcohol series, chemically CH3–OH. It is a highly toxic substance and leads to mental disturbance, blindness, and death when consumed over a period. It is added to industrial alcohol and methylated spirits, to 'denature' the ethyl alcohol and render it undrinkable.

Methylene blue dye-reduction test When the dye methylene blue or resazurin is added to milk, the bacteria present take up oxygen and change the colour of the dye; methylene blue loses its colour; resazurin changes from blue-purple to pink. The speed of the change indicates the bacterial content. Pasteurized milk must not reduce dye in less than half an hour.

Methyl-histidine Derivative of the amino acid, histidine, found almost exclusively in the contractile proteins of muscle (myosin and actin). Useful, among other purposes, as an index of the lean meat content of prepared foods, since it is not present in collagen or other added materials.

Metodo classico sparkling wines made by the methode champenoise.

Meuniere, a la Fish dredged with flour, fried in butter, and served with this butter and chopped parsley; literally, 'in the style of the miller's wife'.

Mezethakia Also orekita; appetizers or hors d'oeuvre; sometimes abbreviated to mezes or mezze.

Micro In general use, very small; as a prefix for units of measurement, one millionth part.

Micro-aerophiles Micro-organisms that can grow in low concentrations of oxygen and so lead to spoilage of foodstuffs unless all oxygen is excluded.

Microbe Microorganism; any organism that can be seen only with the aid of a microscope.

Microbial insecticide Preparation of living microbes - such as bacteria or fungi - pathogenic to specific groups of insects. Regulated as pesticides by the U.S. Environmental Protection Agency; users must follow specific labeling and use guidelines.

Microbiological assay Biological method of measuring compounds such as vitamins and amino acids, using micro-organisms. The principle is that the organism is inoculated into a medium containing all the growth factors needed except the one under examination; the rate of growth is then proportional to the amount of this nutrient added in the test substance.

Microencapsulation Preparation of small particles of solids, or droplets of liquids, inside thin coatings (e.g. beeswax, starch, gelatine, and polyacrylic acid). The microcapsules range from 1 to 100 nm in size. Used to prepare liquids as free-flowing powders or compressed solids, to separate reactive materials, reduce toxicity, protect against oxidation and control the rate of release; used for enzymes, flavours, nutrients, etc.

Microfiltration Filtration under pressure through a membrane of small pore size (0.1-10 μm; larger pores than for ultrafiltration). Used for clarification of beverages and to sterilize liquids by filtering off organisms.

Microgram One-thousandth part of a milligram, and hence 1 millionth part of a gram; symbol μg.

Micrometre One-thousandth of a millimetre, and hence one millionth of a metre; symbol μm.

Micronization Extremely rapid heating with infra-red radiation produced by heating propane on a ceramic tile or with nichrome wire elements. Suggested as an alternative to steam heating or toasting where the shorter heating time is less damaging to the foodstuff.

Micronutrients Vitamins and minerals, which are needed in very small amounts (micrograms or milligrams per day), as distinct from fats, carbohydrates, and proteins which are macronutrients, since they are needed in considerably greater amounts.

Microorganism Microscopic organisms, also called microbes, which include bacteria, viruses, fungi (moulds), algae, protozoans, and yeasts.

Microwave cooking Rapid heating by passing high frequency waves from a magnetron (in the UK 2450 or 896 MHz (million cycles/second)) through the food or liquid to be heated. Water absorbs the microwaves very well, so food with a high water content cooks more rapidly; fat absorbs the energy more slowly, so foods consisting of mixtures of fat and water cook unevenly. The cooking time is short and microwaves do not cause browning, so the food may not develop flavours associated with longer cooking times. Metal containers reflect the microwaves and cannot be used.

Midori Pale-green liqueur made from melons.

Mid-upper-arm-circumference A rapid way of assessing nutritional status, especially applicable to children.

Mien Noodles made from wheat and ground dried beans.

Mignardise Small, dainty dishes.

Mignon Small or dainty; hence filet mignon is small portions of beef fillet.

Mignonette Coarsely ground white pepper.

Milanaise, a la Dish garnished with spaghetti, tomato sauce, and ham or tongue. Also food dipped in egg and a mixture of breadcrumbs and cheese, then fried.

Milchig Jewish term for dishes containing milk or milk products, which cannot be served with or after meat dishes.

Milfoil A common wild plant (Achillea millefolium, or yarrow) with finely divided leaves which can be used in salads or chopped to replace chervil or parsley as a garnish.

Milk The secretion of the mammary gland of animals including cow, buffalo, goat, ass, mare, ewe, and camel, and of human beings. A rich source of vitamins B 2, B 1 2, calcium, and iodine; a source of protein, vitamin A, and vitamin B 1.

Ordinary milk contains 3.9% fat; Channel Islands milk, 5.1%; sheep's milk, 6.0%, buffalo milk, 7.5%; human milk contains 4.1% fat.

Milk baby Infant with iron deficiency anaemia caused by excessive ingestion of milk and delayed or inadequate addition of iron-rich foods to the diet.

Milk-stone Deposit of calcium and magnesium phosphates, protein, etc., produced when milk is heated to temperatures above 60°.

Mille-feuille Cake made of layers of puff pastry split and filled with cream and jam, and iced on top; also puff pastry savoury (literally, 'a thousand leaves').

Millerator Wheat-cleaning machine consisting of two sieves, the upper one retaining particles larger than wheat, the lower one rejecting particles smaller than wheat.

Millet Cereal of a number of species of Gramineae smaller than wheat and rice and high in fibre content. Common millet also known as China, Italian, Indian, French hog, proso, panicled, and broom corn millet, foxtail millet; grows very rapidly, 2-2½ months from sowing to harvest. Protein 10%, fat 2.5%, carbohydrate 73%.

Red, finger, South Indian millet, coracan, or ragi is Eleusine coracana. Protein 6%, fat 1.5%, carbohydrate 75%.

Bulrush millet, pearl millet, bajoa, or Kaffir manna corn is Pennisetum typhoideum or P. americanum; the staple food in poor parts of India. Protein 11%, fat 5%, carbohydrate 69%.

Other species are hungry rice, jajeo millet, Kodo or haraka millet, teff.

Milli As a prefix for units of measurement, one thousandth part; symbol m.

Milling The term usually refers to the conversion of cereal grain into its derivative, e.g. wheat into flour, brown rice to white rice. Flour milling involves two types of rollers: 1. break rolls are corrugated and exert shear pressure and forces which break up the wheat grain and permit sieving into fractions containing varying proportions of germ, bran, and endosperm;

2. Reducing rolls are smooth and subdivide the endosperm to fine particles.

Miltone A toned milk developed in India in which peanut protein is added to buffalo or cow's milk to extend supplies.

Mimosa American name for a mixture of sparkling wine and orange juice, known in the UK as buck's fizz.

Minafen Trade name for food low in phenylalanine for treatment of phenylketonuria.

Minamata disease Poisoning by organic mercury compounds, named after Minamata Bay in Japan, where fish contained high levels of organic mercury compounds after waste water from mercury processing was passed into the estuary.

Minarine Name sometimes given to low-fat spreads with less than the statutory amount of fat in a margarine.

Mince 1. To chop or cut into small pieces with a knife or, more commonly, in a mincing machine or electric mixer.

2. Meat which is finely divided by chopping or passing through a mincing machine; known as ground meat in the USA.

Mince pie Small pastry tarts, filled with mincemeat, traditionally eaten at Christmas.

Minced fish Fish flesh separated from skin and bone in a mechanical bone separator, frequently of the perforated drum and squeezing belt type. Also known as mechanically recovered fish flesh, recovered fish flesh, boneless fish meat, deboned fish flesh. MRM is sometimes used but primarily refers to the beef or poultry industry. Fish or pieces of fish, such as filleting waste or fillet trimmings, are fed from a hopper to be pressed between a moving, flexible belt and the outside of a counter-rotating, perforated, stainless steel drum. Ideally, the material should be fed flesh-side to the drum. The flesh is squeezed through the perforations of the drum and removed from the inside by means of a fixed screw. The coarseness and texture of the mince, the colour of the product, and the yield of flesh, are affected by the diameter of the drum perforations, the pressure applied to the squeezing belt, and the raw material or trimmings being used. For different applications, the perforations may range from about 1 to 10 mm. Under ideal conditions, the skin and bones remain on the outside of the drum and are removed continuously as waste by a fixed scraping blade. Depending on the raw material used and the conditions applied, some skin and bone fragments may pass through with the recovered flesh.

Mineola A citrus fruit.

Mineral salts The inorganic salts, including sodium, potassium, calcium, chloride, phosphate, sulphate, etc. So called because they are obtained by mining.

Mineral water Natural, untreated, spring waters, some of which are naturally carbonated, may be slightly alkaline or salty. Numerous health claims have been made for the benefits arising from the traces of a large number of minerals found in solution. They are normally named after the town nearest the source. Examples are Evian, Malvern, Apollinaris, Vichy, Vittel, Perrier.

Sparkling mineral water may either contain the gases naturally present at the source or may be artificially carbonated. Carbonated beverages are sometimes called minerals.

Mineralocorticoids A general term for the steroid hormones secreted by the adrenal cortex which control the excretion of salt and water.

Minerals, trace Those mineral salts present in the body, and required in the diet, in small amounts: copper, chromium, iodine, manganese, molybdenum, selenium; although required in larger amounts, zinc and iron are sometimes included with the trace minerals.

Minerals, ultra-trace Those mineral salts present in the body, and required in the diet, in extremely small amounts; known to be dietary essentials, although rarely if ever a cause for concern since the amounts required are small and they are widely distributed in foods and water, e.g. cobalt, manganese, molybdenum, silicon, tin, vanadium.

Miners' cramp Cramp due to loss of salt from the body caused by excessive sweating. Occurs in tropical climates and with severe exercise; mining often combines the two. Prevented by consuming salt, perhaps in the form of salt tablets as used by athletes.

Minestra Vegetable soup, thinner than minestrone.

Minestrone Thick mixed vegetable soup with pasta or rice.

Mint Aromatic herbs, Mentha spp., including spearmint, M. spicata; peppermint, M. piperita. Garden mint is spearmint, Mentha spicata. Oil of peppermint is distilled from stem and leaves of M. piperita, and used both pharmaceutically and as a flavour.

Mint jelly Apple jelly flavoured with spearmint, often eaten with lamb.

Mint sauce Mint chopped in vinegar, sometimes with sugar added; traditionally eaten with lamb in Britain.

Mirabelle French spirit prepared by distillation of yellow-green mirabelle plums; strong fruity aroma, not sweetened. Similar to the German quetsch.

Miracle berry The fruit of the West African bush Richardella dulcifica. It contains a taste-modifying glycoprotein that causes sour foods to taste sweet. Hence the name miracle berry.

Miracullin The taste-modifying glycoprotein of the miracle berry.

Mirepoix Mixture of finely chopped carrots, celery, and onions, fried in butter and used to flavour meat dishes.

Miso A Japanese sauce, prepared from autoclaved soya beans mixed with cooked rice and partly fermented with Aspergillus oryzae and A. sojae to form koji. Salt is added to stop further mould growth, bacterial fermentation continues with the addition of Lactobacillus. Barley miso is an alternative to soya beans alone.

Mistelles partially fermented grape juice.

Mitochondrion The subcellular organelles in all cells apart from red blood cells in which the major oxidative reactions of metabolism occur, linked to the formation of ATP from ADP.

Mix To combine ingredients so they are evenly blended.

Mixing bowl All purpose bowl used for combining foods.

Mixiria Process of preserving meat and fish by roasting them in their own fat and preserving in jars covered with a layer of fat.

Mixograph American instrument for measuring the physical properties of a dough, similar in principle to the farinograph.

Ml Millilitre, one-thousandth of a litre.

Mocca Mixture of coffee and cocoa used in bakery and confectionery products.

Mocha 1. Variety of arabica coffee.

2. Flavoured with coffee. 3 In the USA a combination of coffee and chocolate flavouring.

Mock goose North of England; beef heart simmered in stock until tender, then stuffed, wrapped in fatty bacon, and roasted.

Mock turtle soup Gelatinous soup made from calf's head, beef, bacon, and veal; similar to turtle soup, but without the turtle; mock turtle is a calf's head dressed to resemble a turtle.

Mode, a la Beef braised in the classic way; also fruit or other sweet pie served with ice cream.

Moisture Synonymous with water content in the context of proximate composition.

Molasses The residue left after repeated crystallization of sugar; it will not crystallize. Contains 67% sucrose, together with glucose and fructose and raffinose and small quantities of dextrans; 260 kcal /100 g; more than 500 mg iron per 100 g, with traces of other minerals.

Mold Various fungous growths, often causing disintegration of organic matter; fungus that causes mold.

Mole Mexican; sauce made from sweet pepper, avocado, tomato, and sesame, flavoured with aniseed, garlic, coriander, cinnamon, cloves, chilli, and grated chocolate.

Molluscs Marine bivalve shellfish with soft unsegmented bodies; most are enclosed in a hard shell; they include abalone, clams, cockles, mussels, oysters, scallops, whelks, winkles.

Molybdenum A dietary essential mineral, required for a number of enzymes, including xanthine, aldehyde, and pyridoxal oxidases, where it forms the functional part of the coenzyme molybdopterin. Deficiency is unknown, and there are no estimates of requirements. Average intakes are between 50 and 400 ug/day, and are considered to be safe and adequate.

Monellin The active sweet principle, a protein, from the serendipity berry, Dioscoreophyllum cumminsii, 1500-2000 times as sweet as sucrose.

Monkfish A white fish, Lophius piscatorius.

Monoclonal antibody A protein that recognizes and attacks one specific foreign substance invading the body. A monoclonal antibody may be used for early disease diagnosis and therapy.

Monophagia Desire for one type of food.

Monosaccharides Group name of the simplest sugars, including those composed of three carbon atoms, four, five, six, and seven.

Also known as monoses or monosaccharoses. These are the units from which disaccharides, oligosaccharides, and polysaccharides are formed.

Monosodium glutamate The sodium salt of glutamic acid, used to enhance the flavour of savoury dishes and often added to canned meat and soups. Originally called Aginomoto.

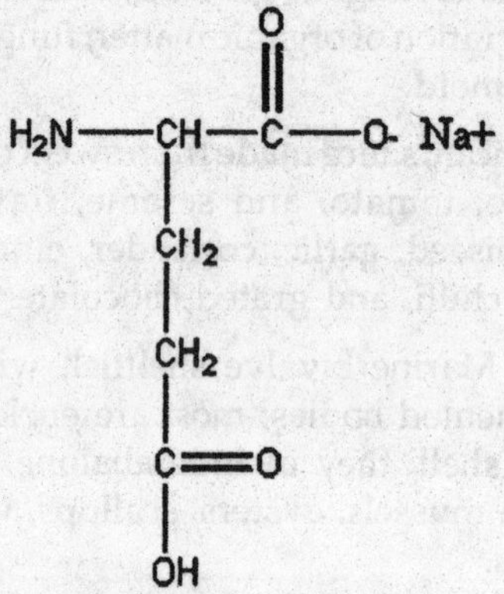

Fig. Monosodium glutamate

Mono-unsaturates Commonly used short term for mono-unsaturated fatty acids.

Monstera Fruit of the Swiss cheese plant, Monstera deliciosa, also known as fruit salad fruit, and 'delicious fruit'.

Montmorency, a la Sweet dishes and cakes which include cherries.

Mooli Long, white radish; oriental variety of Raphanus sativa.

Moonshine American name for illicit home-distilled spirit.

Moorfowl Scottish name for grouse.

Morcilla Spanish; black pudding.

Morel Edible fungus Morchella esculenta, much prized for its delicate flavour.

Moreton bay bug Or Bay lobster, a variety of sand lobster found in Australia.

Mornay, a la Dish served with bechamel cheese sauce.

Mosaic In Down syndrome, this means a a mixture of two type cells in an individual's bodies—trisomy 21 with 47 chromosomes and the normal cell pattern of 46 chromosomes.

Mottled teeth In areas where the drinking water contains fluoride at a level of several parts per million, dull, chalky patches, known as mottling, occur on the teeth. Also known as dental fluorosis. These teeth are relatively free from decay, and lower levels of fluoride, about 1 ppm, reduce decay without causing mottling.

Mould bran A fungal amylase preparation produced by growing mould on moist wheat bran; used as source of starch-splitting enzymes.

Moulds A group of multi-cellular fungi which grow in thread-like strands called hyphae. Moulds can grow on foods and damage them but some are introduced into foods to give added flavour, e.g. in some cheeses.

Moussaka Greek, eastern Mediterranean; minced lamb, potatoes, onions, and aubergine, topped with bechamel sauce and baked.

Mousse Light frothy dish based on white or custard sauce or fruit puree to which eggs, whipped cream and sometimes gelatine are added. May be sweet or savoury, served hot or cold.

Mousseline General term for light, fluffy dish. Mousseline sauce is Hollandaise sauce to which beaten egg-white or whipped cream is added just before serving, to give a frothy texture.

Mozzarella Soft white curd cheese, originally made from buffalo milk, now normally from cow's milk. Mainly used outside Italy as a topping for pizza, although in Italy it is also eaten uncooked. paralysis, and the disease is subject to random periods of spontaneous remission. There is some evidence that supplements of polyunsaturated fatty acids slow the progression of the disease.

Multiculturalism A policy and a set of strategies of State and Commonwealth governments of the 1970s to encourage cultural diversity and tolerance in society.

Muscadet A grape variety widely used for wine making, although not one of the classic varieties; makes light, very dry wines; also known as melon de bourgogne.

Muscat A grape variety widely used for wine making, although not one of the classic varieties; they mostly make perfumed, sweet, white wines.

Muscle The contractile cellular unit of skeletal muscle is the cylindrical fibre, composed of many myofibrils. Chemically, muscle consists of three main proteins, actin, myosin, and tropomyosin, as well as structural proteins such as collagen and elastin. Contraction is achieved by formation of a complex between actin and myosin.

The muscle fibre is surrounded by a thin membrane, the sarcolemma; within the muscle fibre, surrounding the myofibrils, is the sarcoplasm. Individual fibres are separated by a thin network of connective tissue, the endomysium, and bound together in bundles by thicker sheets of connective tissue, the perimysium.

Mush American; cakes of boiled ground rice served cold or fried.

Mushrooms Various edible fungi; correctly the fruiting bodies of the fungi. Altogether some 300 species are sold, fresh or dried, in markets around the world; most of these are gathered wild rather than cultivated.

The common cultivated mushroom, including flat, cup, and button mushrooms is Agaricus bisporus, as is the chestnut or Paris mushroom. Other cultivated mushrooms include: shiitake; oyster mushroom; Chinese straw mushroom.

Some wild species are especially prized, including field mushroom; horse mushroom; parasol mushroom; beefsteak fungus; blewits; wood blewits; cep or boletus; chanterelle; matsutake; puffballs; morels; truffles; wood-ears; yellow mushroom. Many other wild fungi are also edible, but many are poisonous, too.

Mussels Various marine bivalve molluscs, Mytilus edulis, M. californianus. A 100-g portion of boiled flesh is an exceptionally rich source of iodine, a rich source of protein, iron, copper, and selenium; a good source of calcium, vitamin B 2, and niacin; a source of zinc, and contains 200 of sodium; supplies 90 kcal.

Mustard Powdered seeds of black or brown mustard, or white or yellow mustard, or a mixture. English mustard contains not more than 10% wheat flour and turmeric. French mustard: made from dehusked seeds or black or brown seeds with salt, spices, and white wine or unripe grape juice. Bordeaux is black and brown seeds mixed with sugar, vinegar, and herbs. Meaux mustard is grainy and made with mixed seeds. American mustard, mild and sweet, is made with white seeds, sugar, vinegar, and turmeric.

Mutagen Agent that causes biological mutation. Examples include chemicals, radioactive elements and ultraviolet light. See pentachlorophenol.

Mutation Sudden random change in genetic material that may cause that cell and all cells derived from it to look or behave differently.

Mutton Meat from the fully grown sheep, Ovis aries.

Mycoprotein Name given to mould mycelium. used as a food ingredient.

Mycotoxins Toxins produced by fungi, especially Aspergillus flavus under tropical conditions and Penicillium and Fusarium species under temperate conditions. The problem is caused by storage of food under damp conditions which favour the growth of moulds. They include aflatoxin, ochratoxin, patulin, zearalenone, and stigmatocystin.

N

Nan Indian flat bread, an egg dough prepared with white flour and leavened with sodium bicarbonate, normally baked in a tandoor.

Nape Referred to as the flesh exposed at the shoulders of a fish when the head is removed, or that part of the belly flap immediately adjacent to the nape bone, and also called lug. The shoulder is the thickest part of the flesh of a fish, just behind the head on either side of the backbone. The nape bone is the bone at the shoulder of a fish that forms the leading edge of a belly flap; also called lug bone, shoulder bone or collar bone.

Naphthoquinone The chemical ring structure of vitamin K; the various chemical forms of vitamin K can be referred to as substituted naphthoquinones.

Napolitaine, a la Dish garnished with spaghetti, cheese, and tomato sauce.

Naringin A glycoside found in grapefruit, especially in the immature fruit. It is extremely bitter, and dilutions of 1 part in 10 000 parts of water can be detected. Sometimes found in canned grapefruit segments as tiny, white beads. Hydrolysed to the aglucone, trihydroxyflavonone, which is not bitter.

Nasi goreng Dutch, Indonesian; fried rice with chicken, prawns, etc., topped with strips of omelette.

Nasogastric tube Fine plastic tube inserted through the nose and thence into the stomach for enteral nutrition.

Nasturtium Both the leaves and seeds of Tropaeolum officinalis can be eaten; they have a hot flavour. The seeds can be pickled as a substitute for capers, and the flowers can be used to decorate salads.

Natto soya bean fermented using Bacillus natto.

Natural foods A term widely used but with little meaning and sometimes misleading since all foods come from natural sources. No legal definition seems possible but guidelines suggest the term should be applied only to single foods that have been subjected only to mild processing, i.e. largely by physical methods such as heating, concentrating, freezing, etc., but not chemically or 'severely' processed.

Nature-identical Term applied to food additives that are synthesized in the laboratory and are identical with those that occur in nature. naturel, au Plain; uncooked, or very simply cooked.

Navarin ragofit of mutton with potatoes and onions or spring vegetables.

NCHS standards Tables of height and weight for age used as reference values for the assessment of growth and nutritional status of children, based on data collected by the US National Center for Health Statistics. The most comprehensive such set of data, and used in most countries of the world.

Neapolitan ice Block of ice cream in layers of different colours and flavours.

Neat's foot Ox or calf's foot used for making soups and jellies. Now called cow heels.

Neat's-foot oil Oil obtained from the knucklebones of cattle; used in leather working and for canning sardines.

Nebbiolo A grape variety widely used for wine making, although not one of the classic varieties.

Nectarine Smooth-skinned peach. One medium-sized fruit, 150 g weighed with stone, provides 3 g of dietary fibre and is a rich source of vitamin A; a good source of vitamin C; a source of copper; supplies 70 kcal.

Neep Scottish name for root vegetables; now used for turnip.

Negus Drink made from port or sherry with spices, sugar, and hot water.

Nematodes Also called roundworms. Members of a zoological phylum Nematoda widely distributed in nature in water and in soils. They are characterised by having an elongated body with a circular cross section. Many species are parasites, including in man. Fish harbour nematodes parasitic in man, the two most prevalent being the cod worm and the herring worm.

Neohesperidin dihydrochalcone A non-nutritive sweetener, 1000 times as sweet as sucrose; formed by hydrogenation of the naturally occurring flavonoid neohesperidin.

Neomycin Antibiotic isolated in 1949 from Streptomyces fradii; has a broad spectrum of activity, and is not absorbed from the gut; commonly used to treat serious intestinal bacterial infections.

Neroli oil Prepared from blossoms of the bitter orange by steam distillation. Yellowish oil with intense odour of orange blossom.

Nescafe Trade name for a dried, instant coffee. Contains more potassium than any other food, at 4%.

Nesselrode, a la Implies the use of chestnuts; nesselrode pudding is made from chestnuts, custard cream, and candied fruit.

Net dietary protein energy ratio A way of expressing the protein content of a diet or food taking into account both the amount of protein and the protein quality. It is protein energy multiplied by net protein utilization divided by total energy.

Net protein value A way of expressing the amount and quality of the protein in a food; the product of net protein utilization and protein content per cent.

Nettle The young leaves of the stinging nettle, Urtica dioica, can be cooked as a vegetable and used to make nettle beer.

Neural tube defect Congenital malformations of the spinal cord caused by the failure of the closure of the neural tube in early embryonic development. Supplements of folic acid begun before conception reduce the risk.

Neuron A special type of cell used to transmit information within the brain and between the brain and the body.

Neurotransmitter A chemical messenger used by nerve cells to communicate with each other.

New zealand process Drying process applied to meat. It is immersed in hot oil under vacuum when it dries to 3% moisture in about 4 hours. The fat is removed from the dry meat in a hydro-extractor.

Niacin A vitamin; one of the B complex without a numerical designation. It is sometimes referred to as vitamin B 3, and formerly vitamin PP. Deficiency leads to pellagra, photosensitive dermatitis resembling severe sunburn, a depressive psychosis, and intestinal disorders; fatal if untreated. Niacin is the generic descriptor for two compounds in foods which have the biological activity of the vitamin: nicotinic acid and nicotinamide. In the USA niacin is sometimes used specifically to mean nicotinic acid, and niacinamide for nicotinamide.

The metabolic function of niacin is in the coenzymes NAD and NADP, which act as intermediate hydrogen carriers in a wide variety of oxidation and reduction reactions.

Nicotinamide can also be formed in the body from the amino acid tryptophan; on average 60 mg of dietary tryptophan is equivalent to 1 mg of preformed niacin. The total niacin content of foods is generally expressed as mg niacin equivalents; the sum of preformed niacin plus one-sixtieth of the tryptophan. This means that most foods that are good sources of protein are also good sources of niacin: meat, fish, cheese, pulses, eggs. In cereals niacin is largely present as niacytin, which is not biologically available; therefore the preformed niacin content of cereals is generally ignored when calculating intakes. Free niacin is added to white flour and enriched breakfast cereals in many countries.

Niacin toxicity High doses of nicotinic acid have been used to treat hypercholesterolaemia; they can cause an acute flushing reaction, with vasodilatation and severe itching. Intakes of niacin above 500 mg/d can cause liver damage over a period of months; the risk is greater with sustained-release preparations of niacin.

Niacinamide American name for nicotinamide, the amide form of the vitamin niacin.

Niacinogens Name given to protein-niacin complexes found in cereals.

Niacytin The bound forms of the vitamin niacin, found in some foods, particularly cereals. Complexes of niacin with polysaccharides and peptides or glycopeptides; not hydrolysed by intestinal enzymes, so biologically unavailable, but can be liberated by acid or alkaline hydrolysis or by baking the cereal, especially with an alkaline baking powder.

Niceritol A derivative of the vitamin, niacin used in large doses to reduce plasma cholesterol levels.

Niche A job or position which is very suitable for someone, esp. one that they like.

Nickel An ultra-trace mineral; known to be essential for experimental animals, although its function is not known. There is no information on requirements. Metallic nickel is used as a catalyst in the hydrogenation of fats.

Nicoise 1. A garnish for meat with tomatoes, olives, and French beans.

2. Soup garnished with tomato, flageolets, and diced potato.

3. A sauce made from concentrated tomato puree blended with demi-glace.

4. Salade nicoise contains tuna, anchovy, hard boiled egg, tomatoes, lettuce, and olives.

Nicotinamide One of the vitamers of niacin.

Nicotinate, sodium Sodium salt of nicotinic acid; has been used, among other purposes, to preserve the red colour in fresh and processed meats.

Nicotinic acid One of the vitamers of niacin.

Nigerseed Guizotia abyssinica, or nug; grown in India and Ethiopia as a food crop.

Night blindness Nyctalopia. Inability to see in dim light as a result of vitamin A deficiency.

Nim leaf Sweet nim, an aromatic Indian herb with an aroma resembling that of truffles.

Ninhydrin test For proteins and amino acids. Pink, purple, or blue colour is developed on heating the amino acid or peptide with ninhydrin.

Nioigome Perfumed rice.

Nisin Antibiotic isolated in 1944 from lactic streptococci group N; inhibits some but not all clostridia; not used medically. The only antibiotic permitted in Great Britain to preserve specified foods. It is naturally present in cheese, being produced by a number of strains of cheese starter organisms. Useful to prolong storage life of cheese, milk, cream, soups, canned fruits and vegetables, canned fish, and milk puddings. It also lowers the resistance of many thermophilic bacteria to heat and so permits a reduction in the time and/or temperature of heating when processing canned vegetables.

Nitrate Is also used, together with nitrite, for curing meat products.

Nitrates Plant nutrient and natural constituent of plants; the inorganic form of nitrogen used by plants to synthesize protein; found in soils and included in inorganic fertilizer. Nitrate is a natural constituent of crops in amounts sometimes depending on the content of the soil. Also found in drinking water as a result of excessive use of fertilizers.

Health problems can arise because within a day or two of harvesting some crops nitrates are converted into nitrites which can react with the haemoglobin in the blood to produce an oxidized form which cannot transport oxygen. Maximum levels have been established for nitrate levels in certain foods and drinking water.

Nitrites Found in many plant foods, since they are rapidly formed by the reduction of naturally occurring nitrate. Nitrite is the essential agent in preserving meat by pickling, since it inhibits the growth of clostridia; it also combines with the myoglobin of meat to form the characteristic red nitrosomyoglobin.

Nitrogen The amount of basic, nitrogen-containing chemicals distilled from an alkalised extract or suspension of a fishery product. The bases, amines, in the distillate are determined by titration with standard acid. The amines all have one basic nitrogen atom in the molecule and TVB is expressed on a nitrogen basis, typically mg nitrogen/100g of sample. The main component of TVB from very fresh fish is ammonia and, as fish spoils, increasing amounts of trimethylamine (TMA) are present. Small amounts of dimethylamine (DMA) are often present in TVB from spoiled fish. During distillation, nitrogen-containing chemicals in the sample are decomposed by the hot alkaline conditions to ammonia, so that the TVB content is greater than the sum of the bases present in the sample. Elasmobranchs have urea in their muscle tissues which decomposes to ammonia under the typical conditions for TVB measurement. The amount of ammonia from this decomposition depends on distillation conditions, especially temperature and degree of alkalisation, and TVB varies with details of the experimental procedure used to determine it. The lowest, and probably true, value is given by distillation at room temperature from a mildly alkalised sample. There are two main variants of the analytical procedure distillation at boiling point of an aqueous suspension of a sample made alkaline with magnesium oxide; distillation at boiling point of a protein-free extract of the sample with strong alkali. The latter gives a higher result, by around 4 mgN/100g, than the former. TVB is used as a measure of freshness because it increases as fish spoils. But, it does not increase substantially until the fish spoil, typically around 7/8 days in ice for vertebrate fish, and is not suitable for discriminating among freshnesses of fresh fish. The main contribution to changes in TVB during spoilage is the increasing presence of TMA. TVB content is used by some regulatory agencies to confirm fish as being unfit for human consumption.

Nitrogen balance The difference between the dietary intake of nitrogen and its excretion. Healthy adults excrete the same amount as is ingested, and so are in N equilibrium.

During growth and tissue repair the body is in positive N balance, i.e. ingestion is greater than loss and there is an increase in the total body pool of nitrogen. In fevers, fasting, and wasting diseases the loss is greater than the intake and the individual is in negative balance; there is a net loss of nitrogen from the body.

Nitrogen factor, nitrogen content A number used to convert the nitrogen content of a sample to protein content or to fish content. There are two values of relevance to fish technology. One is the factor that converts total nitrogen content to protein content. Conventionally a value of 6.25 is used and the nitrogen content is multiplied by this factor to give the protein content. The other is a value that converts total nitrogen content to fish content. It is the average nitrogen content of flesh of the species of interest. It has a value of around 2.8, and the measured nitrogen content of a sample is divided by this factor to give the fish content of the sample. The latter sample is used by trading standards officers in the determination of the fish content of products.

Nitrogen, metabolic Nitrogen in the faeces derived from internal or endogenous sources, as distinct from nitrogen-containing dietary sources. This nitrogen consists of unabsorbed digestive juices, the shed lining of the gastrointestinal tract and bacteria from the intestine, and continues to be excreted on a proteinfree diet.

Nitrogen-free extract In the analysis of foods and animal feedingstuffs, the fraction that contains the sugars and starches plus small amounts of other materials.

Nitrogenous extractives The nitrogen-containing compounds in a protein-free extract of fish tissue. Typical extractants are dilute solutions of trichloroacetic acid and 80% ethanol. The constituents of extracts of fish muscle include creatinine, amino acids, trimethylamine oxide and, in the case of elasmobranchs, urea.

Nitrosamines Compounds bearing the nitroso group on the N of amines. Found in trace amounts in mushrooms, fermented fish meal and smoked fish, and in pickled foods, where they are formed by reaction between nitrite and secondary amines. They cause cancer in experimental animals, but it is not known whether the small amounts in foods affect human beings, especially since they have also been found in human gastric juice, possibly formed by reaction between amines and nitrites or nitrates from the diet.

Nitrosomyoglobin The red colour of cured meat. It is formed by the reaction of nitric oxide from the pickling salts with myoglobin in the meat. Fades in light to yellow-brown metmyoglobin.

Nitrous oxide A gas used as a propellant in pressurized containers, e.g. to eject cream or salad dressing from containers.

No effect level With respect to food additives, the maximum dose of an additive that has no detectable adverse effects.

Noble rot White grapes affected by the fungus Botrytis cinerea. It spoils the grapes if they are damaged by rain, but if they are ripe and healthy, and the weather is sunny, it causes them to shrivel and concentrates the sugar, so that top quality sweet wines can be made.

Nocino Liqueur made from walnuts steeped in brandy or other spirit.

Nockerln Small oval dumplings served with soups, stews, goulash, etc., known as noques in Alsace. Sometimes known as butternockerln since they are made with butter. Lebernockerln contain fried, finely chopped liver.

Noggin Traditional measure of volume of liquor = 1/4 pint; also known as a quartern.

Noisette 1. Small round individual portions or slices of meat.

2. French liqueur made from hazelnuts.

3. Flavoured or made with hazelnuts.

Nondisjunction A genetic term referring to an event which takes place during cell division where the cell with 46 chromosomes fails to divide properly during cell division and a zygote is created with a total of 47 chromosomes. When non-disjunction occurs, one of the daughter cells of the division will have an extra chromosome, the other daughter cell will have one less.

Non-essential amino acids Those amino acids that can be synthesized in the body and therefore are not dietary essentials.

Non-esterified fatty acids Free fatty acids in the blood, as opposed to triacylglycerols, which are esterified fatty acids; about 10% of the total blood fatty acids, usually 0.5-2.0 mmol/L. The plasma concentration increases in fasting as fatty acids are released from adipose tissue as a metabolic fuel. Also known as unesterified fatty acids or free fatty acids.

Non-pareils The silver beads used to decorate confectionery, made from sugar coated with silver foil or aluminium-copper alloy.

Non-starch polysaccharides Those polysaccharides found in foods other than starches. They are the major part of dietary fibre and can be measured more precisely than total dietary fibre; include cellulose, pectins, glucans, gums, mucilages, inulin, and chitin. The NSP in wheat, maize, and rice are mainly insoluble and have a laxative effect, while those in oats, barley, rye, and beans are mainly soluble and have a bloodcholesterol lowering effect. In vegetables the proportions of soluble to insoluble are roughly equal but vary in fruits. It is recommended that the average UK adult intake of NSP should beincreased from 13 to 18 g per day.

Noodles Type of pasta made with flour and water, sometimes with added egg, the flour being made from various grains such as rice, wheat, buckwheat, and mung bean starch. Made into a vast range of shapes and sizes.

Noodles, cellophane Transparent noodles.

Noradrenaline Hormone secreted by the adrenal medulla together with adrenaline; also a neurotransmitter. Physiological effects similar to those of adrenaline.

Norite Activated carbon used to decolorize solutions.

Normande, a la Dish containing apples or served with Normandy sauce and garnished with shrimps, truffles, crayfish, or mussels.

Normandy pippins Whole peeled and cored dried apples.

Normandy sauce White sauce made with fish stock, butter, and egg.

North kiln See Smoking kiln n-series.

Nouet Small muslin bag containing spices and herbs which is cooked in liquids to impart flavour without leaving solid particles. Also known as mignonette.

Nougat Sweetmeat made from a mixture of gelatine or egg albumin with sugar and starch syrup, and the whole thoroughly aerated. Originated in Montellimar in southern France.

Novadelox Trade name for benzoyl peroxide used for treating flour.

Novel foods Foods and food ingredients consisting of or containing chemical substances not hitherto used for human consumption to a significant extent in the locality in question.

A food or ingredient to which has been applied a process not currently used for food manufacture or which has not been previously marketed and which gives rise to changes that affect its nutritional value or safety.

Noyau Liqueur made from peach, apricot, plum, or cherry kernels, with a flavour of almonds.

NPR Net Protein Ratio, a measure of protein quality.

NPU Net Protein Utilization, a measure of protein quality.

NPV Net Protein Value, a measure of protein quality.

Nubbing Term used in the canning industry for 'topping and tailing' of gooseberries.

Nucellar layer Of wheat, the layer of cells that surrounds the endosperm and protects it from the entry of moisture.

Nucleic acids Polymers of purine and pyrimidine sugar phosphates; two main classes: ribonucleic acid and deoxyribonucleic acid. Collectively the purines and pyrimidines are called bases. DNA is a double-stranded polymer containing the five-carbon sugar deoxyribose. RNA is a single-stranded polymer containing the sugar ribose.

They are not nutritionally important, since dietary nucleic acids are hydrolysed to their bases, ribose and phosphate, in the intestinal tract; purines and pyrimidines can readily be synthesized in the body, and are not dietary essentials.

Nucleoproteins The complex of proteins and nucleic acids found in the cell nucleus.

Nucleosides Compounds of purine or pyrimidine bases with a sugar, most commonly ribose. For example, adenine plus ribose forms adenosine. With the addition of phosphate a nucleotide is formed.

Nucleotides A class of organic compounds whose molecular structure comprises a nitrogen-containing base linked to a sugar and a phosphate group. The nucleotide of interest for fish technology is adenosine triphosphate which is central to many metabolic processes in cells including the reactions that result in contraction of muscles resulting in movement. On the death of the fish, ATP is degraded by a series of reactions, and the first step, loss of ATP to adenosine diphosphate, is linked to the onset of rigor mortis.

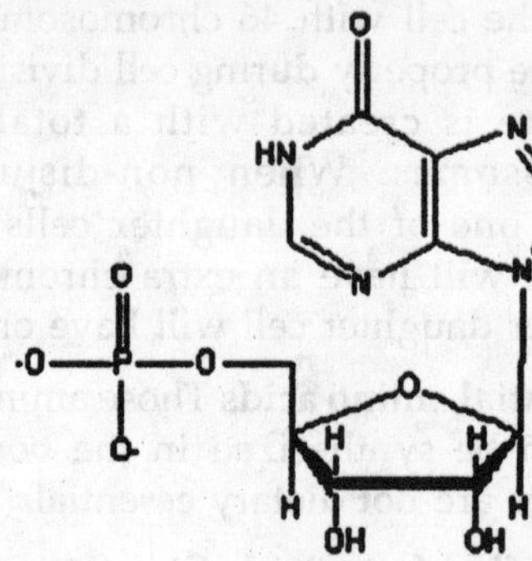

IMP

Fig. Nucleotides

Nun's beads Scottish; small fried pastries filled with cheese.

Nuoc mam Vietnamese, Cambodian; fermented fish sauce. The fish is digested by autolytic enzymes in the presence of salt added to inhibit bacterial growth.

Nutmeg Dried ripe seed of Myristica fragrans; mace is the seed coat of the same species. Both mace and nutmeg are used as flavourings in meat products and bakery goods. Nutmeg contains myristicin, which is toxic in large amounts, and may cause vomiting and hallucinations in moderate excess.

Nutraceuticals Term for compounds in foods that are not nutrients but have beneficial effects.

Nutrasweet Trade name for aspartame.

Nutrient analysis Calculation of the nutrients contained in foods. Analysis may be of a single food or a combination of foods, as in an individual's diet.

Nutrient density A way of expressing the nutrient content of a food or diet relative to the energy yield rather than per unit weight.

Nutrient enemata Rectal feeding can be carried out with nutrient solutions as the colon can absorb 1-2 litres of solution per day; maximum daily amount of glucose that can be given is 75 g, and 1 g of nitrogen, in the form of hydrolysed protein.

Nutrients Essential dietary factors such as vitamins, minerals, amino acids and fatty acids. Metabolic fuels are not termed nutrients so that a commonly used phrase is 'energy and nutrients'.

Nutrients Protein, fat, carbohydrates, vitamins, minerals, and water. These provide chemicals essential to normal growth and development.

Nutrification The addition of nutrients to foods at such a level as to make a major contribution to the diet.

Nutrition facts The uniform list of nutrients provided on all retail food labels. Required by the Food and Drug Administration in the 1993 food labeling regulations.

Nutrition The process by which living organisms take in and use food for the maintenance of life, growth, the functioning of organs and tissues, and the production of energy; the branch of science that studies these processes.

Nutritional recommendations Recommendations comprising nutrient goals, food goals, and dietary guidelines. In addition to reference intakes of nutrients, key recommendations in developed countries are reduction of total fat intake to 30% of energy intake, with a more severe restriction of saturated fats; increase of carbohydrate intake to 55% of energy intake; increased intake of non-starch polysaccharides and reduced intake of salt.

Nutritional status assessment For adults, general adequacy of nutrition is assessed by measuring weight and height; the result is commonly expressed as the body mass index, the ratio of weight (kg) to height2 (m). Body fat may also be estimated, by measuring skinfold thickness, and muscle diameter is also measured.

For children, weight and height for age are compared with standard data for adequately nourished children. The increase in the circumference of the head and the development of bones may also be measured.

Status with respect to individual vitamins and minerals is normally determined by laboratory tests, either measuring the blood and urine concentrations of the nutrients and their metabolites, or by testing for specific metabolic responses.

Nutritional status The condition of the body in those respects influenced by the diet; the levels of nutrients in the body and the ability of those levels to maintain normal metabolic integrity.

Nutritionist One who applies the science of nutrition to the promotion of health and control of disease, instructs auxiliary medical personnel, and participates in surveys.

Nutritive ratio In animal feeding a measure of the value of a feeding ration for growth compared with its fattening value. It is the sum of the digestible carbohydrate, protein, and 2.3 x fat, divided by digestible protein. Value 4-5 for growth, 7-8 for fattening.

Nutritive value index In animal feeding; intake of digestible energy expressed as energy digestibility multiplied by voluntary intake of dry matter of a particular feed, divided by metabolic weight, compared with standard feed.

Nuts Hard-shelled fruit of a wide variety of trees, e.g. almonds, Brazil, cashew, peanut, walnut. A number of nuts are grown specially for their oils.

O

Oats Grain from Avena spp., the three best-known being A. sativa, A. steritis, and A. strigosa.

Obesity Excess body fat that results in adverse health effects. Obesity can be defined as a Body Mass Index of 30 or higher.

Obstipation Extreme and persistent constipation caused by obstruction of the intestinal tract.

Occupational therapy Therapy focusing on the development of fine motor and adaptive skills, such as grasping items, writing, cutting, etc.

Occurrence Aflatoxins often occur in crops in the field prior to harvest. Postharvest contamination can occur if crop drying is delayed and during storage of the crop if water is allowed to exceed critical values for the mould growth. Insect or rodent infestations facilitate mould invasion of some stored commodities.

Aflatoxins are detected occasionally in milk, cheese, corn, peanuts, cottonseed, nuts, almonds, figs, spices, and a variety of other foods and feeds. Milk, eggs, and meat products are sometimes contaminated because of the animal consumption of aflatoxin-contaminated feed. However, the commodities with the highest risk of aflatoxin contamination are corn, peanuts, and cottonseed.

Fungal growth and aflatoxin contamination are the consequence of interactions among the fungus, the host and the environment. The appropriate combination of these factors determines the infestation and colonization of the substrate, and the type and amount of aflatoxin produced. However,. a suitable substrate is required for fungal growth and subsequent toxin production, although the precise factor(s) that initiates toxin formation is not well understood. Water stress, high-temperature stress, and insect damage of the host plant are major determining factors in mould infestation and toxin production. Similarly, specific crop growth stages, poor fertility, high crop densities, and weed competition have been associated with increased mould growth and toxin production.

Aflatoxin formation is also affected by associated growth of other moulds or microbes. For example, preharvest aflatoxin contamination of peanuts and corn is favoured by high temperatures, prolonged drought conditions, and high insect activity; while postharvest production of aflatoxins on corn and peanuts is favoured by warm temperatures and high humidity.

Octave A cask foı wine containing one-eighth of a pipe, about 13 imperial gallons.

Octopus Marine creature with beak-like mouth surrounded by eight tentacles bearing suckers.

Oedema Excess fluid in the body; may be caused by cardiac, renal, or hepatic failure and by starvation.

Oenin An anthocyanidin from the skin of purple grapes.

Oestrogens The female sex hormones; chemically they are steroids, although non-steroidal compounds also have oestrogen activity, including the synthetic compounds stilboestrol and hexoestrol. These have been used for chemical caponi-zation of cockerels and to increase the growth rate of cattle. Compounds with oestrogen activity are found in a variety of plants; collectively these are known as phyto-oestrogens.

Offal With reference to meat, the term includes all parts that are cut away when the carcass is dressed, including liver, kidneys, brain, spleen, pancreas, thymus, tripe, and tongue.(2) With reference to wheat, offal is the bran discarded when milled to white flour.

Off-flavour/odour Sometimes the words are not hyphenated. An unpleasant flavour or odour generated by, chemical, biochemical, or microbiological reactions occurring within the product. Examples are spoilage flavours and odours and cold storage flavours.

Ohmic heating Sterilization by heat generated by passing an electric current through the food or mixture.

Oil, frying, heavy duty Virtually any edible oil or fat can be used to fry foods; heavy duty oils can be used for prolonged periods because they have been processed to resist breakdown by heat and air.

Oils, fixed The triacylglycerols, the edible oils, as distinct from the volatile or essential oils.

Oilseed A wide variety of seeds are grown as a source of oils, e.g. cottonseed, sesame, groundnut, sunflower, soya, and nuts such as coconut, groundnut, and palm nut. After extraction of the oil the residue is a valuable source of protein, especially for animal feedingstuffs, as in oil-seed cake.

Okolehao liqueur made by distillation of fermented extract of the ti tree; also known as oke.

Okra Also known as gumbo, bamya, bamies, and ladies' fingers; the edible seed pods of Hibiscus esculentus. Small ridged mucilaginous pods resembling a small cucumber, grown in South America, the West Indies, and India; used in soups and stews. There are two varieties: gomba are oblong, bamya are round.

Okroschka sour, cold soup made from sour cream, soured milk, and kvass.

Olallie berry Cross between loganberry and youngberry.

Old clothes stew Castilian; dish prepared from left-over cooked meat simmered with a sofrito made from onions, red or green peppers, aubergines, tomatoes, and garlic.

Old fashioned Alcoholic drink made from whisky, sugar, bitters, and soda water.

Oleic acid Mono-unsaturated fatty acid; found to some extent in most fats; olive and rapeseed oils are especially rich sources. By far the most abundant of the unsaturated fatty acids, and beneficial in heart disease.

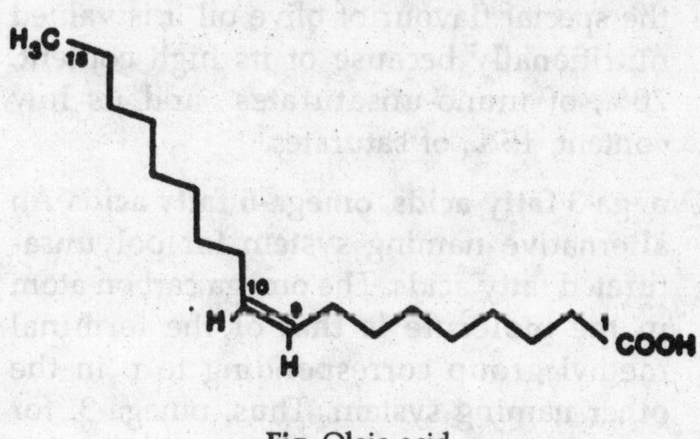

Fig. Oleic acid

Oleoresins In the preparation of some spices such as pepper, ginger, and capsicum, the aromatic material is extracted with solvents which are evaporated off, leaving behind thick oily products known as oleoresins.

Oleovitamin Preparation of fish liver oil or vegetable oil containing one or more of the fat-soluble vitamins.

Oligodynamic Sterilizing effect of traces of certain metals. For example, silver at a concentration of 1 in 5 million will kill Escherichia coli and staphylococci in 3 hours.

Oligopepticles Polymers of 4 or more amino acids; more than about 20-50 are termed polypeptides, and more than about 100 are considered to be proteins.

Oligosaccharides Carbohydrates composed of 3-10 monosaccharide units.

Olive Fruit of the evergreen tree, Olea europea; picked unripe when green or ripe when they have turned dark blue or purplish, and usually pickled in brine. Olives have been known since ancient times. The tree is extremely slow-growing and continues to fruit for many years; there are claims that trees are still fruiting after 1000 years.

Olive oil Pressed from ripe olives, the fruit of Olea europea. Virgin olive oil is not refined and the flavour varies enormously with the locality where it is grown. Other types have been refined to varying extents. Used in cooking, as salad oil, for canning sardines, and for margarine manufacture. Apart from the special flavour of olive oil it is valued nutritionally because of its high content, 70%, of mono-unsaturates and its low content, 15%, of saturates.

Omega-3 fatty acids, omega-6 fatty acids An alternative naming system for polyunsaturated fatty acids. The omega carbon atom in the molecule is that of the terminal methyl group corresponding to n in the other naming system. Thus, omega-3, for example, refers to the position of the first carbon atom counting from the terminal carbon, the third in this case, attached to a double bond.

Omelette Egg whisked and fried. May be plain or filled with mushrooms, bacon, cheese, etc. Spanish omelette is filled with potato.

Omophagia Eating of raw or uncooked food.

Onglet cut of beef corresponding to the top of the skirt.

Onion Bulb of Allium cepa; there are many varieties with white, brown, and red skins.

Onion, Egyptian Allium cepa, proliform group. A type that produces clusters of aerial bulbs which develop shoots to form a multi-tiered plant; the aerial bulbs are cropped.

Onion, everlasting Allium perutile, similar to Welsh onion.

Onion, japanese bunching Allium fistulosum, similar to Welsh onion, but larger.

Onion, spring Young plants of Allium cepa, generally eaten whole as a salad vegetable. Also known as salad onions or scallions.

Onion, welsh The perennial onion, Allium cepa perutile, the leaves of which are cropped, leaving the plant to grow. Similar to, but smaller than, the Japanese bunching onion, Allium fistulosum. Also sometimes used as an alternative name for the leek.

Opaque Preventing light from travelling through, and therefore not transparent or translucent.

Operating characteristics curve A plot showing the probability of acceptance of a batch of stated quality under a given sampling plan. The quality of the batch is defined by the proportion of defectives in the batch in the case of acceptance sampling, or the mean value of properties in the case of sampling by variables. Typically, the probabilities are calculated for set qualities over a range of qualities likely to be obtained for batches using various sampling plans so that sampling plans can be compared for their likelihoods of making wrong decisions - accepting batches that should have been rejected, and rejecting batches that should have been accepted.

Opsomania Craving for special food.

Optic Dispenser attached to bottles of spirits, etc., in bars to ensure delivery of a precise volume.

Optical activity The ability of some compounds to rotate the plane of polarized light because of the asymmetry of the molecule. If the plane of light is rotated to the right, the substance is dextrorotatory and is designated by the prefix (+); if laevorotatory, the prefix is (-). A mixture of the two forms is optically inactive and is termed racemic.

Sucrose is dextrorotatory but is hydrolysed to glucose and fructose, which is more strongly laevorotatory so hydrolysis changes optical activity from (+) to (-); hence, the mixture of glucose and fructose is termed invert sugar.

Oral motor therapy Therapy focusing on the use of the muscles in and around the mouth and face. Oral motor skills are important for learning to eat and talk properly.

Orange butter Chopped whole orange, cooked, sweetened, and homogenized.

Orange Citrus fruit, from the subtropical tree Citrus sinensis. Of nutritional value mainly because of its vitamin C content of 40-60 mg/ 100 g. Blood oranges are coloured by the presence of anthocyanins in the juice vesicles.

Orange, bitter The fruit of the subtropical tree Citrus aurantium; known as Seville orange in Spain, bigaradier in France, melangol in Italy, and khush-khash in Israel. It is used mainly as root stock, because of its resistance to the gummosis disease of citrus. The fruit is too acid to be edible, but is used in the manufacture of marmalade; the peel oil is used in the liqueur curacao; the peel and flower oils and the oils from the green twigs are used in perfumery.

Orangeade Carbonated beverage flavoured with orange juice and orange peel oil.

Organic 1. Chemically the term means a substance containing carbon in the molecule. Substances of animal and vegetable origin are organic; minerals are inorganic.

2. The term organic foods refers to 'organically grown foods', meaning plants grown without the use of pesticides, fungicides, or inorganic fertilizers, and prepared without the use of preservatives. Foodstuffs must be grown on land that has not been treated with chemical fertilizers, herbicides, or pesticides for at least three years. Organic meat is from animals fed on organically grown crops without the use of growth promoters, with only a limited number of medicines to treat disease, and commonly maintained under traditional, non-intensive, conditions.

Organic acids Acids occurring naturally in foods that contain, as do all organic compounds, carbon; e.g. acetic, citric, fumaric, lactic, malic acids. As distinct from the inorganic acids such as hydrochloric and sulphuric acids, they can be metabolized in the body to provide 2.4-3.6 kcal per gram for the various acids; for labelling purposes a value of 2 kcal (8 kJ/ g) is used for all organic acids.

Organic farming Food produced without the use of synthetic chemicals and with concern for the environment.

Organoleptic Relating to the properties of a product that are capable of stimulating one or more senses.

Orientale, a l' Fish, eggs, or vegetables, cooked with tomatoes and flavoured with garlic and saffron.

Ornithine An amino acid that occurs as a metabolic intermediate , but not involved in protein synthesis. Of no nutritional importance, since it is not found in proteins and can be synthesized in adequate amounts in the body.

Ornithine-arginine cycle The metabolic pathway for the synthesis of urea.

Orotic acid An intermediate in the biosynthesis of pyrimidines; a growth factor for some micro-organisms and at one time called vitamin B 13. There is no evidence that it is a human dietary requirement.

Ortanique A Jamaican citrus fruit; a cross between orange and tangerine.

Orthophenylphenol A compound used for the treatment of citrus fruit and nuts after harvesting to prevent the growth of moulds . Diphenyl is also used.

Ortolan Small wild song bird, Emberiza hortulana, still caught in the wild and eaten in parts of Europe, where it is prized for its delicate flavour.

Oryzenin The major protein of rice.

Oslo breakfast A breakfast requiring no preparation, introduced in Oslo, Norway, in 1929 for schoolchildren before classes started. It consisted of rye-biscuit, brown bread, butter or vitaminized margarine, whey cheese, and cod liver oil paste, ? litre of milk, raw carrot, apple, and half an orange.

Osmazome Obsolete name given to an aqueous extract of meat regarded as the 'pure essence of meat'.

Osmophiles Micro-organisms that can flourish under conditions of high osmotic pressure, e.g. in jams, honey, brine pickles; especially yeasts.

Osmosis The passage of water through a semi-permeable membrane, from a region of low concentration of solutes to one of higher concentration. Reverse osmosis is the passage of water from a more concentrated to a less concentrated solution through a semi-permeable membrane by the application of pressure. Used for desalination of sea water, concentration of fruit juices, and processing of cheese whey. The membranes commonly used are cellulose acetate or polyamide of very small pore size, 10-4-10-3 μm.

Osmotic pressure The pressure required to prevent the passage of water through a semi-permeable membrane from a region of low concentration of solutes to one of higher concentration, by osmosis.

Ossein The organic matrix of the bone left behind when the mineral salts are removed by solution in dilute acid. Chemically similar to collagen and hydrolysed by boiling water to gelatine; hence the manufacture of glue from bones, known as ossein gelatine.

Osseomucoid Mucoid substance forming part of the structure of bone.

Osso bucco Veal stew.

Osteocalcin A calcium-binding protein in bone, essential for the normal mineralization of bone. Its synthesis requires vitamin K, and is controlled by vitamin D.

Osteomalacia The adult equivalent of rickets; a bone disorder due to deficiency of vitamin D, leading to inadequate absorption of calcium and loss of calcium from the bones.

Osteoporosis Degeneration of the bones with advancing age due to loss of bone mineral and protein; this is largely a result of loss of hormones with increasing age. Although there is negative calcium balance this is the result of osteoporosis, not its cause, and there is no evidence that calcium supplements affect the progression of the disease. A high calcium intake in early life may be beneficial, since this may result in greater bone density at maturity, but the most important factor is regular exercise to stimulate bone metabolism.

Vitamin D supplements do not improve osteoporosis, and may even exacerbate the condition. Nevertheless, an adequate intake of vitamin D is important for the elderly to prevent the development of osteomalacia.

Ostermilk Trade name for dried milk for infant feeding. Ostermilk No. 1 is half-cream; No. 2 is full-cream.

Otitis media Acute Pain in the ear, hearing loss, bulging and redness of the eardrum, and often a fever. Serious Noninfected accumulation of fluid in the middle ear with no observable symptoms. Chronic Characterized by a perforated eardrum and drainage from the middle ear.

Oude Old Dutch gin.

Outer Semi-rigid carton used, for example, to give additional protection to frozen fish packed in a flexible film, or shatterpacks, or IQF fillets, or a retortable pouch. Sometimes refers to a master carton used to hold a number of smaller packs or blocks, especially of frozen fish such as the laminated block.

Out-grade Category to be discarded in a sorting process.

Ouzo herb liqueur flavoured with anise. Turkish equivalent is raki; Arabic arrack.

Ovalbumin The albumin of egg-white; comprises 55% of the total solids.

Ovaltine Trade name for a preparation of malt extract, milk, eggs, cocoa, and soya, for consumption as a beverage when added to milk. Fortified with thiamin, vitamin D, and niacin.

Oven spring The sudden increases in the volume of a dough during the first 10-12 minutes of baking, due to increased rate of fermentation and expansion of gases.

Overexpression More than a normal amount of a particular chemical is made because there are extra copies of the gene for that chemical

Overrun In ice cream manufacture, the percentage increase in the volume of the mix caused by the beating-in of air. Optimum overrun, 70-100%. To prevent excessive aeration US regulations state that ice cream must weigh 4.5 lb per gallon.

Overweight Excessive accumulation of body fat, but not so great as to be classified as obesity.

Ovcmucin A carbohydrate-protein complex in egg-white, responsible for its firmness. Comprises 1-3% of the total solids.

Ovomucoid A protein of egg-white, 12% of the total solids. It inhibits the digestive enzyme trypsin, but is inactivated by the stomach enzyme pepsin.

Oxalates Salts of oxalic acid.

Oxalic acid A dicarboxylic acid, chemically COOH–COOH. Poisonous in large amounts; present especially in spinach, chocolate, rhubarb, and nuts. The toxicity of rhubarb leaves is due to their high content of oxalic acid.

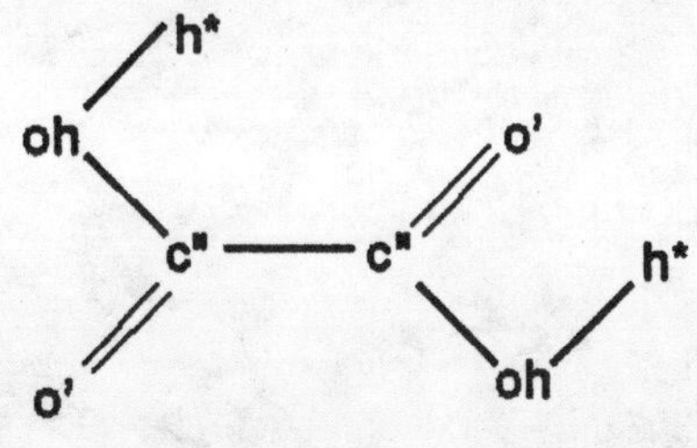

Fig. Oxalic acid

High concentrations of oxalates in the urine can form kidney stones; while most of these oxalates are of endogenous metabolic origin, patients with hyperoxaluria are advised to avoid dietary sources of oxalates.

Oxidation There are formal chemical definitions of oxidation, involving electron transfer and free radicle reactions but, in the context of fish technology, it can be considered as the chemical reaction in which oxygen combines with a compound. Most usually, the oxygen comes from air surrounding the product but, for some reactions, the oxygen can come from oxygen-containing compounds. The oxidation reaction most frequently encountered in fish technology is oxidation of unsaturated fatty acids in the free form or as combined in lipids. Oxidation in this case is only the first step in a series of reactions that result in a variety of reaction products which are often strongly flavoured and give rise to rancid flavours.

Oxo Trade name for a dried preparation of hydrolysed meat, meat extract, salt, and cereal in cube form, used as a drink or gravy.

Oxycalorimeter Instrument for measuring the oxygen consumed and carbon dioxide produced when a food is burned, as distinct from the calorimeter, which

Oxyhaemoglobin Haemoglobin combined with oxygen; the form in which oxygen is transported in the blood. oxymel Medicinal syrup or beverage made from honey and vinegar.

Oxymyoglobin Myoglobin is the muscle oxygen-binding protein; it takes up oxygen to form oxymyoglobin, which is bright red, while myoglobin itself is purplish-red. The surface of fresh meat that is exposed to oxygen is bright red from the oxymyoglobin, while the interior of the meat is darker in colour where the myoglobin is not oxygenated.

Oyster Marine bivalve mollusc, Ostreidae and Crassostrea spp. Oysters are an exceptionally rich source of vitamin B 12 ; a rich source of iron, iodine, selenium, and vitamin D; a good source of protein and niacin; a source of vitamins A, B 1, and B 2.

Oyster crabs Small young crabs found inside oysters, cooked and eaten whole, including the soft shell.

P

Paddy Rice in the husk after threshing; also known as rough rice.

Palatability Pleasant to taste, being acceptable.

Palatable Acceptable, food or drink that tastes good.

Palatinose Isomaltulose. a disaccharide of glucose and fructose.

Palestine soup English, nineteenth century, made from Jerusalem artichokes and named in the mistaken belief that the artichokes came from Jerusalem.

Palm kernel oil One of the major oils of commerce, widely used in cooking fats and margarines; oil extracted from the kernel of the nut of the oil palm, Elaeis guineensis, pale in colour in contrast with 'red' palm oil from the outer part of the nut; about 80% saturates and 15% mono-unsaturates.

Palm oil From outer fibrous pulp of the fruit of the oil palm, Elaeis guineensis. Coloured red because of very high content of ? 1 - carotene and ß-carotene, but these are usually removed to produce a pale oil; 45% saturated and 40% mono-unsaturated, 10% polyunsaturated.

Palm stew West African; prepared by stewing the entire head of nuts of the oil palm Elaeis guineensis, and skimming off the oil as it separates. Although most of the oil is removed, the final dish has a deep red colour because of the high content of carotenes of palm oil.

Palm wine Fermented sap from various palm trees, especially date and coconut palms.

Palm, wild date Phoenix sylvestris, a relative of the true date palm, P. dactylifera, grown in India as a source of sugar, obtained from the sap.

Palmier Cake made from puff pastry sandwiched with cream or jam. Also small sweet biscuits made from puff pastry.

Palmitic acid A saturated fatty acid with sixteen carbon atoms, common in the triglycerides of many animal and vegetable fats.

Palmitoleic acid A mono-unsaturated fatty acid with sixteen carbon atoms, widespread in fats and oils.

Palomino A grape variety widely used for wine making, although not one of the classic varieties.

Pan broil American; to cook food in a pan on top of the stove, dry and with enough oil only to prevent it sticking.

Pan dowdy American; baked apple sponge pudding, served with the apple side up.

Panada Mixture of fat, flour, and liquid mixed to a thick paste; used to bind mixtures such as chopped meat, and also as the basis of souffles and choux pastry.

Panage To coat food with breadcrumbs before frying or grilling. Food so treated is pane.

Panary fermentation Yeast fermentation of dough in bread making.

Pancake Batter fried on a lightly greased griddle. Served plain or filled with sweet or savoury mixture.

Pancetta Spiced pickled bacon, rolled in the form of a long sausage.

Pancreas A gland in the abdomen with two functions: the endocrine pancreas secretes the hormones insulin and glucagon; the exocrine pancreas secretes the pancreatic juice. Known by the butcher as sweetbread or gut sweetbread as distinct from chest sweetbread which is thymus.

A rich source of protein, niacin, and iron; unusually for a meat product, a good source of vitamin C; contains about 20 g of fat of which one-third is saturated, and supplies 350 kcal.

Pancreatic juice The alkaline digestive juice produced by the pancreas and secreted into the duodenum. It contains the inactive precursors of a number of protein digestive enzymes. Trypsinogen is activated to trypsin by enteropeptidase in the intestinal lumen; in turn, trypsin activates the other enzyme precursors: chymotrypsinogen to chymotrypsin, pro-elastase to elastase, procarboxypeptidase to carboxypeptidase. Also contains lipase, amylase, and nucleases.

Pancreatin Preparation made from the pancreas of animals containing the enzymes of pancreatic juice. Used to replace pancreatic enzymes in cystic fibrosis as an aid to digestion.

Pancreozymin Hormone produced by the intestinal mucosa which, together with secretin, stimulates the secretion of pancreatic juice.

Panettone Half bread and half cake; sweet, with candied peel and sultanas; a speciality of Lombardy.

Pangamic acid Chemically the N-di-isopropyl derivative of glucuronic acid. Claimed to be an antioxidant, and to speed recovery from fatigue. Sometimes called vitamin B 15, but there is no evidence that it is a dietary essential, nor that it has any metabolic function.

Panne The fat surrounding the pig's kidneys.

Pannfisch Large fried fishcake.

Panoufle the underpart of the top of a sirloin of beef.

Panthenol The alcohol form of pantothenic acid; biologically active. pantothenic acid A vitamin of the B complex with no numerical designation. Chemically, the ß-alanine derivative of pantoic acid. Required for the synthesis of coenzyme A and of acyl carrier protein.

Dietary deficiency is unknown; it is widely distributed in all living cells, the best sources being liver, kidney, yeast, and fresh vegetables. Human requirements are not known with any certainty; average intakes are between 3 and 7 mg/d, which is therefore considered to be a safe and adequate level of intake.

Experimental deficiency signs in rats include greying of the hair. Experimental deficiency in human beings leads to fatigue, headache, muscle weakness, and gastro-intestinal disturbances.

Pao-tzu Steamed buns.

Papain Proteolytic enzyme from the juice of the pawpaw used in tenderizing meat; sometimes called vegetable pepsin. The enzyme is obtained as the dried latex on the skin of the fruit by scratching it while still on the tree, and collecting the flow. In the tropics meat is often tenderized by wrapping in pawpaw leaves.

The rate of reaction is slow at room temperature, increasing to maximum activity at 80°C and rapidly inactivated at higher temperatures; hence, the papain continues to tenderize the meat during the early stages of cooking.

Papaw Purple fruit of Asiminia triloba, related to the custard apple; distinct from the pawpaw or papaya.

Papin's digester Early version of the pressure cooker. Named after D. Papin, French physicist, 1647-1712; originally invented for the purpose of softening bones for the preparation of gelatine.

Para-amino benzoic acid Essential growth factor for microorganisms. It forms part of the molecule of folic acid and is therefore required for the synthesis of this vitamin. Mammals cannot synthesize folic acid, and PABA has no other known function; there is no evidence that it is a human dietary requirement. Sulphanilamides are chemical analogues of PABA, and exert their antibacterial action by antagonizing PABA utilization.

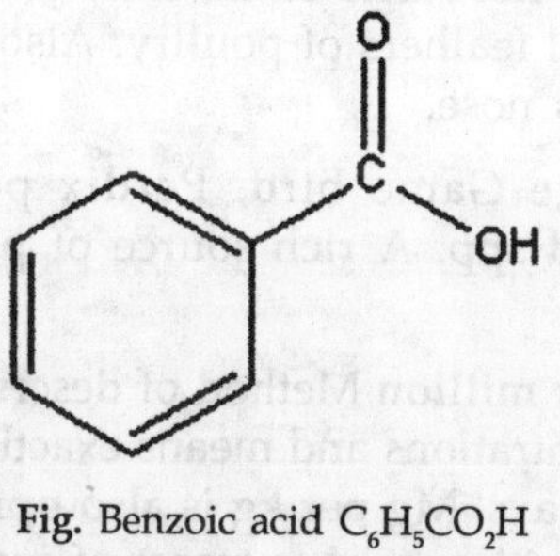

Fig. Benzoic acid $C_6H_5CO_2H$

Parabens Methyl, ethyl, and propyl esters of p-hydroxybenzoic acid used together with their sodium salts as antimicrobials in food. Effective over a wide range of pH; more effective against moulds and yeast than against bacteria.

Paracasein Obsolete name for milk casein after precipitation.

Paraflow Trade name for a plate heat exchanger used for pasteurizing liquids.

Parakeratosis Disease of swine characterized by cessation of growth, erythema, seborrhoea, and hyperkeratosis of the skin; due to zinc deficiency and possibly to changes in essential fatty acid metabolism.

Paralytic shellfish poisoning Caused by shellfish that have accumulated toxins from the dinoflagellate plankton, Gonyaulax spp.

Parasites Organisms that live in or on other organisms, referred to as the host, from which they derive their food. Ectoparasites live on, or attached to, the surface of the host, for example, sea lice and anchor worms; endoparasites live within the host. The latter encompass those such as nematodes and trematodes which cause disease in humans. Some parasites can complete their whole life cycle within one host, but those of public health concern need two or three different hosts to complete the life cycle. The host in which a parasite matures and produces eggs is the definitive host, the others are intermediate hosts.

Parasol mushroom Macrolepiota procera.

Paratha Indian whole-wheat unleavened bread; the dough is rolled and brushed repeatedly with melted butter before cooking on a buttered griddle. Frequently stuffed with spiced potato or other vegetables.

Parathormone Commonly used as an abbreviation for the parathyroid hormone; correctly a trade name for a pharmaceutical preparation of the hormone.

Parathyroid hormone The hormone secreted by the parathyroid glands; four glands situated in the neck near to the thyroid gland but not connected with its function. The hormone is secreted in response to a fall in plasma calcium, and acts on the kidney to increase the formation of the active metabolite of vitamin D, leading to an increase in plasma calcium by increasing intestinal absorption and mobilizing the mineral from bones. It also reduces urinary excretion of phosphate.

Parboil Partially cook. Of special interest in nutrition is the parboiling of brown rice, i.e. steaming rice in the husk before milling. The watersoluble vitamins diffuse from the husk into the grain; when the rice is polished, it contains far more of these vitamins than polished raw rice.

Parch To brown in dry heat.

Parcha Foods cooked en papillote; wrapped in parchment or banana leaf.

Parenteral nutrition Slow infusion of solution of nutrients into the veins through a catheter. This may be partial, to supplement food and nutrient intake, or total, providing the sole source of energy and nutrient intake for patients with major intestinal problems.

Pareve Jewish term for dishes containing neither milk nor meat. Orthodox Jewish law prohibits mixing of milk and meat foods or the consumption of milk products for 3 hours after a meat meal.

Parfait Frozen dessert, similar to a mousse but lighter.

Parillin Highly toxic glycoside from sarsaparilla root; consists of glucose, rhamnose and parigenin.

Parkin Oatmeal, ginger, and treacle cake.

Parlies Scottish; ginger cakes, believed to be so named because they were eaten by members of the Scottish parliament.

Parma ham Smoked, raw ham served in very thin slices.

Parmentier Made or served with potatoes, named after Parmentier, who popularized the potato in France in the eighteenth century. Pommes parmentier are diced and fried; parmentier soup is a leek and potato soup.

Parmesan cheese English name for the hard Italian cheese parmigiana. Made from semi-skimmed cow's milk cooked with rennet, dried for six months, and then coated with a black substance to make the product airtight. When at least 2 years old it is called vecchio; stravecchio is 3 years old; stravecchione, 4 years old. It is hard and usually served on dishes grated; it cooks without becoming sticky.

Parmigiana Made or served with parmesan cheese.

Parnuts An EU term for foods prepared for particular nutritional purposes. Also called dietetic foods.

Parsley Leaves of the herb Petroselinum crispum, P. hertense, or P. sativum. Since it is largely used as a garnish and for flavouring, the nutrients per serving are negligible.

Parsley, hamburg Root of Petroselinum crispum var. tuberosum, grown for its root; similar in appearance to parsnip. A rich source of vitamin C.

Parsnip Root of Pastinaca sativa, eaten as a vegetable.

Parson's nose The small fatty joint holding the tail feathers of poultry. Also known as pope's nose.

Partridge Game bird, Perdix perdix and related spp. A rich source of protein and iron.

Parts per million Method of describing small concentrations and means exactly what the term says. Mg per kg is alsó ppm. Usually used with regard to traces of contaminants and food additives.

Pashka Dessert of curd cheese with cream, almonds, and dried fruit, an Easter speciality.

Passion fruit Also known as parchita, granadilla, and water lemon; fruit of the tropical American vine, Passiflora spp.

Purple or greenishyellow when ripe, it contains watery pulp surrounding small seeds; used in fruit drinks.

Passover The Jewish festival celebrating the Exodus from slavery in Egypt. There was no time to allow bread dough to rise, and unleavened bread was eaten. In commemoration, Jews abstain from leavened bread for the week of Passover, eating matzo instead, and using matzo meal or potato flour for baking. Foods certified free from leavened bread are known as 'kosher for Passover'.

Pasta Dried dough, traditionally made with hard wheat but soft wheat may be added, sometimes with egg and milk. Spinach or tomato may be added to the dough to give a green or red colour. The dough is partly dried in hot air, then more slowly. Sold both completely dry, when it can be stored for a long period, or 'fresh', i.e. less dried and keeping for a week or so only. It is made in numerous shapes: spaghetti is a solid rod about 2 mm in diameter; vermicelli is about one-third this thickness; ravioli, fettucine, and linguini, and a range of twists, spirals, and other shapes. Macaroni is tubular shaped, about 5 mm in diameter; at 10 mm. it is known as zitoni, and at 15 mm fovantini or maccaroncelli. Cannelloni are tubes 1.5-2 cm wide and 10 cm long, stuffed with meat; penne are nibshaped. Lasagne is sheets of pasta. Farfals are ground, granulated, or shredded.

Pasteurisation A form of heat treatment that kills vegetative pathogens and most but not all spoilage microorganisms in foods. Typically, a fish product is heat-treated in water at 70 to 90°C for 5 to 10 minutes, so that a high proportion of spoilage and pathogenic micro-organisms is destroyed, thus extending chilled storage life. Pasteurisation can be used in combination with acid foods to reduce the severity of the heat process and minimise its effects on taste and texture. For example, mussel meats packed, covered in acetic acid, in closed jars, such that the acid content of the pack is not less than 1.8% and the pH is not higher than 4.2, may be processed so that the thermal centre of the jar is raised to 70°C and held at that temperature for 20 minutes, will have a shelf life of about 6 months at ambient temperature.

Pasteurizer Equipment used to pasteurize liquids such as milk, fruit juices, etc. The material is passed continuously over heated plates, or through pipes, where it is heated to the required temperature, maintained for the required time, then immediately cooled.

Pastille Round, flat sweet, often coated with sugar, sometimes medicated.

Pastis Spirit prepared by distilling anise and liquorice as opposed to anise, which is prepared by steeping the herbs in spirit.

Pastourma Greek and Turkish; black-rinded smoked bacon, highly flavoured with garlic; generally fried until crisp.

Pastrami Middle-European; smoked and seasoned beef, also made from turkey. Known in Canada as smoked beef.

Pastry Baked dough of flour, fat, and water. There are six basic types: shortcrust, in which the fat is rubbed into the flour; suet crust, in which chopped suet is mixed with the flour; puff and flaky, in which the fat is rolled into the dough; hotwater crust and choux, in which the fat is melted in hot water before being added to the flour. Phyllo pastry is made from flour and water only.

Suet pastry is raised using baking powder or self-raising flour; puff and flaky and choux pastry are raised by the steam trapped between layers of dough

Pastry brush Used to brush liquid over foods, eg. glazing pastry.

Pasty Individual dish enclosed in pastry, folded over, and baked on a flat tray.

Patatas bravas Spanish; fried slices of potato eaten with a mayonnaise strongly flavoured with garlic.

Pate French for paste; used for pastry, dough, or batter, also for pasta.

Pate Literally, French for a savoury pie, now used almost exclusively to mean a savoury paste of liver, meat, or fish.

Patent ductus arteriosus The ductus arteriosus (DA) is a blood vessel in the developing fetus that diverts circulation away from the lungs and sends it directly to the body. The DA usually closes on its own shortly after birth; it is no longer needed once a newborn breathes on his own. If the DA doesn't close, then a condition called patent ductus arteriosus (PDA) results, which can result in too much blood flow to a newborn's lungs. PDA is common in premature babies.

Patent flour Flour of 30-50% extraction rate; mainly, the endosperm of the grain.

Pathogen Organism able to cause disease in a certain host.

Pathogenic Harmful to the human body.

Pathogens An organism that is capable of causing disease, directly or by excreting a toxin. Examples are viruses, bacteria, parasites, moulds.

Patisserie ornate small cakes and pastries.

Patty Small savoury pie, normally made with short-crust pastry; also small cakes of minced meat or poultry, like croquettes but not dipped in breadcrumbs before cooking.

Patulin Mycotoxin found in juice from fruits infected with one of a variety of moulds. Removed by fermentation and also pasteurization.

Paunching To remove the entrails of rabbits, hares, etc.

Paupiette Small, thinly cut piece of meat wrapped round a filling of forcemeat and braised.

Pave Sweet or savoury cold dish made in a square or rectangular mould.

Pavlova Australian; meringue cake topped with fruit and whipped cream; created in honour of the Russian ballerina Anna Pavlova on her visit to Australia in the 1920s.

Pawpaw Large green or yellow melon-like fruit of the tropical tree Carica papaya, widely grown in all tropical regions. The proteolytic enzyme, papain, is derived from the fruit. A rich source of vitamin C; a good source of vitamin A.

Paysanne, a la Peasant or simple country style. Meat or poultry, usually braised and accompanied by a garnish of mixed vegetables, bacon, etc.

Payusnaya coarse, pressed caviare including skins of ovaries.

Peanut oil From the peanut or groundnut, Arachis hypogaea; 20% saturated, 50% mono-unsaturated, 30% polyunsaturated.

Peanut butter Ground, roasted peanuts; commonly prepared from a mixture of Spanish and Virginia peanuts, since the first alone is too oily and the second is too dry. Separation of the oil is prevented by partial hydrogenation of the oil and the addition of emulsifiers. A source of protein, copper, and niacin.

Peanut Fruit of Arachis hypogaea, also known as groundnut, earth-nut, arachis nut, monkey nut. A rich source of protein, niacin, and vitamins E and B1, a good source of copper and zinc, and a source of protein, vitamin B6, folate, and iron. When dryroasted or roasted and salted some of the vitamin B 1 is damaged and the nuts are only a source of this vitamin.

Pear Fruit of many species of Pyrus; cultivated varieties all descended from P. communis.

Peas, garden or green Seeds of the legume Pisum sativum. Widely available in frozen, dried, and canned forms.

Peas, mushy A dish prepared from processed peas, boiled and mashed to give a smooth texture.

Peas, processed Garden peas that have matured on the plant and are then canned.

Pease pudding A dish prepared from dried peas which are soaked, boiled, mashed, and sieved, traditionally served with baked ham.

Pecan nuts From the American tree Carya illinoensis, a species of hickory nut. A rich source of selenium, vitamin E, and niacin; a good source of vitamin B 1 and zinc; a source of protein and iron.

Pecorino cheese made from ewe's milk.

Pectase An enzyme in the pith of citrus fruits which removes the methoxyl groups from pectin to form water-insoluble pectic acid. The intermediate compounds, with varying numbers of methoxyl groups are pectinic acids. Also known as pectin esterase, pectin methyl esterase, and pectin methoxylase; distinct from pectinase.

Pectic acid Demethylated pectin.

Pectin Is the setting agent in jam; it forms a gel with sugar under acid conditions. Soft fruits, such as strawberry, raspberry, and cherry, are low in pectin; plums, apples, and oranges are rich. Apple pulp and orange pith are the commercial sources of pectin. Added to jams, confectionery, chocolate, and ice cream as an emulsifier and stabilizer instead of agar; used in making jellies, and as an anti-staling agent in cakes. Included in non-starch polysaccharides.

Pectin Plant tissues contain hemicelluloses known as protopectins which cement the cell walls together. As fruit ripens, there is maximum protopectin present; thereafter it breaks down to pectin, pectinic acid, and, finally, pectic acid, and the fruit softens as the adhesive between the cells breaks down.

Pectinase An enzyme present in the pith of citrus fruits, which hydrolyses pectin or pectic acids into smaller polygalacturonic acids, and finally galacturonic acid and its methyl ester. Used to clarify fruit juices. Also known as pectolase, pectozyme, and polygalacturonase.

Pectinic acid Partially demethylated pectin.

Pectins, low-methoxyl Partially demethylated pectins which can form gels with little or no sugar and are therefore used in low-calorie jellies.

Peer Review Analysis of research by a group of professionals of comparable knowledge and expertise in a specific scientific or medical field.

Pekar test A comparative test of flour colour.

Pekmez Turkish; thick jelly made by evaporating grape juice, the basis of Turkish delight and other sugar confectionery.

Pelagic fish Fish that swim near the surface, compared with demersal fish, which live on the sea bottom. Pelagic fish are mostly of the oily type such as herring, mackerel, and pilchard, containing up to 20% oil.

Pellagra The disease due to deficiency of the vitamin niacin and the amino acid tryptophan. Signs include a characteristic symmetrical photosensitive dermatitis, resembling severe sunburn; mental disturbances; and digestive disorders; fatal if untreated.

Most commonly associated with a diet based on maize or sorghum, which are poor sources of both tryptophan and niacin, with little meat or other vegetables.

Pellicle Glossy film formed on the cut surface of the fish during dripping, due mainly to the swelling of protein under the influence of brine and subsequent surface drying in the smoking kiln.

Pemmican Mixture of dried, powdered meat and fat, used as a concentrated food source, e.g. on expeditions. Analysis per 100 g: water, 3 g, protein, 40 g, fat, 45 g; 560 kcal.

Penicillin The first of the antibiotics; found in the culture fluid of the mould Penicillium notatum in 1929. Active against a wide range of bacteria and of great value clinically. Not used as food preservative because of the danger that repeated small doses will increase the development of penicillin-resistant organisms.

Penicillium A genus of moulds; apart from the production of penicillin, several species are valuable in the ripening of cheeses.

Pentachlorophenol Chemical formerly widely used to preserve wood from decay and insect damage, now a restricted-use pesticide. According to the U.S. Environmental Protection Agency, it can be an acute toxin causing such problems as burns and breathing difficulty. It may cause developmental effects such as birth defects and affect male and female reproductive capacity. It can also harm wildlife.

Pentosans Polysaccharides of five-carbon sugars. Widely distributed in plants, e.g. fruit, wood, corncobs, oat hulls. Not digested in the body, and hence a component of non-starch polysaccharides and dietary fibre.

Pentoses Monosaccharide sugars with five carbon atoms. The most important is ribose.

Pentosuria The excretion of pentose sugars in the urine. Idiopathic pentosuria is an inherited metabolic disorder almost wholly restricted to Ashkenazi Jews, which has no adverse effects. Consumption of fruits rich in pentoses can also lead to pentosuria.

Pepper 1. Bell pepper, bullnose pepper, capsicum, paprika, sweet pepper, Spanish name pimiento; fruits of the annual plant Capsicum annuum. Red, yellow, purple, or brown fruits, often eaten raw in salads when green and unripe; very variable in size and shape; some varieties can be spicy but most are non-pungent. One-quarter of a green pepper (45 g) is a rich source of vitamin C; contains 0.4 g of dietary fibre and 0.2 g of fat; supplies 6 kcal (25 kJ).

2. Red pepper, chilli (or chili), small red fruit of the bushy perennial plant Capsicum frutescens. Usually sun-dried and therefore wrinkled. Very pungent, an ingredient of curry powder, pickles, and tabasco sauce. Cayenne pepper is made from the powdered dried fruits. Unripe (green) chillis are also very pungent.

3. Black and white pepper, fruit of the tropical climbing vine, Piper nigrum; the fruits are peppercorns. Black pepper is made from sundried, unripe peppercorns when the red outer skin turns black. White pepper is made by soaking ripe berries and rubbing off the outer skin. Usually ground as a condiment. Green peppercorns are dried or pickled unripe fruit. Pungency due to the alkaloids piperine, piperdine, and chavicine.

4. Japan pepper, black seeds of Zanthoxylum piperitum with a pungent, peppery flavour.

Pepper dulse Red aromatic seaweed, dried and used as a spice in Scotland.

Pepper, melegueta Seeds of the West African tree Amomum melegueta, also known as guinea pepper or grains of paradise.

Peppergrass Peppery-tasting cress, also known as pepperwort and peppermint.

Peppermint A hybrid between M. aquatica and M. spicata. Not used for flavouring dishes but grown for the essential oil which is used in confectionery and medicinally.

Pepperpot Caribbean; stew flavoured with cassareep.

Pepsi-Cola Trade name for a cola beverage. Originally made in 1896 in the USA by Caleb Bradham, druggist.

Pepsin An enzyme in the gastric juice which hydrolyses proteins to give smaller polypeptides, known as peptones: an endopeptidase. Active only at acid pH, 1.5-2.5. Secreted as the inactive precursor pepsinogen, which is activated by acid.

Pepsin digestibility Test used in assessing the quality of fish meal which is meant to mimic to some extent digestion conditions in the stomach of animals. A sample of finely ground meal is treated with pepsin, a proteolytic enzyme from the stomach, under standard conditions, and the amount of protein digested from the sample is a measure of the nutritional availabilty of protein in the fish meal.

Peptidases Enzymes that hydrolyse proteins, and therefore important in protein digestion. Endopeptidases cleave at specific points in the middle of protein molecules; exopeptidases remove amino acids sequentially from either the amino terminal or carboxy terminal.

Peptides Compounds formed when amino acids are linked together through the –CO–NH– linkage. Two amino acids so linked form a dipeptide, three a tripeptide, etc.; medium-length chains of amino acids are known as oligopeptides, longer chains are polypeptides or proteins.

Peptones Small polypeptides that are intermediate products in the hydrolysis of proteins. The term is often used for any partial hydrolysate of proteins as, e.g., bacteriological peptone, which is used as a medium for the growth of micro-organisms.

PER Protein Efficiency Ratio, a measure of protein quality.

Pericarp The fibrous layers next to the outer husk of cereal grains and outside the testa; of low digestibility and removed from grain during milling. The major constituent of bran.

Perigord, a la Dish made or served with truffles and sometimes foie gras.

Perillartine Non-nutritive sweetener derived from perillaldehyde, extracted from shiso oil; 2000 times as sweet as sucrose.

Perishable Having a short shelf-life; food that spoils quickly and needs careful storage.

Peristalsis The rhythmic alternating contraction and relaxation of smooth muscle that forces food through the intestinal tract in peristaltic waves.

Permutit An ion exchange resin.

Pernod Trade name for anise-flavoured liqueur: originated as a substitute for absinthe.

Peroxide Any compound with the peroxy group; atmospheric oxidation of unsaturated fatty acids produces peroxides. Also used to mean specifically hydrogen peroxide.

Peroxide number Or peroxide value; a measure of the oxidative rancidity of fats by determination of the lipid peroxides present.

Peroxide value An index of the extent of oxidation of a lipid in the early stages of oxidation. The initial step of oxidation of lipid is for oxygen to attach to unsaturated bonds in the molecule to form a hydroperoxide, which later breaks down to other products. The peroxide value measures this initial formation of hydroperoxides. Its value increases then decreases as oxidation proceeds, as the hydroperoxides are first formed then decomposed. A common way of measuring peroxide value is to react the lipid with potassium iodide and measure the amount of iodine liberated from the potassium iodide by peroxides in the lipid.

Perrier water Aerated mineral water from Vergèze, France. perry Fermented pear juice analogous to cider from apples. Sparkling perry is sometimes known as champagne perry.

Persimmon Fruit of Diospyros virginiana and D. kaki. Kaki may be eaten raw or cooked; American persimmon develops a sour flavour if cooked.

Pesticides Chemicals used to kill pests.

Pesto Basil and garlic sauce.

PET A light-weight clear plastic with acceptable barrier properties to gas and water vapour.

Petechiae Small, pin-point bleeding under the skin; one of the signs of scurvy.

Petillant Lightly sparkling wines, equivalent to Italian frizzante.

Petit pois Small peas, picked when young; also French for green pea.

Petit sale Salted belly or flank of pork. petits fours Small, rich, sweet cakes and biscuits served after a meal.

pH Formally defined as the negative logarithm of the molar concentration of hydrogen ions in solution. It is a measure of the acidity or alkalinity of a solution. pH 7 is neutral, and the pH of very pure water. A solution with a pH less than 7 is acidic, and one with pH above 7 is alkaline. The pH of fish muscle in live fish is around 7.2; after the death of the animal and resolution of rigor mortis, the typical pH of most species of fish drops to between 6.4 and 6.9. In some species, for example, tunas, salmon, some species of halibut, the pH drops to 6 or lower. The pH of stored fish tends to rise as amines are formed during spoilage.

PH level A number which shows how strongly acid or alkaline a substance is, in a range from 0 to 14.

Below pH 6.5 is acid, above pH 7.5 is alkaline.

Phaeophytin Brownish-green derivative of chlorophyll, due to the loss of the magnesium in acid conditions. The formation of phaeophytin accounts for the colour change when green vegetables are cooked.

Phagomania Morbid obsession with food; also known as sitomania.

Phagophobia Fear of food; also known as sitophobia. phase inversion Cream is an emulsion of fat in water; butter is an emulsion of water in fat. The change from cream to butter is termed phase inversion.

Phaseolin Globulin protein in kidney or haricot bean.

Phaseolunatin Cyanogenic glucoside found in certain legumes, which hydrolyses to glucose, acetone, and hydrocyanic acid; not proven harmful when present in the diet.

Phasin Originally the lectin from the bean Phaseolus vulgaris, now used for non-toxic plant lectins in general.

Pheasant Game bird, Phasianus colchicus and related spp. Total weight 1.5 kg; traditionally sold as a brace, i.e. cock and hen, although now commonly available as single birds; usually hung for 3 days to develop flavour.

Phenol oxidases Enzymes that oxidize phenolic compounds to quinones. For example, monophenol oxidase in mushrooms, polyphenol oxidases in potato and apple are responsible for the development of the brown colour when the cut surface is exposed to air; tyrosinase in plants and animals forms the brown and black pigment melanin.

Phenols Aromatic compounds present in wood smoke that are mainly responsible for reducing bacterial spoilage in smoked fish. Measurement of phenols can serve as a rough check on a smoking process.

Phenylalanine An essential amino acid; in addition to its role in protein synthesis, it is the metabolic precursor of the non-essential amino acid tyrosine. Tyrosine in the diet spares phenylalanine, so reducing the requirement.

Phenylketonuria A genetic disease affecting the metabolism of phenylalanine. Phenylalanine is normally metabolized to tyrosine, catalysed by phenylalanine hydroxylase. Impairment of this reaction leads to a considerable accumulation of phenylalanine in plasma and tissues and metabolism to phenylpyruvate, phen-yllactate, and phenylacetate, collectively known as phenylketones, which are excreted in the urine.

The very high plasma concentration of phenylalanine causes disruption of brain development, and if untreated there is severe mental retardation. All infants are screened for phenylketonuria shortly after birth; treatment is by very strict limitation of phenylalanine intake, only providing sufficient to meet requirements for protein synthesis. Once brain development is complete dietary restriction can be relaxed to a considerable extent, since high concentrations of phenylalanine seem to have little adverse effect on the developed brain. There may, however, be benefits from continuing dietary restriction into adult life, and phenylketonuriac women require extremely careful dietary control through pregnancy to avoid damage to the foetus's developing brain.

Phitosite High-calorie food.

Phonological awareness The ability to hear and use sounds effectively.

Phosphatase test A test for the adequacy of pasteurization of milk. The enzyme phosphatase, normally present in milk, is denatured at a temperature slightly greater than that required to destroy the tubercle bacillus and other pathogens; therefore the presence of detectable phosphatase activity indicates inadequate pasteurization. The test can detect 0.2% raw milk in pasteurized milk.

Phosphate Is also essential for plant growth, hence the use of inorganic phosphate or bone-meal as fertilizer. Bone-meal, is often used as a supplement in human foods but as a source of calcium rather than of phosphate.

Phosphates Salts of phosphoric acid; the form in which the element phosphorus is normally present in foods and body tissues.

Phosphatidic acid Glycerol esterified to two molecules of fatty acid, with the third hydroxyl group esterified to phosphate; chemically diacylglycerol phosphate; intermediates in the metabolism of phospholipids.

Phosphatidylethanolamine A phospholipid containing ethanolamine.

Phospholipids A class of lipids. They are esters of glycerol, as are triglyceride lipids, but in phospholipids only two of the glycerol hydroxy groups are esterified with fatty acids, the third by a compound containing a phosphate group. They are incorporated into cell walls and membranes of structures within the cell. The phospholipid content of fish muscle, lean and fatty fish, is typically in the range 0.5-1.0%. The remainder of the fat content is triglyceride lipid.

Phosphoproteins Proteins containing phosphate, other than as nucleic acids or phospholipids, e.g. casein from milk, ovovitellin from egg yolk.

Phosphoric acid May be one of three types: orthophosphoric acid, metaphosphoric acid, or pyrophosphoric acid.

Phosphorus An essential element, occurring in tissues and foods as phosphate, phospholipids, and phosphoproteins. In the body most (80%) is present in the skeleton and teeth as calcium phosphate; the remainder is in the phospholipids of cell membranes, in nucleic acids, and in a variety of metabolic intermediates, including ATP. The parathyroid glands control the concentration of phosphate in the blood, mainly by modifying its excretion in the urine.

Human dietary needs are always met; a deficiency never occurs in man. The calcium to phosphate ratio of infant foods is, however, important. Phosphate deficiency is common in livestock and gives rise to osteomalacia.

Photosynthesis The synthesis of carbohydrates from carbon dioxide and water by plants in sunlight, with the release of oxygen.

Phrynoderma Blocked pores or 'toad-skin' often encountered in malnourished people.

Originally thought to be due to vitamin A deficiency but possibly due to other deficiencies, and also occurs mildly in adequately nourished people.

Phulouri Fritters made from dried split peas.

Phycotoxins Marine biotoxins that accumulate in fish and shellfish from their diet, as distinct from toxins naturally present.

Phyllo pastry Plain paper-thin pastry made from flour and water, rolled into small balls then tossed in the air and stretched until it forms an extremely thin sheet. Multiple layers are used as the basis for Greek and Middle-Eastern pastry dishes.

Phylloxera An aphid which threatened to destroy the vineyards of Europe in the middle of the nineteenth century. They were saved by grafting susceptible varieties on to resistant American vine rootstock.

Physalin A carotenoid pigment in the fruits of the Cape gooseberry, Physalis spp.

Physical activity level Total energy cost of physical activity throughout the day, expressed as a ratio of basal metabolic rate. Calculated from the physical activity ratio for each activity, multiplied by the time spent in that activity. A desirable PAL for health is considered to be 1.7; the average in the UK is 1.4.

Physical activity ratio Energy cost of physical activity expressed as a ratio of basal metabolic rate.

Physin Name given to a growth factor in liver, needed by rats; found to be vitamin B 12.

Physical therapy Therapy focusing on the gross motor development of an individual, such as crawling, walking, sifting, rolling over. The therapist will help your child overcome physical difficulties related to low muscle tone or weak muscles.

Phytase An enzyme that hydrolyses phytate to inositol and phosphate. Present in yeast, liver, blood, malt, and seeds. If a high level of yeast is used in baking with high extraction flours, some of the phytate is broken down.

Phytate Inositol hexaphosphate, present in cereals, particularly in the bran, in dried legumes and some nuts as both water-soluble salts and insoluble salts of calcium and magnesium. Magnesium calcium phytate is phytin.

It can bind calcium, iron, and zinc to form insoluble complexes; it is not clear how far phytate reduces the availability of these minerals in the diet, especially since there are phytase enzymes in yeast and legumes which may liberate these minerals.

Phytin Magnesium calcium phytate, approximately 12% calcium, 1.5% magnesium, and 22% phosphorus.

Phytoalexins Substances, often harmful to human beings, which increase in plant tissues when they are stressed, as by physical damage, exposure to ultraviolet light, etc.

Phytochemicals Substances plants naturally produce to protect themselves against viruses, bacteria, and fungi. Phytochemicals help protect humans

Phyto-oestrogens Compounds with oestrogen activity found in a variety of plants.

Phytoplankton Minute plants floating in the sea which serve as the basic food for all marine life, since they photosynthesize. The basis of the marine food-chain.

Phytosterol General name given to sterols occurring in plants, the chief of which is sitosterol.

Pica An unnatural desire for foods; alternative words are cissa, cittosis, and allotriophagy. Also a perverted appetite.

Piccalilli Mixture of chopped, brine-preserved vegetables in mustard sauce.

Piccata Small thin slices of veal.

Pickles, Dill Pickles fermented in a mixture of brine, dill weed, mixed spices, and vinegar.

Pickling Also called brining. Vegetables immersed in 5-10% salt solution undergo lactic acid fermentation, while the salt prevents the growth of undesirable organisms. The sugars in the vegetables are converted to lactic acid; at 25°C the process takes a few weeks, finishing at 1% acidity.

Picklizes Pickled mixed vegetables; piccalilli.

Picnic shoulder American name for hand of pork.

Pico As a prefix for units of measurement, one million-millionth part; symbol p.

Pidan Chinese eggs.

Pie Food cooked in a dish and covered with pastry; may be sweet or savoury. Also savoury dishes with a crust of mashed potato.

Piernik Spiced honey cake.

Pierogi Envelopes of dough stuffed with minced meat, cheese, vegetables, or fruit; similar to ravioli.

Pigeon Columbia livia; young about 4 weeks old is a squab.

Pilau Dish of rice cooked in stock, with vegetables, meat, or fish added.

Pilchard Oily fish, Sardina pilchardus; young is the sardine.

Pils Pale type of lager originally made in Czechoslovakia.

Pimentón Powdered, dried red pepper. Pimentón picante is cayenne; pimentón dulce is paprika.

Pimms Trade name; a ready-mixed cocktail based on spirits and flavoured with herbs and liqueurs, normally served as a long drink with ice and soda or lemonade, garnished with fruit, cucumber, or mint. Originally there were four varieties: No. I based on gin, No. 2 on whisky, No. 3 on brandy, No. 4 on rum; only No. 1, now based on vodka, survives.

Piña colada A long drink made from rum, pineapple juice, and coconut milk; also the trade name for a sweet liqueur made from rum, pineapple, and coconut.

Pine nuts Edible seeds of various species of pine cone, especially Mediterranean stone pine, Pinus pinea.

Pineapple Fruit of the tropical plant Ananas sativus, one of the bromeliad family. The fruit contains the proteolytic enzyme bromelain, which has been used to tenderize meat.

Pink Pink discoloration of salted fish, and of salt and brines, caused by halophilic bacteria. Eventually, salted fish attacked by these bacteria go brownish, soften, and develop a foul odour.

Pinot Three varieties of grape used for wine making; pinot noir is one of the 9 'classic' varieties, used especially in champagne; pinot blanc and pinot gris are also widely used.

Pinotage A South African grape variety used for wine making, the result of a cross between the cinsaut and pinot noir varieties.

Pinto bean Phaseolus acutifolius, also known as frijole, Mexican haricot, or tepary bean. Able to grow during drought.

Piononos Caribbean; stuffed plantains.

Pipe Cask for wine; the volume varies with the type of wine, e.g. port, 115 gallons (517 L); Tenerife, 100 gal (450 L); Marsala, 90 gal (418 L).

Piping To force a smooth mixture through a narrow nozzle to form fancy shapes to decorate a dish.

Pipis Edible mollusc, Plebidonas deltoides, widely distributed around Australian coastline.

Piri-piri Seasoning Portuguese; crushed chillies, citrus peel, onion, garlic, pepper, salt, lemcn juice, bay leaves, paprika, pimiento, basil, oregano, and tarragon.

Piri-piri Small red chillies, extremely pungent.

Pirozhki small baked pasties of yeast dough filled with chopped fish, meat, etc.

Pisco South American; brandy made from muscat grapes. A pisco sour is a mixture of pisco with 2 parts lime juice, sweetened to taste.

Pis-en-lit French name for dandelion leaves, eaten raw or cooked. So called because of the diuretic effect of dandelion.

Pissaladière French; savoury tart similar to pizza.

Pistachio Fruit of Pistacchio vera; yellow-green coloured nut. May be roasted and salted or used as flavouring for ice cream and hot, sweet, milk beverage. A source of protein, vitamins B, E, and niacin.

Pisto manchego Spanish; vegetable hash.

Pita Middle-Eastern; unleavened flat bread, baked as an oval or circle, which can be opened up as an envelope.

Pitanga Surinam cherry, Eugenia uniflora or E. michelii; small, round fruit, deeply ribbed, cherry-like with a single stone.

Pitch Lake pudding Caribbean; dark rum and chocolate mousse.

Pits Stones from cherries, plums, peaches, apricots. Oil extracted from these pits is used in cosmetics, pharmaceuticals, for canning sardines, and as table oil. The press cake left behind contains amygdalin.

Pitting Removing the stones from cherries, olives, etc.

Pitz Swiss raw tomatoes stuffed with minced apple and celeriac.

Pizza Savoury tart on a base of yeast dough, traditionally cooked in a wood-burning oven. The topping varies with the region and may contain tomatoes, cheese, salami, or seafood.

Plaice Flat fish, Pleuronectus platessa.

Plaki Fish baked or braised with vegetables.

Plancha, A la Spanish; cooking 'on the plate'. The plancha is a thick iron or aluminium plate, traditionally incorporated into the stove, an oiled hot plate or griddle.

Planking Cooking meat or fish on a small board of oiled hardwood.

Plankton Minute organisms, both plant and animal, drifting in the sea, which serve as the basic foodstuffs of marine life; the basis of the marine food chain.

Plansifter A nest of sieves mounted together so that material being sieved is divided into a number of fractions of different size. Widely used in flour milling.

Plantain Adam's fig; variety of banana with higher starch and lower sugar content than dessert bananas, picked when the flesh is too hard to be eaten raw, and therefore cooked. Some varieties become sweet if left to ripen, others never develop a high sugar content.

Plaque, Dental The growth of bacteria that adhere tightly to teeth, leading to the development of caries.

Plasmid A circular piece of DNA found outside the chromosome in bacteria. Plasmids are the principle tool for inserting new genetic information into microbes or plants.

Plate count To estimate the number of bacteria in a sample, it is poured on to an agar plate, when each bacterial cell multiplies to produce a colony of bacteria which is visible to the naked eye. A count of the number of colonies gives the number of bacteria in the portion of the sample that was taken. Pasteurized milk contains about 100,000 bacteria/mL; good quality raw milk contains less than 500,000/mL.

Platyhelminths Members of the zoological phylum Platyhelminthes, a group of soft-bodied, usually much flattened invertebrates. A trivial name is flatworms referring to their shape. The phylum includes the trematodes and cestodes, parasites of man and animals transmitted by fish or crustacean shellfish.

Pliofilm Trade name for varieties of rubber hydrochloride, the first transparent wrapping paper that could be heat-sealed.

Plonk Colloquial term for cheap, indifferent wine; originated in Australia.

Ploughman's lunch Basically cheese, bread, and pickle, now an English pub lunch; cheese with salad and crusty bread.

Pluck Butchers' term for heart, liver, and lungs of an animal.

Plum Fruit of numerous species of Prunus. Common European plums are.

Plum sauce Chutney-like sauce made from plums, apricots, chilli, vinegar, and sugar.

Plumcote American fruit, a cross between plum and apricot.

Pneumatic conveying Transfer of material in powder form by means of air currents. Applied to flour, sugar, etc.

Pneumatic dryers The material is dried almost instantaneously in a turbulent stream of hot air, which also acts as a conveyor system. Applicable to powdered, granular, and flaky materials: used for starch, mashed potatoes, cereals, flour, and powdered soups.

Poach To cook for a short time in a shallow layer of liquid kept at a temperature just below the boiling point.

Polenta Traditional Italian porridge made from maize meal, often with cheese added. May be further cooked by baking or frying. Also the Italian word for coarsely ground maize meal, as hominy grits is the American term.

Polonaise, a la Dishes made with soured cream, beetroot, and red cabbage.

Polony smoked pork and veal sausage, ready to slice and eat; also known as bologna.

Polycyclic aromatic hydrocarbons A complex group of aromatic hydrocarbons containing 2 to 6 benzene rings that may have aliphatic and/or alicyclic hydrocarbon substituents. The parent compounds range from naphthalene to anthanthrene. They are found in many foods as environmental contaminants at low concentrations, including fish and shellfish, for which selected total 3 to 5 ring compounds in the edible parts are generally within the range from 0.1 to 50 mg/kg from non-industrialised areas. Lipid rich tissues may have concentrations up to 10 times that in lean muscle. PAHs are also found in fisheries by-products such as fish meal, fish oil and supplements derived from fish oils.

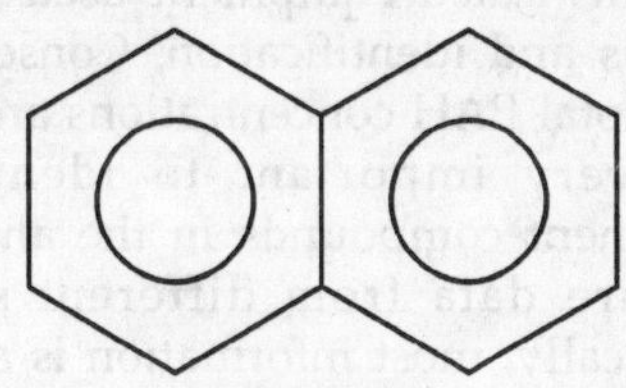

Naphthalene
$C_{10}H_8$

Fig. Polycyclic aromatic hydrocarbons

In fish and shellfish exposed to environmental sources of PAHs such as industrially polluted sediments or spilled petroleum and petroleum products, the concentrations may be enhanced considerably by up to 1000 fold, particularly if exposure is acute. Once the source is removed the contaminants are slowly lost by the living animal. In crustaceans and bivalve molluscs the bioconcentration factors with respect to water for the higher ring PAHs are generally in the range 3000 6000, compared with fish which are generally less than 500. PAHs are produced during pyrolysis, and so methods of cooking and preparation of food such as barbecuing, grilling, broiling and frying tend to increase the concentration of PAHs. PAHs are invariably found in smoked fish products at concentrations up to about 10 times or more of those in unsmoked products depending on the degree of exposure to smoke. They are deposited from the smoke in which more than 200 PAHs have been identified at concentrations and combinations that depend on the type of wood and the temperature of smouldering. Some of the PAHs are human carcinogens, co-carcinogens, teratogens and mutagens, and so selected ompounds are subject to surveillance in foods as part of food safety and environmental programmes. About 16 two to five ring PAHs have been identified as priority environmental pollutants. Most of these have been assessed in a range of smoked fisheries products, but the PAHs selected depend on the availability and type of sophisticated equipment used for trace analysis and identification. Consequently, when total PAH concentrations are quoted it is very important to identify the component compounds in the analysis to compare data from different sources. Historically, most information is available on the presence of benzo[a]pyrene in fisheries products which, together with dibenzo[a,h]anthracene, represents the most potent of the probable human carcinogens. Sometimes the total carcinogenic PAHs measured are expressed as benzo[a]pyrene equivalents. The highest concentrations of PAHs can be expected in heavily smoked traditional products; up to 10 times more benzo[a]pyrene may be deposited during hot smoking compared with cold smoking and up to 10 times more benzo[a]pyrene may be deposited on the product in a traditional smokehouse compared with a modern kiln in which the smoke is recirculated. The use of liquid smoke products in which the PAHs have been removed or reduced is effective in reducing the PAH concentrations of smoke flavoured foods.

Polydextrose, modified Glucose polymer prepared by heating glucose and sorbitol with citric acid. It is more resistant to enzymic digestion than normal polysaccharides and 60% is excreted in faeces undigested, so providing only about 1 kcal (4 kJ)/g; hence, termed 'nons-weetening sucrose replacement', or bulking agent.

Polygalacturonase Enzyme that digests part of the pectin that forms plant cell walls. Polygalacturonase causes ripening fruit to become soft.

Polymerase chain reaction Multiplying a particular DNA segment in repeated cycles. The "copies" made in a previous cycle are used as "originals" or templates in the next cycle. For example, PCR enables forensics experts to do DNA testing on very small blood samples.

Polymorphism The ability to crystallize in two or more different forms. For example, depending on the conditions under which it is solidified, the fat tristearin can form three kinds of crystals, each of which has a different melting point, namely, 54, 65, and 71°C.

Polymyxins Antibiotics isolated from Bacillus polymyxin; polymyxin A is aerosporin. They are polypeptides, active against coliform bacteria; apart from clinical use, they are of value in controlling infection in brewing.

Polyphagia Excessive or continuous eating.

Polyphosphates Polymerised phosphates are prepared by heating orthophosphates. The degree of polymerisation varies with production conditions, but the mixture often used in fish processing is predominately tripolyphosphate. Generally approved as a food additive and is used as a dip for fillets to reduce dripin prepacked fish and from thawed fish. Also added to surimi. Polyunsaturated fatty acids The lipids of marine organisms are characterised by the presence of polyenoic fatty acids containing

16 to 22 carbon atoms with 2 to 6 conjugated double bonds which originate from unicellular marine phytoplankton. Of particular interest, because of their essential benefits to human nutrition and health, are the long chain polyunsaturated fatty acids, eicosapentaenoic acid (EPA), which has 20 carbon atoms in the chain and 5 double bonds, and docosahexaenoic acid (22:6) or (DHA). The double bonds have the cis configuration which confers greater instability and lower melting point than the trans configuration. Thus, the correct chemical notation is cis-5,8,11,14,17-eicosapentaenoic acid and cis-4,7,10,13,16,19-docosahexaenoic acid. These acids are usually the major PUFA constituents of fish lipids and fish oils, and belong to the n-3 series in which the first double bond is at the third carbon atom along the chain from the terminal methyl group. They are considered to be important inter alia in relation to cardiovascular disease, inflammatory diseases and cancer, and DHA is important for normal post-natal development of cerebral and visual tissues. There is a wide range of fish oil supplements available for retail sale containing approximately 12% DHA and 18% EPA.

Polysaccharides Complex carbohydrates formed by the condensation of large numbers of monosaccharide units, e.g. starch, glycogen, cellulose, dextrins, inulin. On hydrolysis the simple sugar is liberated.

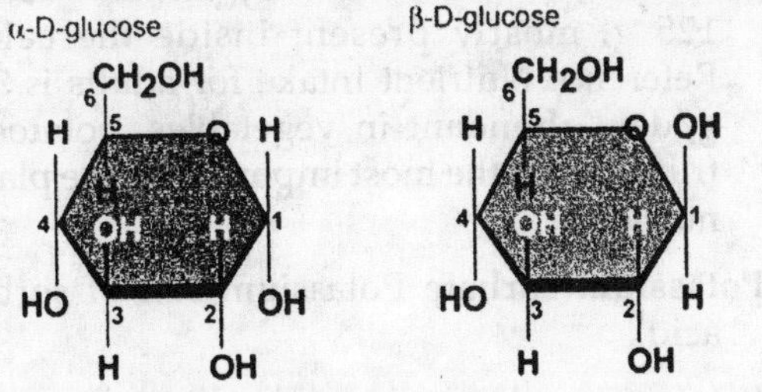

Fig. Polysaccharides

Polysaccharose Alternative name for polysaccharides.

Polyunsaturates Commonly used short term for polyunsaturated fatty acids.

Pomace Residue of fruit pulp after expressing juice: also applied to fish from which oil has been expressed.

Pombe African beer prepared from millet; also known as kaffir beer. Commonly associated with iron overload, since it is frequently brewed in iron drums, and alcohol greatly increases the absorption of iron.

Pomegranate The fruit of the subtropical tree Punica granatum. The juice is contained in a pulpy sac surrounding each of a mass of seeds; the outer skin contains tannin and is therefore bitter. The sweet juice is used to prepare grenadine syrup for alcoholic and fruit drinks.

Pomelo Fruit of Citrus grandis, from which the grapefruit is descended; also called shaddock, after Captain Shaddock, who introduced it into Barbados in the sixteenth century.

Pomes Botanical name for fruits such as apple or pear, formed by the enlargement of the receptacle which becomes fleshy and surrounds the carpels.

Ponderal index An index of fatness, used as a measure of obesity: the cube root of body weight divided by height. Confusingly, the ponderal index is higher for thin people, and lower for fat people.

Ponderocrescive Foods tending to increase weight: easily gaining weight; the opposite of ponderoperditive.

Ponderoperditive Stimulating weight loss.

Pone bread Colloquial name for corn bread in the southern states of the USA.

Pontefract cake A round, flat sweetmeat made from liquorice, originally in Pontefract in England; also called pomfret.

Poonac The residue of coconut after the extraction of the oil.

Poor man's goose Casserole dish of liver and potatoes.

Popcorn Variety of maize that expands on heating, also the name of the fluffy white mass so formed.

Pope's eye The small circle of fat in the centre of a leg of pork or mutton.

Popover Individual batter pudding; small Yorkshire pudding.

Poppadom Thin roasted or fried crisps made from lentil flour; may be spiced.

Poppy seed Seeds of the opium poppy, Papaver somniferum, used mixed with honey in cakes, and as a flavouring on the crust of bread and rolls. Also called maw seed.

Popsicle American name for ice lolly.

Porcine somatotropin Pig growth hormone; a protein hormone produced in the pituitary gland of pigs. Like BGH/BST, PST can also be produced by inserting a copy of the gene for PST into laboratory bacteria. When purified from pituitary glands or from bacterial cultures, PST injected into hogs can increase the leanness of pork and improve pork quality.

Pork Meat from the pig, Suidae spp., eaten fresh, as opposed to bacon and ham which are cured. By far the richest of all meat sources of vitamin B 1, a 150 source of protein, niacin, vitamin B 12, copper, and selenium; a good source of vitamins B 2 and B 6, zinc, and iron.

Porphyra Red seaweed used to make laverbread.

Porridge Oatmeal cooked in water or milk as a breakfast dish; originally Scottish. Also similar thick soups made with other cereals.

Port Fortified wines from the upper Douro valley of north-east Portugal. Mostly aged in wood and bottled when ready for drinking; vintage port is aged in wood for 2 years, then in the bottle for at least 10; late bottled vintage is aged less than 6 years. Crusted port is blended from quality vintages bottled young, and develops a sediment (crust) in the bottle. Ruby port is young; old tawny is aged for 10 or more years; fine old tawny is a blend of young and old wines. Tawny port is aged in wood, vintage in the bottle. White port is made from white grapes; generally served chilled as an aperitif.

Porterhouse American term originally for large beef steak cut from the rear end of sirloin and including some fillet: served with beer in porterhouses. Now a thick slice cut from the wing-rib of beef.

Portuguese, a la Dishes made with tomato, onion, and garlic.

Poskitt index Index of fatness in children; per cent of expected weight for age.

Posset Drink made from hot milk curdled with ale or wine, sometimes thickened with breadcrumbs and spiced. Formerly used as a remedy for colds; popular in the later Middle Ages.

Postum, instant Trade name for a preparation of bran, wheat, and molasses consumed as a beverage.

Pot herbs Herbs for culinary use. Originally vegetables such as cabbage whose leaves were boiled for food.

Pot roasting A method of cooking joints of meat in a saucepan or casserole with fat and a small amount of liquid.

Potage Thick soup.

Potassium An essential mineral, widespread in nature; the human body contains about 125 g, mostly present inside the cells. Reference Nutrient Intake for adults is 3.5 g/day; abundant in vegetables, potatoes, fruit. One of the most important of the plant nutrients.

Potassium sorbate Potassium salt of sorbic acid.

Potato crisps Flavoured, thin slices of potato, deep-fried and eaten cold, sometimes as an accompaniment to meals, more commonly as a snack. Called chips in the USA.

Potato flour Dried potato tuber.

Potato starch Also called farina. Prepared from potato tuber and widely used as a stabilizing agent when gelatinized by heat.

Potato, Irish The 'ordinary' potato, tuber of Solanum tuberosum.

Potato, sweet Tubers of the herbaceous climbing plant Ipomoea batatas, known in Britain before the Irish potato. The flesh may be white, yellow, or pink; the leaves are also edible.

Pot-au-feu large deep earthenware casserole or marmite; also a traditional dish of meat and vegetables.

Poteen Irish name for illicit home-distilled spirit.

Pottage Thick, well-seasoned meat or vegetable soup, usually containing barley or other cereal or a pulse.

Pottle Traditional English wine measure; ½ gallon.

Poularde A neutered hen bird.

Poulette, a la Fricassee of cooked meat or poultry in rich white sauce, garnished with onions or garlic.

Poultry General term for farmyard birds kept for eggs and/or meat; chicken, duck, goose, guinea fowl, pigeon, and turkey.

PPT The Planning and Placement Team meeting is where the IEP for your child is determined and agreed upon.

Prairie oyster Traditional cure for a hangover; a raw egg with Worcestershire sauce and brandy; the egg is swirled with the liquid but the yolk remains intact.

Praline 1. Confection of nuts and partially caramelized sugar, often used as a centre for chocolates.

2. In France, a sugar-coated almond.

Prawns Shellfish of various tribes of suborder Macrura. Large fish of species of Palaemonidae, Penaeidae, and Pandalidae are prawns; smaller ones are shrimps. A rich source of protein, niacin, and selenium; a good source of calcium. The deep-water prawn is Pandalus borealis; common pink shrimp is Pandalus montagui; brown shrimp is Crangon spp.

PRE Protein Retention Efficiency, a measure of protein quality.

Prebiotics Non-digestible oligosaccharides that support the growth of colonies of certain bacteria in the colon. They include derivatives of fructose and galactose, and lead to the growth of bifidobacteria, so changing and possibly improving the colonic flora. Probiotics and prebiotics are sometimes termed synbiotics. They are considered to play a role as functional foods.

Precursors, enzyme Inactive forms of enzymes, activated after secretion; also called pro-enzymes or zymogens.

Predictive microbiology A branch of microbiology that studies the use mathematical modelling techniques to predict the growth of microorganisms in foods or in other substrates.

Pregnancy, nutritional needs Pregnant women have slightly increased energy and protein requirements compared with their needs before pregnancy, although there are metabolic adaptations in early pregnancy which result in laying down increased reserves for the great stress of the last trimester, and high requirements for iron and calcium. These increased needs are reflected in the increased reference intakes for pregnancy.

Premier jus Best-quality suet prepared from oxen and sheep kidneys. The fat is chilled, shredded, and heated at moderate temperature. When pressed, premier jus separates into a liquid fraction and a solid fraction.

Pre-sale Mutton or lamb from sheep raised on salt marshes; prized for its flavour.

Preservation Any process used to slow or stop the progress of spoilage. It allows easier distribution and transport and the food can be stored for longer before use.

Preservative Additives that extend the storage lives of foods either by inhibiting growth of microorganisms and hence delaying spoilage, or by inhibiting chemical reactions that produce unwanted sensory changes in the product. An example of the latter in the context of fish processing is the use of sulphites to inhibit development of black spot in chill stored crustaceans. Preservatives are not used, or permitted in many countries, for preservation of chill- or frozen- stored raw fish or fish products, but are permitted for use in some semi-preserves such as marinades and in dried fish.

Pressure cooking By cooking under pressure, it is possible to achieve a higher temperature, and therefore boil or steam foods faster than in an open pan.

Pretzels Hard, brittle biscuits in the shape of a knot, made from flour, water, shortening, yeast, and salt. Also called bretzels.

Prevention Enterobacteria are heat-sensitive and will be killed by thorough heating. Raw or undercooked foods and cross-contamination, when cooked material comes into contact with raw produce or contaminated materials, are the main causes of infection. Proper

Prickly pear Fruit of the cactus Opuntia spp., also called Indian fig, barberry fig, and tuna; an important part of the diet in certain areas of Mexico.

Primary production Plants and animals produced by the agricultural and fisheries sector.

Primeurs early forced vegetables and fruits. Also used for young, early wine.

Princesse, a la Dish with a garnish of asparagus tips and truffles or noisette potatoes.

Printanière, a la Spring-style dish containing, or garnished with, small, young vegetables.

Prion The infective agent(s) believed to be responsible for KreutzfeldJacob disease, kuru and possibly other degenerative diseases of the brain in human beings, scrapie in sheep, and bovine spongiform encephalopathy (BSE). They are simple proteins, and unlike viruses do not contain any nucleic acid. Transmission occurs by ingestion of infected tissue.

Probiotics Preparations of live micro-organisms added to food, claimed to be beneficial to health by restoring microbial balance in the intestine. The organisms commonly involved are lactobacilli, bifidobacteria, streptococci, and some yeasts and moulds, alone or as mixtures.

Procea Trade name of a white bread with slightly increased protein content.

Processed food A food that has been altered from its natural state. Examples of processing include cooking, canning, milling, and the addition of chemicals, such as preservatives.

Processing Treating a food in such a way as to change its nature and properties in order to preserve it, to improve its eating quality or to make useful ingredients.

Processing specification One of the documents that contributes to the quality assurance programme of a processing plant or company. It is the formalised description of necessary procedures for the manufacture of a product of the required quality. It can be used of itself or in conjunction with an end-product specification of a product. A processing specification can describe those processing steps that are important for the quality of the end-product when it is not easy to specify the necessary quality attributes, or to monitor them, or measure them in the end-product. An example would be in the production of a smoked product. The conditions for brining and smoking would be specified -amount of salt or brine strength, brining time, time and temperature of smoking, smoke density, nature of wood used as source of smoke - to achieve the right flavour and texture in the end-product.

Pro-enzymes Inactive precursors of enzymes, activated after secretion; also called zymogens.

Profiterole Originally a light cake baked on hot ashes and filled with cream; now a small case of choux pastry, baked, filled with cream, and served with chocolate sauce. Also small rounds of choux pastry used as a garnish for clear soups or consommes.

Progoitrins Substances found in plant foods which are precursors of goitrogens.

Proline A non-essential amino acid.

Promega Trade name for mixture of long-chain marine fatty acids: eicosapentaenoic and docosohexaenoic acids.

Pronutro Protein-rich baby food developed in South Africa; made from maize, skim-milk powder, groundnut flour, soya flour, and fish protein concentrate with added vitamins.

Proof spirit An old method of describing the alcohol content of spirits; originally defined as a solution of alcohol of such strength that it will ignite when mixed with gunpowder. Proof spirit contains 57.07% alcohol by volume or 49.24% by weight in the UK. In the USA it contains 50% alcohol by volume. Pure alcohol is 175.25°proof UK or 200°proof USA.

Spirits were described as under or over proof; a drink 30° over proof means that 100 volumes contains as much alcohol as 130 volumes of proof spirit; 30° under proof means that 100 volumes contains as much alcohol as 70 volumes of proof spirit.

Nowadays alcohol content is usually measured as % alcohol by volume.

Pro-oxidant A substance that promotes oxidation of lipids. Salt is an example in fish technology.

Propionates Salts of propionic acid, CH 3 CH 2 COOH, a normal metabolic intermediate. The free acid and salts are used as mould inhibitors, e.g. on cheese surfaces, and to inhibit rope in bread and baked goods (E-280-283).

Proprioceptive Proprioceptive information is that which arises for one's own movements, the sensations generated from muscles and joints.

Propyl gallate An antioxidant, E-310.

Prosparol Trade name for an emulsion containing 50% vegetable fat, 405 kcal (1.7 MJ)/100 g; used as a concentrated source of energy.

Prosthetic group Non-protein part of an enzyme molecule; either a coenzyme or a metal ion. Essential for catalytic activity. The enzyme protein without its prosthetic group is the apo-enzyme and is catalytically inactive. With the prosthetic group, it is known as the holo-enzyme.

Protein A molecule composed of amino acids linked together in a particular order specified by a gene's DNA sequence. Proteins perform a wide variety of functions in the cell; these include serving as enzymes, structural components, or signaling molecules.

Protein content It is not easy to measure the true protein content of fish products. There are a large number of proteins in fish tissues, and it is not possible to have a reasonably simple procedure to determine them as a group of constituents. What is normally given as the protein content of fish products - and of foods generally - is the 'crude protein' content. This is based on the fact that all proteins contain nitrogen and measurement of total nitrogen is an estimate of the amount of protein in a sample. The nitrogen content is commonly measured by the Kjehldahl method. The crude protein content is not a very accurate measure of the true protein content of fish products for two principle reasons. One is that not all of the nitrogen-containing compounds in the product are proteins; about 85% is true protein. However, much of the non-protein material consists of amino acids, which are constituents of proteins and, more-or-less, nutritionally equivalent Thus, this error in the context of nutritional labelling is not all that important. The other reason is that the nitrogen content is converted to protein content by dividing it by the nitrogen content of protein, The nitrogen content of a protein is not the same for all proteins and a crude protein content is conventionally obtained by multiplying the total nitrogen content by 6.25. This factor is higher than the true value for muscle proteins as a group, and protein content estimated in this way overestimates slightly the true protein content.

Protein score A measure of protein quality based on chemical analysis.

Protein solubility The amount of protein in a sample of fish muscle that can be brought into solution under specified conditions of solvent and procedure. It is used especially as an index of damage to fish muscle structure and proteins incurred by frozen storage. Its value decreases with frozen storage at rates that depend on storage temperature; the lower the temperature, the slower the rate. Some 20-25% of the proteins in fish muscle can be solubilised by simple extraction of minced fish muscle with water.

A further 65-70% can be solubilised by gently stirring or blending minced fish muscle in 0.5% salt solution at slightly alkaline pH, about 7.5. The solubilised fraction is separated from the insoluble fraction, usually by centrifugation, the protein content of each is measured, and the protein in the soluble fraction expressed as a percentage of the total.

Protein-energy malnutrition A spectrum of disorders, especially in children, due to inadequate feeding. Marasmus is severe wasting and can occur in adults; it is the result of a food intake inadequate to meet energy expenditure. Kwashiorkor affects only young children and includes severe oedema, fatty infiltration of the liver, and a sooty dermatitis; it is likely that deficiency of antioxidant nutrients and the stress of infection may be involved. The name kwashiorkor is derived from the Ga language of Ghana to describe the illness of the first child when it is weaned on the arrival of the second child.

Emaciation, similar to that seen in marasmus, occurs in patients with advanced cancer and AIDS; in this case it is known as cachexia.

Protein-energy ratio The protein content of a food or diet expressed as the proportion of the total energy provided by protein. The average requirement for protein is about 7% of total energy intake; average Western diets provide about 14%.

Proteins, conjugated Proteins that include a non-protein prosthetic group, e.g. haemoglobin and cytochromes contain haem; many oxidative enzymes contain a prosthetic group derived from vitamin B 2; many proteins are conjugated with carbohydrates or fatty acids.

Proteolysis The breaking down of proteins to their consitituent amino acids by enzymes. Proteins can be roken down to amino acids by the action of acids and alkalis, but this process is better referred to as hydrolysis. Organs in the gut of fish, for example, the pyloric caeca, excrete proteolytic enzymes, which assist in the digestion of food, and muscle tissue contains proteases called cathepsins.

Proteoses Partial degradation products of proteins; soluble in water. The stages of breakdown are protein ? proteoses ? peptones ? polypeptides ? oligopeptides ? amino acids.

Prothrombin Protein in plasma involved in coagulation of blood.

Prothrombin time An index of the coagulability of blood based on the time taken for a citrated sample of blood to clot when calcium ions and thromboplastin are added.

Protopectinase The enzyme in the pith of citrus fruits which converts protopectin into pectin with the resultant separation of the plant cells from one another.

Provencale, a la Containing olive oil, garlic, and often tomato. Normally flavoured with mixed provencale herbs.

Proving The stage in bread making when the dough is left to rise.

Provitamin A substance that is converted into a vitamin, such as 7-dehydrocholesterol, which is converted into vitamin D, or those carotenes that can be converted to vitamin A.

Proximate analysis Analysis of foods and feedingstuffs for nitrogen, ether extract, crude fibre and. ash, together with soluble carbohydrate calculated by subtracting these values from the total. Also known as Weende analysis, after the Weende Experimental Station in Germany, which in 1865 outlined the methods of analysis to be used.

Proximate composition The main constituents of a product, and their amounts in unit mass or volume. In the case of products without added components, proximate composition typically comprises water, protein, fat, and ash, expressed as percentage of the weight. These components usually sum to over 100%, typically around 101-102%, in fish muscle tissue. This is mainly due to the way protein content is measured, which slightly overestimates the true protein content. Packaged products for retail sale will list the proximate composition, usually not including ash, on the package. Packaged products with added ingredients will include the carbohydrate content in the list as appropriate. Carbohydrate content is usually calculated by difference, that is, by subtracting the sum of water, protein, fat and ash, as percentages, from 100.

Prunelle French name for sloe; also a liqueur made from sloes, similar to sloe gin.

Prunus Genus of plants including plums, peaches, nectarines, cherries, and almonds.

Pruteen Trade name for microbial protein produced by growing bacteria, Methylophilus methylotrophus, on methanol; 70% protein on dry weight.

P/S ratio The ratio between polyunsaturated and saturated fatty acids. In Western diets the ratio is about 0.6; it is suggested that increasing it to near 1.0 would reduce the risk of atherosclerosis and coronary heart disease.

Psychrophiles, Psychrophilic or **Psychrotrophic bacteria** Bacteria with optimum growth temperatures in the region 10-15°C and capable of growth down to 0°C. The bacteria growing on fish spoiling in ice are predominately psychrophilic.

Psyllium Plantago psyllium, also known as plantago or flea seed. Small, dark reddish-brown seeds which form a mucilaginous mass with water, taken medicinally to assist the passage of intestinal contents.

Ptomaines Loosely used name for amines formed by decarboxylation of amino acids during putrefaction of proteins: putrescine from arginine, cadaverine from lysine, muscarine in mushrooms, neurine formed by dehydration of choline. They have an unpleasant smell and were formerly thought to cause food poisoning.

Ptyalin Obsolete name for salivary amylase.

Pudding A baked or steamed sponge or suet dish, usually sweet and served as a dessert, but also savoury suet puddings. Also milk puddings, made by baking rice, semolina, or sago in milk.

Puff pastry Prepared from alternating layers of fat and dough; upon baking steam accumulates between the dough layers and causes them to expand, forming large spaces between thin layers of pastry.

Puffballs Edible wild fungi; mosaic puffball Calvatia caelata, and giant puffball C. gigantea, are normally eaten while still relatively small and fleshy. Much prized for their delicate flavour.

Puffed wheat Trade name of a breakfast cereal prepared by heating wheat grains under pressure and then rapidly releasing the pressure, when the superheated steam in the grain suddenly expands, so puffing or 'exploding' the grain.

Pulmonary atresia This heart defect is the complete absence of the pulmonic valve and may also include an absence of the main blood vessel that runs between the right ventricle and the lungs.

Pulmonary stenosis The pulmonic valve is stiffened and has a narrowed opening. It does not open and close properly, which increases strain on the heart because the right ventricle has to pump harder to send blood out to the lungs.

Pulpeta Caribbean; meat loaf, served hot or cold.

Pulque Sourish beer produced in Central and South America by the rapid natural fermentation of aquamiel, the sweet, mucilaginous sap of the agave. Contains 6% alcohol by volume.

Pulses Name given to the dried seeds of legumes such as peas, beans, and lentils. In the fresh, wet form they contain about 90% water, but the dried form contains about 10% water and can be stored.

Pumpernickel Dense, sour-flavoured black bread made from rye, originating from Germany. Name said to be derived from Napoleon's remark that it was 'pain pour ma Nicole'.

Pumpkin A gourd, fruit of Cucurbita pepo.

Punch A drink, hot or cold, made from wine and spirit, lemons and other fruit, spices and sugar.

Punchnep Welsh; mashed potato and turnip with small holes punched into the top, into which cream is poured.

Puree Fruit, vegetable, meat, or fish that has been pounded or sieved to give a smooth, finely divided pulp. Also a soup made by sieving vegetables with the liquor in which they were cooked.

Puri Unleavened whole-wheat bread prepared from a butter-rich dough, shaped into small pancakes and deep-fried in hot oil.

Purim fritters Traditional Jewish fritters eaten at Purim, similar to French toast.

Purines Nitrogenous compounds that occur in nucleic acids and their precursors and metabolites; inosine, caffeine, and theobromine are also purines. They are not dietary essentials; both dietary and endogenously formed purines are excreted as uric acid.

Sweetbread is rich in purines, as is fish roe; there are moderate amounts in sardines and anchovies, lesser amounts in other fish and meat; little in vegetables, fruits, and cereals.

Purl Old English winter drink; warmed ale with bitters and brandy or milk, sugar and spirit.

Pursindah Roast tender fillets of meat, cut into thin strips and beaten flat; normally cooked on skewers over charcoal.

Putromaine Any toxin produced by the decay of food within the body.

Putty fish A defect in salted fish in which the flesh, more so in thicker parts of the fish, has a soft and plastic texture. It happens when thick fish are salted at high temperature so that parts of the fish spoil before the salt has penetrated to prevent spoilage.

Pyloric caeca Hollow finger like pouches in bony fish, in which proteolytic digestive enzymes are secreted, attached to the intestine near to the bottom of the stomach.

Pyrimidines Nitrogenous compounds that occur in nucleic acids: cytosine, thymidine, and uracil.

Pyrophosphates Salts formed by the condensation of two molecules of orthophosphate.

Pyruvic acid An intermediate in the metabolism of carbohydrates, formed by the anaerobic glycolysis of glucose.

Q

Quality The totality of the features and characteristics of a product that bear on its ability to satisfy a given need, In the case of a fishery product, quality is the aggregate of the properties that influence the acceptability of the product for the consumer. These properties are predominately the particular properties that influence the sensory properties of the product - appearance, flavour/odour, texture. Merchants and processors must take account of these properties when marketing or processing fishery products, but other properties related to the processing of the products will be taken into account, for example, size, condition, presentation - whole, headed and gutted - frozen or not, season, fat content, fishing ground. These properties might influence the sensory properties, but are not directly apparent to the consumer. Merchants and processors might prepare a specification of quality for buying or selling products.

Quality assurance All the activities and functions concerned with the attainment and maintenance of quality. Quality assurance refers to the management systems in a company for attaining and maintaining quality, and not to the operating systems. It will include such matters as recruitment and training of staff, documenting specifications, codes of practice and operating procedures, preparing HACCP plans, developing and maintaining records of operational practices, and of quality control for management purposes and for inspection by customers and regulatory authorities.

Quality control All activities and functions concerned with the attainment of quality. It is an active process in which quality controllers monitor handling, processing, storage and distribution systems, and intervene as necessary by selecting appropriate raw materials, modifying the process and withdrawing out-of-specification material to ensure the end-products meet the required specifications for quality.

Quality Index Method A demerit points system for rating the quality of fish products. A system was first developed in New Zealand, but it has been adopted in recent years by laboratories in Europe and Scandinavia. The method has been developed mostly for raw, chilled fish, but systems exist for frozen fish.

Quamash Or camash; starchy roots of Camassia quamash, the staple food of West Coast native Americans.

Quark When made from skim milk this is a very low-fat cheese originating from Germany; contains 80% water; a good source of protein and vitamin B 12, a source of vitamin B 2, with only a trace of fat. Also available in some countries as '20% fat on dry matter'; '40% fat on dry matter', being a good source of protein and a source of vitamin B 2.

Quart Imperial measure of volume, equal to ¼ imperial gallon or 2 pints.

Quart, reputed The traditional 'bottle' of wine or spirits; approximately ? imperial quart, or 26? fl oz. Reputed pint is 13? fl oz.

Queen cake Individual, small, light, rich cakes containing dried fruit.

Queen of puddings Pudding made from custard and breadcrumbs, flavoured with lemon rind and vanilla, topped with jam or sliced fruit and meringue.

Quenelle Dumplings made from finely pounded meat or fish.

Quercitin A flavone, found in onion skins, tea, hops, and horse chestnuts. Not known to be a dietary essential or to have any function in the body, but sometimes called vitamin P.

Querns Pair of grinding stones used for pulverizing grain. The lower stone was slightly hollowed and the upper stone was rolled by hand on the lower one.

Quetsch Plum brandy, prepared by distillation of fermented plums. Similar to slivovitz and mirabelle. Sometimes called zwetschgenwasser or zwetschenwasser.

Quiche Savoury egg custard tart in a pastry case containing a wide variety of vegetable, meat, or fish fillings. Speciality of Alsace and Lorraine in France.

Quick breads Baked goods such as biscuits, muffins, popovers, griddles, cakes, waffles, and dumplings, in which no yeast is used, but the raising is carried out quickly with baking powder or other chemical agents.

Quick freezing Rapid freezing of food by exposure to a blast of air at a very low temperature. Unlike slow freezing, very small crystals of ice are formed which do not rupture the cells of the food and so the structure is relatively undamaged.

A quick-frozen food is commonly defined as one that has been cooled from a temperature of 0°C to -5°C or lower, in a period of not more than 2 hours, and then cooled to -18°C.

Quillaja Or soapbark; the dried bark of the shrub Quillaja saponaria, which contains saponins and tannins. Used to produce foam in soft drinks, shampoos, and fire extinguishers.

Quince Pear-shaped fruit of Cydonia oblongata, with flesh similar to that of the apple; sour but strong aromatic flavour when cooked; rich in pectin and used chiefly in jams and jellies; used to be known as 'the apple and the vine'.

Quince, japanese Fruit of the ornamental shrub Chaenomeles lagenaria, hard, sour and aromatic, used in preserves and jellies.

Quinine Bitter alkaloid extracted from the bark of the cinchona tree, used to treat or prevent malaria and in aperitif wines, bitters, and tonic water.

Quinoa Glutinous seeds of the South American plant Chenopodium album, used in Chile and Peru to make bread.

Quinquina bitter vermouths based on partially fermented grape juice and quinine; trade names include Byrrh, Dubonnet, and St Raphaël.

Quorn Trade name for mycoprotein from the mould Fusarium gramineanum.

R

Rabbit 1 Lepus cuniculus; both wild and farmed rabbits are eaten.

2 Original form of rarebit.

Raffinose Trisaccharide of fructose, glucose and galactose, found in cotton seed, sugar-beet molasses, and Australian manna; also known as melitose or melitriose. Has 23% of the sweetness of sucrose. It is not digested.

Rainbow sugar Crystals of coffee sugar, coloured.

Raisin Dried seedless grapes of several kinds. Valencia raisins from Spanish grapes; Thompson seedless raisins produced mainly in California from the sultanina grape. Raisins are also produced in Australia and South Africa. Raisin oil is extracted from the seeds of muscat grapes, which are removed before drying them to yield raisins. The oil is used primarily to coat the raisins to prevent them sticking together, to render them soft and pliable and less subject to insect infestation.

Raita Yoghurt with chopped cucumber, onion, garlic, and spices.

Rancid, rancidity Rancid describes the harsh, somewhat bitter, flavour and odour of oxidised fats.

Randomized controlled trial A research study that uses two or more groups of research subjects. The treatment group receives the experimental treatment while the control group receives either a placebo or current standard treatment.

Rape Brassica napus, also known as cole, coleseed, or colza. Grown for its seed, as source of oil for both industrial and food use. Varieties low in erucic acid are termed '0' or single low; varieties also low in glucosinolates are termed '00' or double low, both these being undesirable constituents of ordinary rapeseed. Oil is very rich in monounsaturates (60%), contains 33% polyunsaturates, and only 7% saturates.

Rasgulla Dessert of small balls of milk curd, ground almond, and semolina boiled in syrup.

Rasher Slice of bacon or ham.

Raspberry Fruit of Rubus idaeus. A rich source of vitamin C; a source of folate and copper.

Raspings Finely ground breadcrumbs, used for coating rissoles, fishcakes, etc.

Rastrello Sharp-edged spoon used to cut out the pulp from halved oranges or other citrus fruit.

Ratafia 1. Flavouring essence made from bitter almonds.

2. Small macaroon-like biscuits flavoured with almonds.

3. Almond-flavoured liqueur.

Ratatouille Casserole or stew of aubergine, onions, peppers, courgettes, and tomatoes, originally from Provence in southern France.

Ravigote sauce 1. Salad dressing containing pounded hardboiled egg, highly flavoured with garlic and chopped herbs

2. Sauce made from ravigote butter melted in wine and vinegar added to a veloute sauce; served with boiled poultry.

Ravioli Square envelope of pasta stuffed with minced meat or cheese.

Raw sugar Brown unrefined sugar,96-98%pure, as imported for refining. Contaminated with mould spores, bacteria, cane fibre, and dirt.

Ray Cartilaginous fish, Raja spp.

RDA Refers to both Recommended Dietary Allowance and U.S. Recommended Daily Allowance. Dietary intake goals for healthy people in 16 different age and gender groups based on a consensus of the Food and Nutrition Board of the Institute of Medicine of the National Academy of Sciences. RDAs serve as the basis for the U.S. Recommended Daily Allowances, the Food and Drug Administration's standards for required nutrition labeling of foods. US RDAs are more widely used than RDAs.

RDI Reference Daily Intake.

Ready brek Trade name for instant, hot, oatmeal breakfast cereal.

Rebaudioside Very sweet substance extracted from the leaves of Stevia rebaudiana; 400 times as sweet as sucrose.

Receptive language The child's ability to understand what is being said by other people.

Rechauffe Made-up dish of reheated food; literally 'reheated'.

Reciprocal ponderal index An index of adiposity; height divided by cube root of weight.

Recombinant DNA Technique of isolating DNA molecules and inserting them into the DNA of a cell. This technique includes taking copies of genes from one organism and inserting them in another organism. The two organisms can be totally unrelated. Recombinant DNA has a variety of uses, such as studying how genes work, and producing medicines such as human insulin and other commercial products.

Recycled To use again.

Red colours Amaranth, carmoisine, cochineal, erythrosine, poncho, red.

Red cooking Chinese method of cooking; meat or poultry is first stir fried, then simmered in broth or water with soy sauce, herbs, and spices.

Red herrings Herrings that have been well salted and smoked for about ten days. Bloaters are salted less and smoked for a shorter time; kippers lightly salted and smoked overnight. Also called Yarmouth bloaters.

Red tide Sudden, unexplained increase in numbers of toxic organisms in the sea which cause fish and shellfish feeding on them to become seasonally toxic.

Redcurrants Fruit of Ribes sativum; the UK National Fruit Collection contains 78 varieties.

Reduce The process of boiling a mixture in an uncovered pan to evaporate surplus liquid and give a more concentrated product.

Reducing sugars Sugars that are chemically reducing agents, including glucose, fructose, lactose, pentoses, but not sucrose.

Reduction The opposite of oxidation; chemical reactions resulting in a gain of electrons, or hydrogen, or the loss of oxygen.

Reference intakes Amounts of nutrients greater than the requirements of almost all members of the population, determined on the basis of the average requirement plus twice the standard deviation, to allow for individual variation in requirements and thus cover the theoretical needs of 97.5% of the population.

Reference intakes for energy are based on the average requirement, without the allowance for individual variation. Used for planning institutional catering and assessing the adequacy of diets of groups of people, but not strictly applicable to individuals.

Reference man An arbitrary physiological standard; defined as a man aged 25, weighing 65 kg, living in a temperate zone of a mean annual temperature of 10°C, performing medium work, and assumed to require an average daily intake of 3200 kcal.

Reference woman An arbitrary physiological standard; defined as a woman aged 25, weighing 55 kg, living in a temperate zone of a mean annual temperature of 10°C, engaged in general household duties or light industry, and assumed to require an average daily intake of 2300 kcal.

Reform sauce Espagnole sauce enriched with port and redcurrant jelly; originated at the Reform Club in London as an accompaniment for breaded and fried lamb cutlets.

Refractive index Measure of the bending or refraction of a beam of light on entering a denser medium. It is constant for pure substances under standard conditions. Used as a measure of sugar or total solids in solution, purity of oils, etc.

Refractometer Instrument to measure the refractive index.

Refreezing Freezing fish again after it has been thawed or partially thawed as in tempering.

Refreshing A process used by cooks when preparing vegetables. After the vegetables have been cooked, cold water is poured over them to preserve the colour; they are then reheated before serving.

Refrigerant The working fluid of a refrigeration system that directly or indirectly extracts and carries away heat from a product or space being cooled. The primary refrigerant in a system takes up the heat in an evaporator or heat exchanger and discharges it in a condenser. The secondary refrigerant is the intermediate fluid in the system, cooled in a heat exchanger by the primary refrigerant, and used to actually cool the product or space.

Refrigerated sea water Sea water in fixed tanks chilled by mechanical refrigeration, especially at sea installed in the hold, for quickly and uniformly chilling pelagic fish, in particular, by immersion. In a full tank the proportion of sea water to fish should be between 1:3 and 1:2 and the water is circulated by pump to maintain a uniform temperature and prevent temperature stratification. The sea water in the tank should be pre-cooled before fish are loaded. The sea water can be lowered to about -1°C without freezing the fish to obtain the maximum benefit, and the catch can be cooled more rapidly than by stowage in ice. The refrigeration capacity of an RSW system should be sufficient to cool the fish to 0°C within a few hours. Some species take up unacceptable amounts of water and salt when kept in sea water, the eyes become cloudy and the gills are bleached as the blood is leached out. Consequently, stowage in RSW is usually limited to a maximum of 3 to 4 days. RSW systems should be designed specifically to suit both the vessel and the fishery.

Register of accredited nutritionists A register of those with approved qualifications in nutrition maintained by the British Nutrition Society and Institute of Biology.

Regulatory standards Standards That are incorporated into legislation or are issued by a regulatory body.

Rehfuss tube A small-diameter tube with a slotted metal tip for removing samples of food from the stomach after a test meal.

Reichert-meissl value Measure of the volatile fatty acids in fats.

Reine, a la Dish based on chicken.

Relative humidity The ratio of the amount of moisture or water vapour in the air to the total moisture that the air could hold at that temperature, i.e. saturated with water vapour, is the relative humidity. It is usually expressed as a percentage and abbreviated to rh. Air saturated with water vapour has a relative humidity of 100% and when only half saturated the rh is 50%.

Relative protein value A measure of protein quality.

Release agents Substances applied to tinned or enamelled surfaces or plastic films to prevent the food adhering, e.g. fatty, acid amides, microcrystalline waxes, petrolatums, starch, methyl cellulose.

Relish Culinary term for any spicy or piquant preparation used to enhance flavour of plain food. In the UK, a thin pickle or sauce with a vinegar base; in the USA it includes finely chopped fruit or vegetables with a dressing of salt, sugar, and vinegar, sometimes eaten as a first course.

Remoulade sauce Mayonnaise with added mustard and finely chopped capers, gherkins, parsley, chervil, and anchovy fillet.

Remove Obsolete term for the main course of dinner.

Renal failure Inability of the kidneys to excrete waste, as a result of kidney disease. Especially in advanced cases, as well as haemodialysis, treatment includes feeding a very low protein diet, so as to minimize the amount of nitrogenous waste that must be excreted.

Rendering The process of liberating the fat from the cells that constitute the adipose tissue. Dry rendering, heating the fat dry, or wet rendering, when water is present.

Rennet Extract of calf stomach; contains the enzyme chymosin which clots milk. Used in cheese making and for junket.

Rennet, vegetable The name given to proteolytic enzymes derived from plants, such as bromelain and ficin, as well as biosynthetic chymosin. Used for the preparation of 'vegetarian' cheeses.

Rentschlerizing Sterilizing by treatment with ultraviolet light, named after Dr H. C. Rentschler, who developed the lamp.

Renverse Turned out of a mould, as for a cream or jelly.

Respiratory quotient Ratio of the volume of carbon dioxide produced when a substance is oxidized, to the volume of oxygen used. The oxidation of carbohydrate results in an RQ of 1.0; of fat, 0.7; and of protein, 0.8.

Respite care Service provided to families of children who require extraordinary forms of care so that the family can take holidays, etc. and have some relief from the duty of caring for the child.

Restoration The addition of nutrients to replace those lost in processing, as in milling of cereals.

Retailer Sells goods or services to the consumer for personal use.

Reticulocyte Immature precursor of the red blood cell in which the remains of the nucleus are visible as a reticulum. Very few are seen in normal blood as they are retained in the marrow until mature, but on remission of anaemia, when there is a high rate of production, reticulocytes appear in the bloodstream.

Retinoid receptors Intracellular proteins that bind the vitamers of vitamin A, resulting in the modulation of gene expression, and hence the basis of the actions of the vitamin other than in vision.

Retinoids Compounds chemically related to, or derived from, vitamin A, which display some of the biological activities of the vitamin, but have lower toxicity; they are used for treatment of serious skin disorders and some cancers.

Retinol Vitamin A.

Retort In food technology, an autoclave.

Retsina Greek dry white wine, with a strong resinous flavour due to the addition of pine resin as a preservative.

Reuben sandwich American; salt beef with cheese and sauerkraut, on rye bread, served hot. Originated in 1956 as the winner of a competition for novel sandwiches; the origin of the name is unclear.

RFLP RFLP mapping detects patterns in DNA that can indicate the presence of a gene for a trait. Both RFLP and polymerase chain reaction (PCR) analysis can be used in DNA fingerprinting for genealogical studies and forensics.

Rhamnose A methylated pentose sugar; it has 33% of the sweetness of sucrose, and is widely distributed in plant foods.

Rheology It involves the measurement of textural properties by instrumental, not sensory, methods.

Rheology The study of flow properties of materials. These properties include viscosity, plasticity, hardness, toughness, elasticity, i.e. all are properties that have a bearing on the texture of fish products.

Rhizome Botanical term for swollen stem that produces roots and leafy shoots.

Rhizopterin Obsolete name for folic acid.

Rhodopsin The pigment in the cone cells of the retina of the eye, also known as visual purple, consisting of the protein opsin and retinaldehyde, which is responsible for the visual process. In rod cells of the retina the equivalent protein is iodopsin.

Rhubarb Leaf-stalks of the perennial plant, Rheum rhaponticum. Has a high content of oxalate.

Ribena Trade name for a preparation of blackcurrant juice; extremely rich source of vitamin C, 200 mg per 100 g, so a 300 mL glassful, diluted 1:6 would be a very rich source of the vitamin.

Ribose A pentose sugar which occurs as an intermediate in the metabolism of glucose; especially important in the nucleic acids and various coenzymes: occurs widely in foods.

Ribosome Site of protein synthesis in the cytoplasm. selection and breeding - Manipulating microbes, plants or animals, and choosing desirable individuals or populations as breeding stock for new generations.

Ribotide Trade name for a mixture of the purine derivatives, disodium inosinate and disodium guanylate, used as a flavour enhancer for savoury dishes.

Rice Grain of Oryza sativa; major food in many countries. Rice when threshed is known as paddy, and is covered with a fibrous husk comprising nearly 40% of the grain. When the husk has been removed, brown rice is left. When the outer bran layers up to the endosperm and germ are removed, the ordinary white rice of commerce or polished rice is obtained.

Rice, unpolished American term; rice that has been under-milled in that the husk, germ, and bran layers have been partially removed.

Rice, wild Also known as zizanie, Tuscarora rice, Indian rice, and American wild rice; Zizania aquatica, native to eastern North America, grows 12 feet high; it has a long, thin, greenish grain; little is grown and it is difficult to harvest, so it is strictly a gourmet food. Higher in protein content than ordinary rice at 14%.

Richelieu, a la Dish made with Richelieu sauce or garnished with mushrooms, artichoke bottoms, or stuffed tomatoes, etc.

Ricin A lectin in the castor oil bean.

Ricing Culinary term: cutting into small pieces about the size of rice grains.

Ricinoleic acid A hydroxylated mono-unsaturated fatty acid in castor oil.

Rickets Malformation of the bones in growing children due to deficiency of vitamin D, leading to poor absorption of calcium. In adults the equivalent is osteomalacia.

Rickey Long drink made from liqueur or spirit with ginger ale and fresh fruit.

Ricotta Cheese made originally from the whey left by cheese making, but now often with milk added.

Riesling One of the 9 'classic' grape varieties used for wine making; the wines have a flowery aroma when young.

Riffle flumes Washing equipment consisting of stepped channels along which the product being washed is carried in a flow of water; stones and grit are retained on the steps.

Rigor mortis Rigor mortis is the Latin term for stiffness of death; the stiffening of an animal body after death. Because it is a Latin term it is conventional to italicise 'rgor mortis'; 'rigor', exists as a term meaning stiffness and is not usually italicised.

Typically, vertebrate fish start to go into rigor 8-24 hours after death, but this period can be shorter or longer, and rigor is resolved after a further 1-3 days. Time into, time in, and strength of rigor varies with several factors, the main ones being species, nutritional status, amount of exercise by the fish immediately before death, and the ambient temperature of the fish.

Rijstaffel Dutch, Indonesian; meal consisting of a variety of different dishes served at the same time.

Rillauds Potted pork, goose, or rabbit made from very small pieces, flavoured, seasoned, and cooked very slowly in stock. Rillettes and rillons are modifications of the procedure.

Rioja Wines from the Rioja region of Spain; white or red. Traditionally matured in oak casks, so have a strong 'oaky' flavour.

Ripening Ripening or maturing of small fatty species such as anchovy and herring with salt, sugar, and sometimes spices, in barrels, in order to obtain characteristic changes in texture and flavour, is a complex process which takes several months. It is thought to involve degradation of proteins and lipids in the fish muscle by enzymes from the digestive tract, especially from the pyloric caeca, and to be seasonally dependent particularly in herring. The products are semi-preserves and must be kept chilled.

Risk populations All people are believed to be susceptible to hemorrhagic colitis, but young children and the elderly appear to progress to more serious symptoms more frequently.

Risotto Rice dishes cooked with meat, poultry, vegetables, etc.

Rissole Small round cake of cooked minced meat, bound with mashed potato, coated in egg and breadcrumb, and fried. Originally enclosed in an envelope of thin pastry before frying. rissole Baked or fried and well-browned.

RNA Ribonucleic Acid. A single-stranded nucleic acid that contains the sugar ribose. Many forms of RNA, including messenger RNA, transfer RNA, and ribosomal RNA, are involved in protein synthesis.

Roast Originally meant to cook meat over an open fire on a spit; now refers to cooking in. an enclosed oven, and so is 'dry heating'. With meat the juices are squeezed out and evaporate on the surface, producing the Maillard complex characte-ristic of roasted meat.

Robert sauce Piquant sauce containing onion, white wine, vinegar, mustard, and espagnole or demi-glace sauce; served with meat.

Robertsonian translocations Centric fusions is the most frequent translocations where the two short arms of both chromosomes and the center of one of them are lost.

Rocambole Mild variety of garlic, Allium scordoprasum, also called sand leek.

Rock Traditional English seaside sugar confectionery, made by pulling melted sugar. It is white, with an outer coloured layer and an inner ring of coloured sugar that spells out the name of the town.

Rock cake Plain buns containing fruit and spice, baked in small heaps on a tin or baking sheet.

Rock eel, salmon Alternative name for dogfish.

Rocket Cruciferous plant, Eruca sativa, with small spear-shaped leaves and peppery taste, eaten raw in salads or cooked. Also called arugula, rucola, Italian cress.

Roe Hard roe is the eggs of the female fish. Soft roe is from the male fish, also known as milt. Hard roe of sturgeon and lumpfish are used to make caviare and mock caviare.

Rokelax Scandinavian term for smoked salmon.

Roll-an closure Aluminium or lacquered tinplate cap for sealing on to narrow-necked bottles with a threaded neck. The unthreaded cap is moulded on to the neck of the bottle and forms an air-tight seal.

Roller dryer The material to be dried is spread over the surface of internally heated rollers and drying is complete within a few seconds. The rollers rotate against a knife that scrapes off the dried film as soon as it forms. There is little damage to nutrients by this method; for example, roller-dried milk is not scorched, but there is some loss of vitamins B, and C, more than in spray drying.

Roller mill Pairs of horizontal cylindrical rollers, separated by only a small gap and revolving at different speeds. The material is thus ground and crushed in one operation. Used in flour milling.

Rollmop Filleted uncooked herring pickled in spiced vinegar.

Roly-poly Sweet suet pudding filled with jam.

Romaine French and American name for cos lettuce.

Romesco sauce Spanish; pungent red sauce made from small red peppers, tomatoes, almonds, olive oil, garlic, and cayenne pepper.

Rooibos tea Fermented leaves of the South African bush Aspalathus liniearis. Contains a unique polyphenol, aspalathin, which becomes red during preparation and produces a reddish herbal tea; free from caffeine and theaflavin.

Root beer American; non-alcoholic carbonated beverage flavoured with extract of sassafras root and oil of wintergreen.

Rope Spore-forming bacteria occurring on wheat and hence in flour. The spores can survive baking and then are present in the bread. Under the right conditions of warmth and moisture the spores germinate and the mass of bacteria convert the bread into sticky, yellowish patches which can be pulled out into ropelike threads, hence the term 'ropy' bread. The bacterial growth is inhibited by acid substances. Can also occur in milk, called long milk in Scandinavia.

Ropy brine A defect in pickling brines that contain sugars. Bacteria can convert the sugars to a polysaccharide giving a stringy, slimy consistency to the brine.

Roquefort Green-blue marbled French cheese made in Roquefort-surSoulzon from cow's milk; ripened in limestone caves where the mould, Penicillium roquefortii is present and the cheese thus inoculated.

Rosado Portuguese, Spanish; rose wines.

Rosato Rose wines.

Rose hips The fruit of the rose; a rich source of vitamin C. Rose-hip syrup is extract of rose hips with added sugar.

Rose Pink-coloured wines, either made from red grapes, allowing the skin to remain in the fermentation for only 12-36 hours, or by mixing red and white wines. Known as blush wines in the USA.

Rose water Fragrant water made by distillation or extraction of the essential oils of rose petals. Used in confectionery.

Rose-gottlieb test For fat in milk, by extracting the fat with solvent.

Rosella Caribbean plant grown for its fleshy red sepals, used to make drinks, jams and jelly. Also known as sorrel, flor de Jamaica.

Rosemary A bushy shrub, Rosmarinus officinalis, cultivated commercially for its essential oil, used in medicine and perfumery. The leaves are used to flavour soups, sauces and meat.

Rösti potatoes are partly baked or boiled, then sliced and fried in hot fat to form a cake with a golden crust.

Rotary louvre dryer A drying process whereby hot air passes through a moving bed of the solid inside a rotating drum.

Rotisserie Method of cooking which developed from the traditional rotating spit above an open fire; the food is rotated - while roasting, so bastes itself.

Rotmos Swedish; mashed boiled potatoes and turnips.

Rouille French sauce made from red chillies, garlic, olive oil, and potatoes or breadcrumbs, served with fish dishes.

Fermentation allows ruminants to obtain nourishment from grass and hay which cannot be digested by monogastric animals.

Roundfish Sometimes written Round fish. May refer to a whole, ungutted fish, or to fish that have a roughly round cross section of the body, such as cod, as compared with a flatfish such as plaice.

Rub in To incorporate butter or margarine into flour using fingertips until the mixture resembles breadcrumbs.

Rumpbone Cut of meat: in the USA it is called the aitchbone, in the UK, the loin or haunch.

Rusk 1. Sweetened biscuit or piece of bread or cake crisped in the oven, especially as food for young children when teething.

2. Cereal added to sausages and hamburgers.

Russe, a la Containing beetroot and/or sour cream.

Russian tea Tea without milk, served with lemon. Traditionally it is not sweetened, but is drunk by sucking through a crystal of sugar held between the teeth.

Rutabaga American name for swede.

Rutin A citrus bioflavonoid found in buckwheat, the leaves and petioles of *Rheum* species, and the fruit of the Fava D'Anta tree, as well as other sources.

It can combine with cations, supplying nutrients from the soil to the cells in plants. In humans, it attaches to the iron ion Fe^{2+}, preventing it from binding to hydrogen peroxide and creating a highly reactive free radical that may damage cells. It is also an antioxidant, and therefore plays an important role in inhibiting some cancers.

Rutin also strengthens the capillaries, and therefore can reduce the symptoms of haemophilia. It also may help to prevent a common unpleasant-looking venous edema of the legs. Rutin, as ferulic acid, can reduce the cytotoxicity of oxidized LDL cholesterol and lower the risk of heart disease.

Rye Rye is a grass grown extensively as a grain and forage crop. It is a member of the wheat tribe and is closely related to barley and wheat. Rye grain is used to make flour, feed, beer, some whiskies and most vodkas. It can also be eaten whole, either as boiled rye berries, or by being rolled, similar to rolled oats.

S

Saccharases Enzymes that hydrolyse sugars to liberate their constituent monosaccharides.

Saccharic acid The dicarboxylic acid derived from glucose.

Saccharin A synthetic chemical, benzoic sulphimide, 550 times as sweet as sucrose. Soluble saccharin is the sodium salt. It has no food value, but is useful as a sweetening agent for diabetics and slimmers. Discovered in the USA in 1879.

Sachertorte Austrian; chocolate sponge cake with rich chocolate icing and whipped cream.

Saddle The whole back of the animal, from the end of the loin to the best end of the neck.

Safe and adequate intake Where there is inadequate scientific evidence to establish requirements and reference intakes for a nutrient for which deficiency is rarely, if ever, seen, the observed levels of intake are assumed to be greater than requirements, and thus provide an estimate of intakes that are safe and adequate to meet needs.

Safflower oil Vegetable oil extracted from the seeds of Carthamus tinctoria, 75% polyunsaturated fatty acids.

Saffron Deep orange-red powder from the stigmata of the saffron crocus, Crocus sativus; 1 g requires the stigmata of 1500 flowers and yields about 50 mg of extract. Used as natural dyestuff and spice. Very soluble in water.

Sage Leaf of the Dalmatian sage, Salvia officinalis, of the mint family; fragrant and spicy and an important herb used in the kitchen for flavouring meat and fish dishes and in poultry stuffing. Other sages differ in flavour from the Dalmatian variety.

Sago Starchy grains prepared from the pith of the swamp sago and the sugar palm; almost pure starch and sugars, free from protein.

Saithe Also known as coley and coal fish, Polachius virens. Apart from being eaten cooked, it is smoked, salted, and dyed red when it is similar to smoked salmon.

Sake Japanese wine made from rice. Cooked whole rice grains are fermented with a yeast-like fungus culture for 10-14 days and stored in wooden barrels. Contains about 17% alcohol, by volume. Normally drunk warm and traditionally served in conical ceramic cups.

Salad Originally derived from the Latin sal for salt, meaning something dipped into salt. Now normally a dish of uncooked vegetables; either a mixed salad or just one item. In France it can mean a small, hot, savoury dish, e.g. of chicken liver, etc.

Salami Type of sausage speckled with pieces of fat and flavoured with garlic; originally Italian.

Saliva Secretion of the salivary glands in the mouth: 1-1.5 L secreted daily. A dilute solution of the protein mucin and the enzyme amylase, with small quantities of urea, potassium thiocyanate, sodium chloride, and bicarbonate.

Salmagundi Old English dish consisting of diced fresh and salt meats mixed with hard-boiled eggs, pickled vegetables, and spices, arranged on a bed of salad.

Salmonella A genus of Gram negative, carbohydrate fermenting, non-spore forming bacteria of faecal origin.

They can survive and may grow well in factory environments. The majority of strains are regarded as potentially pathogenic to humans. They cause an infective type of food poisoning, i.e. they multiply in the body. Consequently, only a small numbers of cells need to be present in the contaminated food. Salmonellosis is the illness caused by Salmonella bacteria. The illness is typical of a gastrointestinal infection - onset 12-36 hours after consumption of the food, with symptoms of diarrhoea, abdominal pain, fever, nausea, vomiting. The organisms are not indigenous to the aquatic environment, but can be found there following contamination with animal, including human, faeces. They are not normally present in fish and shellfish and are not often found in fresh fish products, but can be found in molluscan shellfish taken from waters, especially inshore waters, that have become contaminated with sewage. The Salmonellae along with other bacteria and viruses become concentrated in the gut of these filter-feeding molluscs. Fish from unpolluted waters do not harbour the bacteria, but can become carriers if caught or harvested from polluted waters, and products after harvesting can become contaminated in unsanitary surroundings.

Salpicon various mixtures of chopped meat, fish, or vegetables in a sauce, used as stuffing or filling.

Salsify Long, white, tapering root of the biennial plant Tragopogon porrifolius. Black salsify is very similar, the hardy perennial, Scorzonera hispanica.

Salt Defined chemically, salts are ionic compounds in which the cation comes from a base and the anion comes from an acid, e.g. sodium chloride, calcium chloride. Solutions of these salts, or brines, are used for brine freezing fish such as tuna in a brine immersion freezer. Sodium chloride is also known as common salt. It is the source of a basic taste. Salt curing is a traditional method of preserving both white fish and fatty fish. Three main types of salt are used commercially, rock or mined salt from underground deposits, evaporation of strong brines pumped from deep mines, and solar salt by solar evaporation from sea water or inland salt lakes. For modern food products, it is preferable to use purified, manufactured, vacuum dried salt that can be produced from the types described above. Salt is used for dry salting fish or, in solution as a brine, for brining fish prior to smoking, drying, canning or some other form of processing that requires a brining stage. Solid salt is also used for pickle curing, the traditional way of preserving fatty fish in airtight barrels. The purity of the salt used for curing affects the curing process and the characteristics of the cured product, particularly in dry salting.

Salt concentration For food safety reasons, particularly to consider the potential growth of pathogenic microorganisms in smoked fish products intended to be eaten without further cooking, it is important to know the salt concentration in the product as distinct from the salt content. In smoked fish, salt concentration is the percentage weight of salt in the water phase of the product. Both water content and salt content must be measured. The percentage salt concentration is calculated as salt content divided by salt content plus water content multiplied by 100.

Salt content The weight of salt in a given weight of fish, for example in a smoked fish product, usually expressed as g salt/100 g fish, or as a percentage and, importantly, distinguished from the salt concentration.

Salt-boiled fish Cooking in brine may be used as a preliminary treatment before further processing, such as canning, or for the initial handling of some crustaceans such as shrimp by boiling in sea water. In southeast Asia, salt-boiled fish is an important commercial process and the amount of salt used determines the shelf life of the product. Boiling inactivates enzymes and kills most but not all non-spore forming micro-organisms and alone gives no more than 1 or 2 days shelf life at the ambient temperature. Storage up to 3 months under tropical conditions can be achieved if the fish are cooked in salt until there is no free water left in the bottom of the pan, the surface of the fish in the pot is covered with a thick layer of salt, and the pot is sealed with paper. Similarly cooked and covered in a glass jar with an effective closure and sealing ring can extend the storage life up to 9 months.

Salt-free' diets Diets low in sodium, for the treatment of hypertension and other conditions. Most of the sodium of the diet is consumed as sodium chloride or salt, and hence such diets are referred to as salt-restricted or low-salt diets, or sometimes 'salt-free', to emphasize that no salt is added to foods in preparation or at the table. Since foods naturally contain sodium chloride, a truly salt-free diet is not possible. It is the sodium and not the chloride that is important.

Foods low in salt: sugar, flour, fruit, green vegetables, macaroni, nuts. Medium salt: chicken, fish, eggs, meat, milk. High salt: corned beef, bread, ham, bacon, kippers, sausages, cheese.

Salting Addition of salt or salting may only be required to flavour a product prior to further processing such as smoking or canning. This is carried out by brining or dry salting for a short time. In salt curing the fish is in contact with a high concentration of salt for a long time. The aim is to increase the concentration of solutes in the flesh, remove water and lower the water activity, but long-term preservation is only possible if the salt concentration in the flesh reaches concentrations corresponding to a saturated brine and a water activity of 0.75. Salt cured products remain important in many parts of the world. Salting may be dry salting in which fish, usually split or as sides, are stacked or covered in solid salt and the brine formed is allowed to flow away, or wet salting by immersion in a strong brine or pickle, i.e. pickling. However, a common traditional method, particularly for herring and other fatty fish, is to bury the fish in dry salt in air tight barrels or containers so that the fish subsequently becomes immersed in the liquid pickle formed by solution of the salt in the liquid extracted from the fish. The liquid pickle is termed the blood pickle. A lighter cure using 80 to 90% saturated brine is used for herring for the matjes cure to give a soft texture and characteristic flavour. It has a lower salt content and has to be kept chilled because it spoils readily at higher temperatures. Fish may be salted whole and ungutted, partially gutted, gutted and split, or as sides, fillets, fillet pieces or as minced fish, depending on size of fish and product required. Only small fatty fish such as anchovy are salted whole because the skin is a barrier to salt penetration such that whole larger fish spoil at the centre before salt can reach the centre. Scoring of fillets and sides of larger fish is used to aid salt penetration. Generally, pickling in concentrated brines for long periods is used for long term preservation of fatty fish. Immersion in a full air tight container restricts the presence of oxygen, retards oxidation reactions and development of rancidity, although some rancidity is essential to the development of the characteristic flavour of the product. Some products depend on the presence of proteolytic enzymes in the guts, or part of the guts, to develop the characteristic flavour, texture and colour of the particular product by the complex process of maturing during pickling, e.g. matjes herring, and also depends on the seasonal condition of the fish before processing. Split white fish, opened flat and laid in layers between layers of salt, such that the liquor formed is allowed to drain away is called kench curing. This can be done onshore or at sea on the fishing grounds. Onshore, the fish are restacked a number of times to produce an even cure in the wet stack, and then dried in the open in the wind and sun, or artificially in a drier or drying kiln, to produce a dry and white product ranging from 10 to 30% moisture. Typically, fish produced to a biscuit cure after drying have a water content of 25% and a salt content of 26%. The salted and dried product is known as a heavy cure or hard cure, and has a long shelf life. However, at relative humidity of air above 76%, solid salt, and therefore dried salt fish, absorbs water and, under these conditions, salted and dried fish spoils. Dry salting of split, white fish carried out in tubs in which the fish are weighted to keep them immersed for 2 to 3 days in the brine liquor formed and then dried in the sun or in drying kilns is termed a light cure or Gaspe cure. The product is less heavily salted, has a cheesy taste and amber coloured translucent flesh.

Saltlicks An adequate intake of sodium is necessary to all animals. Grass is relatively poor in sodium, and its high potassium content induces excretion of sodium in the urine. This loss causes a craving for sodium which is satisfied by licking on natural outcrops of salt-rich rock or artificial salt crystals.

Saltpetre Potassium nitrate.

Sambuca liqueur flavoured with liquorice and elderberry. Traditionally served with coffee beans in the glass, and set alight.

Samosa deep-fried stuffed pancakes, rolled into a cone or folded into an envelope. The filling is normally spiced; it may be meat or vegetables.

Samp Coarsely cut portions of maize with bran and germ partly removed.

Samphire 1 Rock samphire, St Peter's herb, succulent plant of cliffs and salt marshes; grows on coastal rocks, and has fleshy aromatic leaves which may be eaten raw, boiled, or pickled.

2. Marsh samphire, Salicornia spp., grows in salt marshes, is salty, and is eaten cooked as a vegetable.

Sample, Sampling A sample is a set of items taken from a larger set of items which is intended to reflect the properties of the larger set. Sampling is the process of taking a sample. The properties of items in the sample are measured, and the properties of the larger set are deduced from the properties of the sample.

Sampling plan An element of the practice of statistical quality control especially in the examination of batches of product. It is the set of instructions for sampling and testing the batch, and coming to a decision about the fate of the batch as a result of the sampling and testing. A typical sampling plan would have a series of actions. First, the batch for decision making is defined. This should be a set of items of similar characteristics, for example, the products prepared in a shift, or contained in a delivery.

The decision rules will state the number of items to be taken as a sample, and how they should be drawn from the batch. The sample units are tested according to a stated test protocol and the results of the tests compared against a specification for the product. In attribute sampling, each item in the sample is defined as defective or effective, that is, meeting the specification or, in three class sampling plans, being in a marginal class, on the basis of the test protocol. The sampling plan will state the maximum number of items in the sample that may be defective, or maximum numbers in each of defective and marginal classes, in the case of three class sampling plans, for the batch as a whole to be accepted. These are the decision rules. In sampling by variable sampling plans, the properties are measured on a continuous scale and the mean values obtained for the sample items are compared with the specification. A batch will be accepted or rejected according to how the mean values compare with control values in the specification.

Sanatogen Trade name for a preparation of casein and sodium glycerophosphate for consumption as a beverage when added to milk.

Sancoche Caribbean; casseroled salt pork, beef, and vegetables.

Sand cake Sponge or madeira cake made with ground rice, cornflour, or potato flour and matzo meal instead of wheat flour, flavoured with grated lemon rind; a traditional Jewish Passover delicacy.

Sandwich spread Finely chopped vegetables in salad cream, as a filling for sandwiches.

Sandwich Two slices of bread enclosing a filling. Invention attributed to the 4th Earl of Sandwich who spent long periods at the gaming table and carried a portable meal of beef sandwiched with bread.

Sanecta Trade name for aspartame.

Sangaree Caribbean; drink made from spiced and sweetened port and brandy, diluted with crushed ice.

Sangiovese A grape variety widely used for wine making, although not one of the classic varieties.

Sanitation The application of procedures for achieving and maintaining the cleanliness of premises and equipment, especially in the context of preventing contamination of food products by noxious agents.

A sanitary programme for a fish processing plant will include the location and design of premises to reduce the risk of contamination. To facilitate cleaning, it will include the design and construction of equipment to ensure they can be cleaned and disinfected. It will include the control of pests, the disposal of solid and liquid wastes, provision of facilities for personnel hygiene, provision of facilities for the safe and separate storage, of raw materials, finished products, and materials used in the processes. The management of the plant must institute procedures for maintaining the sanitation of premises and equipment by regular programmes of cleaning and disinfection.

Sanka Trade name for decaffeinated instant coffee.

Sapodilla Fruit of the sapodilla tree; size of a small apple, rough-grained, yellow to greyish pulp. Chicle, the basis of chewing gum, is made from the latex of the tree.

Saponification Hydrolysis of fat into its constituent glycerol and fatty acids by boiling with alkali. The fatty acids will be present as the sodium salts or soaps.

Saponins Group of substances that occur in plants and can produce a soapy lather with water. Extracted commercially from soapwort or soapbark and used as foam producer in beverages and fire extinguishers, as detergents, and for emulsifying oils. Bitter in flavour.

Saran Generic name for thermoplastic materials made from polymers of vinylidene chloride and vinyl chloride. They are clear, transparent films used for wrapping food; resistant to oils and chemicals; can be heat-shrunk on to the product.

Sarcolactic acid Obsolete name for (+) lactic acid, found in muscle, as distinct from the optically inactive lactic acid found in sour milk. Also known as paralactic acid.

Sarcosine An intermediate in the metabolism of choline, chemically N-methylglycine. Found in relatively large amounts in starfish and sea urchins, used as an intermediate in the synthesis of antienzyme agents in toothpaste.

Sardine Young pilchard Sardina pilchardus; commonly canned in oil, brine, or tomato paste. Norwegian canned sardines are salted and smoked before canning; French are salted and steamed.

Saridele Protein-rich baby food developed in Indonesia; extract of soya bean with sugar, calcium carbonate, thiamin, and vitamins B 12 and C.

Sarsaparilla 1. Flavour prepared from oil of sassafras and oil of wintergreen or oil of sweet birch.

2. Roots of a South American plant; both used to flavour the beverage called sarsparilla.

Sassafras American tree with aromatic bark and leaves. The root is used to make root beer and the young leaves are powdered to make file powder, an essential flavouring of gumbo. Sassafras oil from the root-bark is used medicinally and as a flavour in beverages, but banned in some countries because of its toxicity.

Sate Indonesian, Malaysian; marinated lamb, pork, or chicken, grilled on wooden skewers and served with a peanut sauce.

Satiety The sensation of fullness after a meal.

Saturated With reference to fatty acids means that the carbon chain carries its full complement of hydrogen atoms, as compared with mono-unsaturated and polyunsaturated. Where the nutrient content of foods is described in these entries the total amount of fat is listed, together with the proportion of saturates. Where mono-unsaturates are sufficiently important this is also quoted and the polyunsaturates can be calculated by difference from the total. Where the amounts of mono-unsaturates are small the entry lists the proportion of polyunsaturates and the mono-unsaturates can be calculated by difference from the total.

Saturates Commonly used short term for saturated fatty acids.

Sauce Used to flavour, coat, or accompany a dish, or may be used in the cooking to bind ingredients together; may be sweet or savoury. Thick sauces may be: 1. roux sauces based on flour heated with fat;

2. Thickened with starch or modified starch;

3. Thickened with egg;

4. Thickened by reduction.

Sauce, table Condiment or relish eaten with food; there are two main types: brown sauce, a blend of herbs, spices, and vegetables in vinegar, and tomato sauce or ketchup.

Sauerbraten Beef marinated for several days in wine vinegar, then braised and served with sour cream.

Sauerkraut Prepared by lactic fermentation of shredded cabbage. In the presence of 2-3% salt, acid-forming bacteria thrive and convert sugars in the cabbage into acetic and lactic acids, which then act as preservatives.

Sausage casings Natural casings are made from hog intestines for fresh frying sausages, and from sheep intestines for chipolatas and frankfurters. Skinless sausages are prepared in cellulose casing, which is then peeled off.

Sausage Chopped meat, mostly beef or pork, seasoned with salt and spices, mixed with cereal and packed into casings made from the connective tissue of animal intestines or cellulose.

There are six main types: fresh, smoked, cooked, smoked and cooked, semi-dry, and dry. Frankfurters, Bologna, Polish, and Berliner sausages are made from cured meat and are smoked and cooked. Thuringer, soft salami, mortadella, and soft cervelat are semi-dry sausages. Pepperoni, chorizos, dry salami, dry cervelat are slowly dried to a hard texture. In the UK pork sausages must be 65% meat and beef sausages 50% meat.

Sausage meat Mixture of meats as used to make sausages, sausage rolls, pies, and stuffing.

Sausage toad Sausages baked in batter, also known as toad-in-the-hole.

Saute To toss in hot fat, from the French sauter, to jump. Saute potatoes are usually boiled, cut into slices, and cooked in a little fat until lightly browned.

Sauvignon blanc One of the 9 'classic' grape varieties used for wine making; the wines are distinctively aromatic and sometimes smokey.

Saveloy Highly seasoned smoked sausage; the addition of saltpetre gives rise to the bright red colour. Originally a sausage made from pig brains.

Savory Herb with strongly flavoured leaves used as seasoning in sauces, soups, and salad dishes. Summer savory is an annual, Satureja hortensis; winter savory is a perennial, Satureja montana. The plants are cut down at flowering time and dried for later use.

Savoury Foods with a savoury or salty flavour; may be small titbits eaten with the fingers or a savoury dish served after the sweet course and before the dessert at a formal banquet.

Savoy biscuits Small sponge fingers.

Savoy Variety of cabbage with crimped leaves; said to have a more delicate flavour than ordinary cabbage.

Sawdust Tiny particles of wood produced during sawing that are burned to generate smoke for fish smoking.

Saxin Trade name for saccharin.

Scaffolding Modeling of a learning support activity with cues and prompts, primarily verbal and/or manual. The cues/prompts are dropped as the child becomes proficient.

Scald 1. To pour boiling water over a food to clean it, loosen hairs, or remove the skin of fruit, tomatoes, etc.

2. To heat milk almost to boiling point, to retard souring or to make clotted cream.

3. Defect occurring in stored apples; the formation of brown patches under the skin, with browning and softening of the tissue underneath. Due to accumulation of gases given off during ripening.

Scale Apparatus for measuring the weight of an object.

Scallion Small onion which has not developed a bulb, widely used in Chinese cooking; also used for shallots and spring onion.

Scalloped dishes Food baked in a scallop shell or similar small container, usually combined with a creamy sauce, topped with breadcrumbs, and surrounded by a border of piped potato.

Scalloping A way of decorating the double edge of the pastry covering of a pie by making horizontal cuts with a knife close together round the edge of the pie, then, using the back of the knife, pulling the edge up vertically at regular intervals to form scallops.

Scallops Marine bivalve molluscs, Pectinidae spp.; Queen scallop is Chamys opercularis.

Scampi Shellfish, Norway lobster or Dublin Bay prawn, Nephrops norvegicus;

Scavengers Materials that remove gases from packaging including oxygen scavengers, ethylene scavengers, carbon dioxide scavengers and water vapour absorbents.

Schilling Test A test of vitamin B 12 nutritional status and absorption.

Schmaltz Jewish name for dripping, especially from chicken or goose.

Schnitzel Cutlet or escalope of veal or pork. Normally coated in egg and breadcrumbs and fried, then served with various garnishes.

Bismarck schnitzel is garnished with plovers' eggs, mushrooms, truffles, and tomato sauce; holsteiner schnitzel with fried egg, capers, olives, and anchovies; rahmschnitzel with paprika-flavoured cream sauce; zigeuner schnitzel with tomato sauce, mushrooms, and smoked ox tongue.

Schweitzer kraut Chopped cooked marigold leaves.

Scombrotoxicosis, scombroid poisoning, scombroid fish poisoning, scombrotoxin poisoning An illness associated with the consumption of spoiled samples of some species of fish, particularly, but not exclusively, of scombroid fish, hence the name. World-wide, it is considered to be one of the most common causes of seafood poisoning. It is usually a mild illness with rapid onset of a variety of relatively non-specific symptoms - rashes on the face, nausea, vomiting and/or diarrhoea, headache, palpitations - though a particular patient might not show all these symptoms. Sometimes symptoms are confused with those of a Salmonella infection or food allergy. Fish and products that have been implicated in such food poisoning incidents have often, but not always, been associated with high concentrations of histamine. Because of this, and the similarity between the symptoms of scombrotoxicosis and those of histamine poisoning, the illness is, or was, often referred to as 'histamine poisoning' or histamine fish poisoning'. Studies have demonstrated that the aetiology of the illness is not as simple as just ingestion of large amounts of histamine. It has been postulated that other toxic agents present in the fish may cause similar symptoms, or that there are other compounds formed in the spoiling fish which of themselves can cause the illness or interact with the histamine to potentiate the effects of histamine. Nevertheless, high histamine contents in fish are sufficiently associated with the illness for histamine content to be a indication of likely toxicosis, and it is used in quality control and mandatory inspection of species with a history of causing scombrotoxicosis.

Scum Undesirable accumulation of impurities as a film floating on the surface of brine.

Seafood A general term to include crustaceans and shellfish, sometimes also fish.

Seafood stick Characteristically pink and white sticks of cooked fish, eaten cold in salads and buffet meals, mainly made from cod or other white fish and containing little or no shellfish. Also known as crab sticks or fish sticks.

Sear To brown meat quickly in a little fat before grilling or roasting. The term is sometimes used when vegetables are browned in fat before being used in soup or sauce making.

Seasoning Normally used to mean salt and pepper, but may include any herbs, spices, and condiments added to a savoury dish.

Seaweed Marine algae of interest as food include Irish moss, laver bread and kelp, which are eaten to some extent in different communities and serve as a mineral supplement in animal feed.

Secco Dry wines.

Seco Portuguese, Spanish; dry wines.

Secretin A peptide hormone secreted by the intestinal mucosa which stimulates pancreatic secretion.

Sedoheptulose A seven-carbon sugar which is an intermediate in glucose metabolism by the pentose phosphate pathway.

Seed cake A sponge or madeira cake containing caraway seeds, mace, and nutmeg. Tipperary seed cake contains orange-flower water and caraway seeds, but no mace or nutmeg.

Seek kababs Minced meat croquettes shaped over an iron skewer and cooked rapidly by grilling.

Seitz filter A filter disc with pores so fine that they will not permit passage of bacteria; thus solutions filtered through a Seitz filter emerge sterile.

Sekt Sparkling wine, usually dry, made by tank fermentation, not the methode champenoise.

Selenium A dietary essential mineral, which is part of enzymes glutathione peroxidase and thyroxine deiodinase. Through its role in glutathione peroxidase it acts as an antioxidant, and to some extent can compensate for vitamin E deficiency. Similarly, vitamin E can compensate for selenium deficiency to some extent.

Requirements are of the order of 50 μg/day; in parts of New Zealand, Finland, and China soils are especially poor in selenium and deficiency occurs. In China selenium deficiency is associated with Keshan disease.

Rich sources include: fish and shellfish, mung (dahl) and red kidney beans, Brazil nuts, bread, kidney, lentils, liver, pork, rabbit, veal.

Selenium is toxic in excess; mild selenium intoxication results in production of foul-smelling hydrogen selenide, which is excreted on the breath and through the skin. Intakes above 450 μg/day are considered hazardous.

Seltzer Effervescent mineral water, originally from Niederselters, Germany. Trade name of effervescent headache cure.

Semillon One of the 9 'classic' grape varieties used for wine making, especially the great sauternes, graves, and white bordeaux. Traditionally called riesling in some parts of Australia.

Semi-preserves Fish products, especially those preserved in salt and acetic acid, that have a longer storage life than the untreated fish, typically 1 to 2 months in chilled storage.

Semi-seco Spanish; medium dry wines.

Semolina The inner, granular, starchy endosperm of hard or durum wheat; used to make pasta and semolina milk pudding.

Sensory' and 'organoleptic' should not be considered as synonyms. Use of 'sensoric' and 'sensorial' instead of 'sensory' is deprecated. The term 'subjective' should not be used describe sensory methods, as distinct from instrumental methods considered to be 'objective', because sensory analysisencompasses both objective and subjective methods. Objective methods are those in which experienced and trained assessors evaluate defined and described properties of products without any contribution of the personal likes and dislikes of the assessor; subjective methods are those using untrained assessors, as in market research, to give personal opinions of products. Sensory analysis is widely and extensively used in evaluating the properties of fish and fishery products in quality control and in research.

Sensory integration therapy Therapy focusing on integrating all the senses of the body. It identifies the importance of external stimulation as one part of the early intervention and educational therapy.

Sensory panel A group of persons brought together to participate in a sensory evaluation test. The preferred term for a member of a panel is 'assessor' rather than 'judge', 'panellist', 'taster' or 'subject'.

Serendipity berry Or Nigerian berry, fruit of the West African plant Dioscoreophyllum cumminsii. It has an extremely sweet taste; the active principle is a protein, monellin.

Serine A non-essential amino acid.

Serotonin A hormone that acts as a neurotransmitter, transmitting nerve signals from one neuron to another.

Serum Clear liquid left after the protein has been clotted; the serum from milk, occasionally referred to as lacto-serum, is whey.

Blood serum is blood plasma without the fibrinogen. When blood clots, the fibrinogen is converted to fibrin, which is deposited in strands that trap the red cells and form the clot. The clear liquid that is exuded is the serum.

SES Socio-economic status.

Sesame A tropical and subtropical plant, Sesamum indicum. Known as sim-sim in East Africa, benniseed in West Africa, gingelly and til in Asia. Seeds are small and, in most varieties, white; used whole in sweetmeats, in stews, and to decorate cakes and bread, and for extraction of the oil, which is used as a seasoning. The seeds contain 60% oil, of which 15% is saturated and 45% polyunsaturated. It has a strong nutty flavour and is used more as a seasoning than a cooking oil.

Sfogato Minced meat casserole.

Sfumatrice Machine for obtaining the oil from the peel of citrus fruit. Based on the principle that the natural turgor of the oil sacs forces out the oil when the peel is folded.

Shallot Bulb of the plant Allium escalonium related to the onion, with essentially the same flavour but less pungency; each plant has a cluster of small bulbs rather than the single large bulb of the onion.

Shandy A mixture of lemonade and beer, originally shandy-gaff. Ginger beer may be used for ginger beer shandy. Contains 1% alcohol.

Sharon fruit Alternative name for the sweet persimmon or kaki.

Sharples centrifuge Continuous high-speed centrifuge, consisting of a vertical cylinder. Used to separate liquids of different densities or to clarify by sedimenting solids.

Shatterpack A block of frozen fillets or portions of fish interleaved by inserting sheets of paper or flexible film such as polyethylene between the fillets or portions in a pack before freezing so that the frozen pack can be divided readily while still frozen by giving it a light blow or knock.

Shchi Green vegetable soup, sometimes called Russian cabbage soup.

Shearling 15-18-month-old sheep.

Sheer Creamed-rice dessert.

Shelf life The words are sometimes hyphenated. A concept in food science and technology applied to indicate the length of time a food product can be stored and remain acceptable, or not pose a health risk to the consumer. It stems from the wish or requirement to indicate the storage lives of products during retail display and distribution. Shelf life in this context need not be the same as storage life. For example, retail packs of frozen foods are labelled with shelf lives based on date of final manufacture and packing. This means that a product prepared from fish frozen just a few days previously will have the same indicated shelf life as one prepared from fish frozen years previously, and might have been stored under inappropriate conditions. There are other difficulties in trying to measure and indicate shelf lives or storage lives of fishery products. A major one is that there are no generally accepted criteria for defining end of shelf life. There are broadly three sets of criteria those based on risk of harm to the consumer by growth of pathogenic bacteria or development of toxins; those based on wholesomeness, that is, fit for consumption; and those based on changes that render the product unacceptable to the consumer. The first two are explicit or implicit in food regulations; the last is the criterion, additional to the first two, adopted by retailers who wish to ensure consumers like, as distinct from not disliking, their products.

Shelfing Stowage in the fishroom at sea of loose fish laid neatly in a single layer on a bed of ice on a shelf, preferably with ice also on top of the fish.

Shellfish, edible A wide range of marine molluscs and crustacea.

Shepherd's pie Dish made from minced lamb or mutton with a crust of mashed potato, baked. Cottage pie is made with minced beef.

Sherbet 1. Arabic name for water-ice, also known by French name, sorbet, and the Italian name, granite. Used to be served between courses during a meal to refresh the palate.

2. Originally a Middle-Eastern drink made from fruit juice, often chilled with snow. Modern version is made with bicarbonate of soda and tartaric acid with sugar and flavours. Sherbet powder is the same mixture in dry form.

Sherry Fortified wines from the southwest of Spain, around Jerez and Cadiz. Matured by the solera process rather than by discrete vintages; each year 30% of the wine in the oldest barrel is drawn off for bottling and replaced with wine from the next oldest; this in turn is replaced from the next barrel, and so on.

In order of increasing sweetness, sherries are: fino; manzanilla; amontillado; oloroso; amoroso or cream.

Sherry-type Wines are also produced in other countries, including South Africa, Cyprus, and Britain and may legally be described as sherry as long as the country of origin is clearly shown.

Shigella A genus of rod-shaped bacteria in the family Enterobacteriaceae. They are responsible for the disease shigellosis, Fish from unpolluted waters do not harbour the bacteria, but can become carriers if caught or harvested from polluted waters, and products after harvesting can become contaminated in unsanitary surroundings.

Shiitake Or Black Forest mushroom, Lentinula edodes.

Shir To bake food in a small shallow container or ramekin dish.

Shoo-fly pie American; pie with sweet filling made from molasses, brown sugar, and flour.

Shortbread Sweet biscuit baked with a high proportion of butter to flour. Sometimes called shortcake in the USA.

Shortcake American; rich cake with a filling of fruit and cream.

Shortening Soft fats that produce a crisp, flaky effect in baked products. Lard possesses the correct properties to a greater extent than any other single fat. Shortenings compounded from mixtures of fats or prepared by hydrogenation are still called lard compounds or lard substitutes.

Unlike oils, shortenings are plastic and disperse as a film through the batter and prevent the formation of a hard, tough mass.

Short-term memory Ability to remember things immediately after learning them.

Showarma Arabic name for döner kebab.

Shred To slice a food such as cheese or raw vegetables into very fine pieces, which often curl as they are cut.

Shredded wheat Trade name for a breakfast cereal prepared from wheat grains. Two biscuits (44 g) are a source of protein, niacin, iron, and zinc; provide 4 g of dietary fibre; supply 140 kcal.

Shrimp Small shellfish, Paleamonidae and Pandalidae spp, Crangon crangon, and Pandalus montagui.

Shrub A bottled cordial made from various fruits, spirits, and sugar.

Shucking Removal of molluscan shellfish meat from the shell to yield the shucked meat.

Sialogogue Substance that stimulates the flow of saliva.

Sialorrhoea Excessive flow of saliva.

Siderosis Accumulation of the iron-protein complex, haemosiderin, in liver, spleen, and bone marrow in cases of excessive red cell destruction and on diets exceptionally rich in iron. It is common among Bantu; apparently owing to intakes of about 100 mg of iron daily from iron cooking pots and pombe.

Sieve/strainer Used for sift dry ingredients such as flour or straining liquid off foods such as pasta and rice.

Sift To place dry ingredients such as flour and salt through a sieve to aerate and remove lumps.

Silage A liquefied product prepared by acid hydrolysis of fish or fish wastes. Used as a feed for animals.

Formic acid or a mixture of formic and propionic acids are commonly used at 3-4% of the weight of fish or wastes. These silages do not need to be neutralised before use as feeds and are not corrosive to mild steel containers. Mineral acids, for example sulphuric acids, can also be used. They are much cheaper than formic and propionic acids but are corrosive to mild steel and need neutralising before use as feeds.

Silica gel A drying agent.

Silicones Organic compounds of silicon; in the food field they are used as antifoaming agents, as semi-permanent glazes on baking tins and other metal containers, and on non-stick wrapping paper.

Silver The silver beads used to decorate confectionery. Present in traces in all plant and animal tissues but not known to be a dietary essential, and has no known function, nor is enough ever absorbed to cause toxicity.

Silverside Boned joint of beef cut from the top part of the hind leg, commonly boiled or salted, although it can be roasted.

Simmer A method of cooking in water slightly below boiling point.

Simnel cake Fruit cake with a layer of almond paste on top and sometimes another baked in the middle. Originally baked for Mothering Sunday, now normally eaten at Easter.

Simplesse Trade name for fat substitute made from milk or egg-white. Consists of 0.1-2.0 μm diameter particles which create the feeling of creaminess in the mouth but cannot be used for cooking. Yields 1.3 kcal /g, compared with 9 kcal /g for fats.

Singe To pass a plucked bird quickly over a flame to burn off the down.

Singing hinnie Large, round, currant scone baked on a girdle or hotplate, usually cut in half, buttered, and eaten hot. The name comes from the hissing noise it makes as it cooks.

Single-cell protein Collective term used for biomass of bacteria, algae, and yeast, and also moulds, of potential use as animal or human food.

Sippet A small piece of bread, fried or toasted, served as a garnish to a mince or hash.

Sirloin Upper part of the hind loin of beef.

Sitapophasis Refusal to eat as expression of mental disorder.

Sitology Science of food.

Sitomania Mania for eating, morbid obsession with food; also known as phagomania.

Sitophobia Fear of food; also known as phagophobia.

Sitosterol The main sterol found in vegetable oils, similar in structure to cholesterol; may reduce the absorption of cholesterol from the intestinal tract and has been used in treatment of hyperlipidaemia.

Skate Cartilaginous fish, Raja undulata.

Skillet Frying pan.

Skim To remove fat from the surface of stock, gravy, stews, etc., or scum from jams, etc., while they are cooking.

Skinfold thickness Index of subcutaneous fat and hence body fat content. Measured at four sites: biceps, triceps , subscapular, supra-iliac. Rapid surveys often involve only biceps.

Skink Originally an essence or extract, now a thick soupstew of meat and vegetables. skirlie Scottish; cakes of oatmeal and onion fried in pork dripping. Sold commercially in a skin, known as mealie pudding.

Skorthalia Sauce with garlic, lemon, and nuts.

Slacked fish Thawed fish as it may be termed in the USA.

Sleep apnea Temporary stopage of breathing during sleep

Slid Traditional UK name applied to a mixture of young herrings and young sprats when canned, since they are caught together and cannot be separated on a commercial scale. When fresh or frozen the mixture is termed whitebait.

Slimming clubs Groups which provide help and support for overweight and obese people trying to lose weight, including nutritional counselling and provision of clear, easy-to-follow, weight reducing diets. The most successful groups are often those in which the counsellors were once obese people themselves, and understand the problems involved in losing excess weight.

Sling Drink made from gin and fruit juice.

Slivaner A grape variety widely used for wine making, although not one of the classic varieties.

Slivovitz East European; distilled spirit made from fermented plums; similar to German quetsch and French mirabelle. Some of the stones are included with the fruit and produce a characteristic bitter flavour from the hydrocyanic acid.

Sloe gin Liqueur made by steeping wild sloes in gin or neutral spirit. Known in France as prunelle.

Sloe Wild plum, fruit of the blackthorn with a sour and astringent flavour; almost the only use for it is for the preparation of sloe gin.

SLP Speech and Language Pathologist. A person who is qualified to diagnose and treat speech, language, and voice disorders.

Smetana Thin, sour cream, originally Russian.

Smoke point The temperature at which the decomposition products of frying oils become visible as bluish smoke. The temperature varies with different fats, ranging between 160 and 260°C.

Smoking kiln There are basically two types of commercial kiln, the traditional kiln, sometimes called a smokehouse, a batch operation that depends on natural convection, and the mechanical kiln with forced convection and temperature control which can be designed as a batch or continuous operation. The smoke is usually generated by burning wood, wood chips, or sawdust obtained from hardwood or softwood or some combination of these. In some regions of the world, wood is burned instead of sawdust, and this produces a hotter fire with less smoke so that the fish is charred rather than smoked. The wood or sawdust must be dry, not mouldy, and free from preservatives. Wood smoke is a mixture of aliphatic and aromatic vapours, and condensed tarry droplets, and it has a very complex composition comprising among other components, acids, phenols, aldehydes and ketones, alcohols, and aliphatic and aromatic hydrocarbons. The droplets represent the visible component of the smoke, and the concentration of the visible smoke particles, the smoke density or thickness of the smoke, is used as an indication of the concentration of the invisible smoke vapours in a kiln. The composition and characteristics of the smoke depend on the type of wood, the water content of the wood, the temperature to which it is heated to produce the smoke, and the manner in which it is heated. Any preservative effect of the smoke is probably largely due to the presence of a range of phenolic compounds, formaldehyde and acids.

In a traditional kiln, the movement of smoke and warm air depends entirely on natural draught, usually by means of a chimney, as in the chimney kiln, where the smoke rises by convection through fish or fillets hung above a fire of smouldering sawdust. The results are very variable and there is very little element of control. It is also known as a tall kiln or vertical smokehouse. Various designs have been built such as the London kiln or smaller north kiln, both used mainly for cold smoking, or those designed mainly for hot smoking. In a mechanical kiln, the smoke is moved over the fish vertically or horizontally by means of a fan. The temperature of the smoke can be controlled much more closely because the smoke is generated outside the kiln in a separate smoke producer, designs of which vary.

The smoke can be distributed much more evenly over the fish so that the fish are all smoked and dried consistently at the same rate. The Torry kiln, originally developed in the 1930s at Torry Research Station, is a batch kiln that can be built to a range of capacities from 100 kg to 1200 kg/4 hours. The temperature, humidity and speed of a constant horizontal stream of smoke can be controlled. It is one of a number of widely used successful designs suitable for both cold and hot smoking. The smoke producer is the apparatus in which the wood is burnt to produce the smoke. It can range from a simple hearth with dampers controlling the supply of air, to an automatic smoke producer in which the wood is burned continuously and the combustion temperature is controlled. Typically, sawdust is fed continuously from a hopper onto an electrically heated plate or firebed, but smoke can also be produced by blowing hot air or super-heated steam through a bed of sawdust, or combustion can be achieved by application of friction to a block of wood. The smoke density in the kiln can be measured using a smoke density meter or smokemeter usually designed to measure the degree of light scattered by the smoke. In modern kilns the production and intensity of the smoke, and the temperature and humidity are computer controlled, the smoke is recirculated, weight loss can be monitored, emissions are controlled, and the consumption of sawdust or wood shavings is about 10% of consumption in the traditional smokehouse.

Smoking The process of applying wood smoke to impart a smoky or smoke flavour to, and to partially dry a fish, or part of a fish such as a fillet, or shellfish meat, to produce a smoked fish product, and also to extend the shelf life of the product under some conditions. In many parts of the world, preservation is still the main purpose of smoking. Any preservative effect of the smoke itself is probably largely due to the presence of a range of phenolic compounds, nitrites and formaldehyde. The principal preservative effect is due to the lowering of the water activity as a consequence of the presence of salt and the degree of drying. There are three principal parts to the whole process, salting or brining, deposition of smoke, and drying. The process can be carried out commercially on a large scale, on an artisanal scale, or domestically, but the further information given here refers to commercial production.

The fish are first prepared by washing, heading and/or gutting for some products, and splitting, or filleting, depending on the products. The prepared fish are either brined in a salt brine, that may contain a permitted dye or colour to achieve the desired yellow/orange/brown colour in some smoked fish products, or dry salted. Most fish are brined but dry salting is used for some salmon, herring and hot smoked products. After brining the fish may be left dripping or draining for several hours before smoking begins. For some products such as sides of salmon, some recipes require the fish to be dried for a period before smoking. The smoked fish is produced in a traditional or mechanical smoking kiln. In this stage the chemical components of smoke which impart the flavour, colour and some preservative effects are deposited on the fish and the fish are dried to some degree. Alternatively, the smoke flavour can be added as a mixture of natural or synthetic flavorous chemicals. For example, so called liquid smoke is a solution or condensate of flavour components of wood smoke from which the polycyclic aromatic hydrocarbons have been removed or reduced. It is sometimes called a smoke solution and is used as a smoke dip, or as a spray. The spray is suitable for application in electrostatic smoking in which the droplets are electrically charged in a high voltage electric field and can be rapidly deposited on the fish or fillet. The smoking time is the length of time a product requires in the kiln to achieve the desirable flavour, colour and degree of drying, to which must be added the required for brining, dripping, cooling and so on to give the total processing time. The degree to which smoke constituents are taken up by the fish depends on the wetness of the surface and the concentration of the smoke constituents in the vapour. There are two principal smoking processes. In cold smoking, the smoke temperature does not exceed 30°C and the fish are uncooked. Fish from temperate waters require a temperature just above 30°C for cooking to begin, especially when salt is present. Cold smoked products are intended to be cooked before consumption, but cold smoked salmon is a notable exception. In hot smoking, the smoke temperature is raised to at least 70°C at some stage during smoking, so that the fish are cooked during the process, and the product can be eaten without further cooking. In some products the temperature may be raised in stages as high as 95°C. Special attention needs to be given to food safety requirements for products that are intended to be eaten without further cooking. Traditionally, fish was subject to heavy smoking in which there was prolonged exposure to smoke to preserve it, often in combination with salting and prolonged drying. Cold smoking for a prolonged period until the fish is hard, mainly as a result of drying is known as hard smoking. Typically, the salt content is 10 to 15% and the fish loses 20 to 30% by weight during the process. Traditional smoked products are still produced and remain important in some countries, but they tend to be too salty, too dry and have too strong a smoky flavour for the mass market in developed countries. In modern smoked products for these markets, salt and smoke are used primarily to flavour the products, and mild and light smoking processes are used. These milder processes contribute only a limited preservative effect resulting in only a small extension of shelf-life compared with that of the raw fish and in some cases should be regarded having much the same shelf life as unsmoked fish. If such a product, e.g. cold smoked salmon, is intended to be eaten without further cooking, the salt concentration in the water phase of the product and proper refrigeration of the product below 5°C along the whole chill chain from producer to consumer are critical factors for product safety. The minimum salt concentration in the water phase of the product should be 3.5%, as measured in the thickest part of the fillet.

Smut Group of fungi that attack wheat; includes loose or common smut and stinking smut or bunt.

Snail The small snail eaten in Europe is Helix pomatia; giant African snail is Achatima fulica.

Snibbing Topping and tailing of gooseberries.

Snow Mixture of sweetened fruit pulp with whisked egg-white, normally served with sponge fingers or biscuits, or on a bed of sponge cake soaked in fruit juice.

Soapstock In the refining of crude edible oils the free fatty acids are removed by agitation with alkali. The fatty acids settle to the bottom as alkali soaps and are known as soapstock or 'foots'.

Sobresada Spanish; spiced sausage containing pork, tripe, paprika and cayenne pepper; large in diameter and soft in texture, so it can be spread like a pate. A firmer variety, which requires cooking, is longaniza.

Soda bread Made from flour and whey, or buttermilk, using sodium bicarbonate and acid in place of yeast. Common in Ireland. soda water Artificially carbonated water, also known as club soda; if sodium bicarbonate is also added, the product is seltzer water.

Sodium bicarbonate Also known as baking soda or bicarbonate of soda, liberates carbon dioxide when in contact with acid. Used as a raising agent in baking flour confectionery. A small pinch of sodium bicarbonate preserves the green colour in cooked vegetables. Also helps to reduce acidity when stewing sour plums or rhubarb.

Sodium chloride Common salt, the commonest form in which sodium is consumed.

Sodium controls The retention of fluid in the body, and reduced retention, aided by low-sodium diets, is required in cardiac insufficiency accompanied by oedema, in certain kidney diseases, toxaemias of pregnancy, and hypertension. Low-sodium diets are usually about 2.3 g and can be as low as 0.6 g. To improve the palatability of such diets, 'salt' mixtures are available, containing potassium and ammonium chlorides together with citrates, formates, phosphates, glutamates, as well as herbs and spices. Such mixtures may be contra-indicated in conditions where potassium intake also has to be restricted.

Sodium Mineral used by the body to regulate blood pressure. Sodium chloride is the chemical name of salt, which is 40% sodium by weight. Salt is the most common source of sodium in the diet. Diets high in sodium increase the risk of high blood pressure in some adults. High blood pressure increases the risk of death from heart attack or stroke.

Sofrito Spanish; sauce prepared from onions, tomatoes, garlic, and other vegetables fried gently in oil.

Soft drinks Term applied to non-alcoholic drinks, usually fruit juice or fruit-flavoured, but also a variety of carbonated beverages. Various concentrations and preparations are termed squash, crush, and cordial, which usually require dilution before drinking; others are ready to drink. In the

USA cider means unfermented apple juice, while the fermented product is hard cider.

Softwood Wood from coniferous trees such as pine; the sawdust produces a resinous smoke when burned. It is less suitable for imparting a smoky flavour to fish than that from hardwood, but the smoke from softwoods generally colours the fish more quickly. Wood treated with preservatives must not be used for smoking.

Sol A colloidal solution, i.e. a suspension of particles intermediate in size between ordinary molecules and coarse particles. A jelly-like sol is a gel.

Solanaceae Family of plants including aubergine, Cape gooseberry, potato, tomato.

Solanine Heat-stable toxic compound, found in small amounts in potatoes, and larger and sometimes toxic amounts in sprouted potatoes and in skin when they become green through exposure to light. Causes gastrointestinal disturbances and neurological disorders; 20 mg solanine per 100 g fresh weight of potato tissue is the upper acceptable limit.

Soluble Something that is soluble is able to be dissolved to form a solution. For example, instant coffee is soluble.

Solution A liquid into which a solid has been mixed and has dissolved.

Somatic cells Cells other than sex or germ cells. High somatic cell counts found in milk indicate mastitis in dairy cows. Low somatic cell counts indicate better quality milk and net a premium price from dairy processors.

Sometribove Generic name used by the U.S. Food and Drug Administration to refer to commercial bovine growth hormone/bovine somatotropin. Posilacþ is Monsanto Company's trade name under which it sells sometribove.

Sorbic acid An organic acid found naturally in some fruits. It is a permitted food additive as a preservative and is used as the sodium or potassium salt for this purpose in some fishery products, especially dried products.

Sorbistat Trade name for sorbic acid and its potassium salt.

Sorbitol A six-carbon sugar alcohol found in plums, apricots, cherries, and apples and manufactured from glucose. Although it is metabolized in the body, yielding the same amount of energy as other carbohydrates, 4 kcal.

Sorghum Sorghum vulgare, S. bicolor; cereals that thrive in semi-arid regions and provide important human food in tropical Africa, Central and North India, and China. Sorghum produced in the USA and Australia is used for animal feed. Also known as kaffir corn, guinea corn, jowar, Indian millet, and millo maize. The white-grain variety is eaten as meal; the redgrained has a bitter taste and is used for beer; sugar syrup is obtained from the crushed stems of the sweet sorghum. A rich source of protein, vitamin B 1, niacin, and iron; a good source of zinc; a source of vitamin B 2.

Sorghum syrup The concentrated juice from a sweet variety of sorghum.

sorrel A common wild plant; the leaves have a strong acid flavour, and are cooked together with spinach or cabbage, used to make soup, and used in salads.

Sorrel syrup Caribbean; syrup prepared from rosella flowers. soubise Dish flavoured with onion or garnished with onion puree; soubise sauce is a white sauce containing onion.

Sorting Separating into defined categories of species, sex, length, weight, either at sea or onshore, or grading to sort into defined categories of some aspect of quality such as size or freshness. If used, the terms sorting and grading should be qualified to indicate the type of sorting or grading, e.g.

species sorting or size grading, to avoid misinterpretation. Sorting is often used in the context of defining the out-grade to be discarded because, the weight, or the degree of spoilage, freezer-burn, parasitic infestation, colour, blemishes, damage, taint, belly-burst, or some other quality aspect is unacceptable. Sorting into various types of categories may be carried out by eye, by smelling or tasting. It may be carried out mechanically for size by means of inclined roller graders of various designs, or by automatically making a physical measurement such as weighing, or by scanning, for example, by computer aided vision, laser or infrared, all of which can be linked electronically to control sorting flaps on a conveyor system.

Souffle Light, fluffy, baked dish of eggs, flour, and butter with various fillings; may be sweet or savoury.

Soul food Afro-Caribbean term for food with traditional or cultural links, having emotional significance.

Sous vide Usually refers to composite foods intended for catering outlets, e.g. fish or shellfish with vegetables and sauce, sealed in a vacuum pack and pasteurised. These products are often given an extended shelf life at chilled temperatures of 2°C or less.

Souvlaki, souvlakia kebabs, especially shishkebab. sow thistle A wild plant; the young leaves can be used in salads, and the roots can be roasted or boiled, and eaten in the same way as salsify. Also known as milk weed or hare's lettuce.

Soy sauce A condiment prepared from fermented soya bean, commonly used in China and Japan. Traditionally the bean, often mixed with wheat, is fermented with Aspergillus oryzae over a period of 1-3 years. The modern process is carried out at a high temperature or in an autoclave for a short time.

Soya A bean important as a source of both oil and protein. The protein is of high biological value, higher than that of many other vegetable proteins, and is of great value for animal and human food. When raw it contains a trypsin inhibitor, which is destroyed by heat. Native of China, where it has been cultivated for 5000 years; grows 60-100 cm high with 2-3 beans per pod. Contains indigestible sugars, stachyose and raffinose, which can cause flatulence. A good source of protein and iron; a source of niacin and calcium.

Soybean milk Extract of the soya bean.

Spanish toxic oil syndrome Widespread disease in Spain during 1981/2, with 450 deaths and many people chronically disabled, due to consumption of an oil containing aniline-denatured industrial rapeseed oil, sold as pure olive oil. The disease appears to be unique and the precise cause is unknown.

Spans Trade name for non-ionic surface agents derived from fatty acids and hexahydric alcohols. They are oil-soluble, in contrast to Tweens which are water-soluble or well-dispersible in water. Used in bread as crumb-softeners to improve doughs, cakes, and biscuits, and as emulsifiers.

Spareribs Pork ribs with most of the surrounding meat removed.

Spätzle Very small noodles made by forcing noodle dough through a colander directly into boiling water.

Species Group of organisms with common or similar characteristics and capable of inter breeding.

Species identification Determination of the species of a fish or part of a fish to establish its authenticity. Identification might be required to support enforcement of regulations, for example restrictions on fishing for a particular species or enforcement of regulations relating to naming of fish offered for sale, or in testing for compliance with specifications. If the whole fish, or substantial parts of it, are available then identification is usually a matter of taxonomy. Identification becomes difficult when parts without clear distinguishing features, for example fillets, are only available. Chemical methods can be used to determine species. Frequently used procedures are electrophoresis or iso-electric focussing of proteins. These procedures are not always successful with heat processed fish as the extreme heat processing renders the proteins insoluble in the simple aqueous solution used for raw fish.

Stronger solvents can be used to solubilise the proteins but, in this case, there is less differentiation among species. More recently, methods based on identification and comparison of DNA fragments have been developed.

Specific heat capacity The quantity of heat required to raise the temperature of the unit weight of a substance by one unit.

Specific-purpose foods Foods produced for a specific purpose such as, to supply military personnel, for space explorations, to supply food for areas of temperature extremes or to supply high-protein biscuits for famine areas, and more recently, foods developed for a more specific health purpose in the general population.

Spectrophotometer Instrument that measures the amount of light absorbed at any

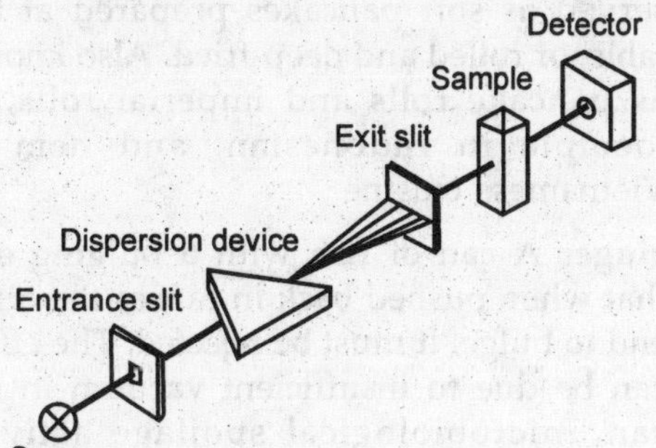

Fig. Spectrophotometer

particular wavelength, which is directly related to the concentration of the material in the solution. Used extensively to measure substances that have specific absorption in the visible, infrared, or ultraviolet range, or can react to form colour derivatives.

Speech therapy Therapy focusing on improvement of oral motor skills, and to learn both receptive and expressive language.

Spelt Coarse type of wheat, mainly used as cattle feed. spent wash Liquor remaining in the whisky still after distilling the spirit. A source of unidentified compounds that are growth factors for chicks. When dried it is known as distillers' dried solubles.

Sphingomyelins Complex phospholipids found in cell membranes, and especially in brain and nerve tissue; composed of the base sphingosine plus fatty acids, phosphoric acid, and choline.

Spices Distinguished from herbs in that part, instead of the whole, of the aromatic plant is used: root, stem, or seeds. Originally used to mask putrefactive flavours. Some have a preservative effect because of their essential oils, e.g. cloves, cinnamon, and mustard.

They are consumed in amounts too small to provide any nutrients except possibly for curry powder, which may contain 7.5-10 mg of iron in a 10-g portion.

Spina bifida Congenital neural tube defect due to developmental anomaly in early embryonic development. Supplements of folic acid, begun before conception, reduce the risk.

Spinach Leaves of Spinacia oleracea.

Spirits Beverages of high alcohol content made by distillation of fermented liquors, including brandy, gin, rum, vodka, whisky; usually 40% alcohol by volume. A standard measure was formerly one-fifth or one-sixth of a gill; now 25 mL. spirit, silent Highly purified alcohol, or neutral spirit, distilled from any fermented material.

Spirometer Apparatus used to measure the amount of oxygen consumed from which to calculate the energy expended.

Spirulina Blue-green alga which can make use of atmospheric nitrogen to form proteins; eaten for centuries round Lake Chad in North Africa, and in Mexico. Many health claims are made, but are negated by the small amounts eaten. spit Thin metal bar on which meat, poultry, or game is roasted in front of an open fire, and rotated during cooking; now also inside an oven or grill.

Spleen Gland near the stomach with main function of destroying 'worn-out' red blood cells and recycling the iron. As a food it is called melts; A rich source of iron, vitamins B 2, niacin, and C.

Splitting Production of a split fish by cutting partly through a gutted fish along its length from throat to vent or tail and laying it open to expose some or all of the backbone prior to smoking; sometimes the head is removed. Splitting may be done in different ways depending on the product, for example, along the belly as for a finnan or along the back as for a kipper.

Spoilage The process of rendering a stored food unfit for consumption, particularly by the action of microorganisms such as bacteria, moulds, and yeasts. The spoilage process in fish and shellfish produces unpleasant odours and flavours and adversely affects appearance and texture. The rate of spoilage is much reduced by chilling, but even under well chilled conditions, the first signs of spoilage are detected in white fish, for example, after about 7 days and the fish is unfit for consumption after about 15 days. Apart from temperature, pH affects the rate. Low pH fish such as tunas spoil more slowly than typical demersal fish. Small, fatty, pelagic fish such as sprat and herring spoil more rapidly than typical demersal fish. Other, non-microbiological, reactions, such as the hydrolysis and oxidation of oils in fatty fish, contribute to the off-flavours of spoiled fish and can be included in the general concept of spoilage. Reactions that produce off-flavours, but are not associated with microbiological spoilage, for example production of off-flavours in frozen stored fish, should not be referred to as spoilage.

Sponge Light-textured cake made from self-raising flour beaten with butter, eggs, and sugar.

Spores In relation to bacteria they are the resting state; they are thickwalled and highly resistant to damage by heat. Under suitable conditions they germinate to produce bacteria. Not all bacteria can form spores; the so-called spore-bearers are a health hazard because they can survive pasteurization and sterilization.

Spotted dick Steamed or boiled suet pudding containing currants and sultanas.

Sprat Small oily fish, Sprattus sprattus, fresh or frozen; young are canned as brisling.

Spray dryer Equipment in which the material to be dried is sprayed as a fine mist into a hot-air chamber and falls to the bottom as dry powder. The period of heating is very brief and so nutritional and functional damage are avoided. Dried powder consists of hollow particles of low density; widely applied to many foods and pharmaceuticals.

Spring rolls Chinese; pancakes filled with quick-fried vegetables and meat; may be served as soft pancakes prepared at the table, or rolled and deep-fried. Also known as pancake rolls and imperial rolls, as loempia in Indonesian, and nem in Vietnamese cuisine.

Springer A can of fish with a bulging end that when pushed back in causes the other end to bulge. It must be rejected. The effect can be due to insufficient vacuum in the can, microbiological spoilage activity occurring post-processing, or internal corrosion in the can.

Spruce beer Branches, bark, and cones of black spruce boiled for several hours, then put in a cask with molasses, hops, and yeast, and allowed to ferment.

Squid Marine creature with elongated body and eight arms, Loligo and Illex spp.

Stabilisers Substances which allow food compounds which do not mix well to be mixed and stay in a homogeneous state.

Stabismus Imbalance in the muscles of the eye so that they cross and cannot focus properly.

Stachyose Tetrasaccharide sugar composed of two units of galactose and one each of fructose and glucose. Not hydrolysed in the human digestive tract and passes to the large intestine, where it is fermented by bacteria. Present in soya beans and some other legumes; gives rise to the flatulence commonly associated with eating beans.

Stack burn In canning, an undesirable effect of over-processing caused by the stacking together of processed cans while still hot.

Staling Starch has a crystalline structure which is lost during baking. Subsequently the starch recrystallizes, i.e. it retrogrades and, in the instance of bread, the crumb loses its softness and the bread goes stale.

Standard A specification that is issued by a regulatory body, standards agency, or an international agency, such as Codex Alimentarius, closely linked with governments.

Staphylococcus aureus Species of food poisoning bacteria. Widely distributed on human skin, hair and in the nose and throat of healthy people, and the cause of boils and a variety of skin and wound infections. It can contaminate and grow on foods under favourable conditions and some strains may produce heat stable enterotoxins that cause abdominal pain, sickness and diarrhoea in 6 to 24 hours after consumption. It is used as an indicator of general hygiene and good food handling practices where less than 100 organisms/g in a sample is considered satisfactory, but greater than 1000/g may indicate the potential presence of enterotoxins and the risk of food poisoning. Enterotoxins from prior contamination may still be present after heat processing despite the absence of viable S. aureus cells.

Starch equivalent A measure of the energy value of animal feedingstuffs; the number of parts of pure starch that would be equivalent to 100 parts of the ration as a source of energy.

Starch, modified Starch altered by physical or chemical treatment to give special properties of value in food processing, e.g. change in gel strength, flow properties, colour, clarity, stability of the paste. Acidmodified starch results from acid treatment that reduces the viscosity of the paste.

Starter Culture of bacteria used to inoculate or start growth in a fermentation, e.g. milk for cheese production, or butter to develop the flavour.

Steak A slice of varying thickness cut from across a fish at right angles to, and including a piece of, the backbone.

Steam baking In baking an even temperature is maintained in the oven by means of closed pipes through which steam circulates. This is sometimes erroneously taken to mean that the bread is baked in live steam.

Steamed buns Steamed dough buns containing a savoury meat and vegetable stuffing, also known as pao-tzu.

Steaming Method of cooking food in the steam above boiling water; there is less loss of nutrients into the cooking water than with boiling. Steaming is also carried out above 100°C by means of pressure cookers.

Stearic acid Saturated fatty acid with eighteen

carbon atoms; is present in most animal and vegetable fats.

Steatorrhoea Excretion of faeces containing a large amount of fat, and generally foul-smelling. May be due to lack of bile, lack of lipase in the digestive juices, or defective absorption of fat. Treatment is by feeding low-fat diet.

Steep The process of leaving a food to stand in water, either to soften it or to extract its flavour and colour. Also the preparation of fruit liqueurs by steeping fruit in spirit.

Steer Bull castrated when very young; if castrated after reaching maturity, it is known as a stag.

Stelk Irish; potatoes boiled, then mashed with milk in which spring onions have been simmered. Also called thump.

Stercobilin One of the brown pigments of the faeces; formed from the bile pigments, which, in turn, are formed as breakdown products of the haemoglobin of old red blood cells.

Stereo-isomerism Occurs when compounds have the same molecular and structural formula, but with the atoms arranged differently in space.

Sterile Free from all living micro-organisms.

Sterilisation A process in which foods are treated to kill all forms of micro-organisms and spores. Foods can be sterilised with high temperature treatment or with ionising radiation.

Sterility, commercial Canned foods that are not sterile but which will not spoil during storage, because of the high acid content of the food, or the presence of pickling salts, or a high concentration of sugar.

Steroids Chemically, compounds that contain the cyclopentenophenanthrene ring system. All the biologically important steroids are derived metabolically from cholesterol; they include vitamin D, and hormones including the sex hormones and the hormones of the adrenal cortexsitosterol.

Sterols Alcohols derived from the steroids; including cholesterol, ergosterol in yeast, sitosterol and stigmasterol in plants, and coprosterol in faeces.

Stevia leaves Leaves of the Paraguayan shrub, Stevia rebaudiana, the source of stevioside and rebaudioside, also known as yerba dulce.

Stevioside Naturally occurring glucoside of steviol, a steroid derivative, which is 300 times as sweet as sucrose. Isolated from leaves of the Paraguayan shrub, yerba dulce, the same source as rebaudioside.

Stevix Trade name for mixture of the sweet glycosides extracted from stevia leaves.

Stew Meat and vegetables cooked together, also known as hotpot. Two main types: brown stew in which meat, vegetables, and flour are fried together before stewing, and white stew in which the ingredients are not fried first. Irish stew is a thin white stew.

Stewing Slow cooking in an enclosed pan at a temperature below boiling point, about 90°C. Such slow cooking is useful for low-quality meat , since it slowly breaks down the connective tissue to gelatine and so softens the meat.

Stickwater The aqueous fraction from pressing cooked fish in the manufacture of fish meal. Contains amino acids, vitamins, and minerals, and is either added to animal feed or mixed back with the fish meal and dried. Also known as fish solubles.

Stilboestrol Synthetic substance with potent activity as female sex hormone; the first non-steroid compound developed to have oestrogen activity. Formerly widely used both clinically and for chemical caponization of cockerels and to stimulate the growth of cattle.

Stilton Semi-hard, creamy white or blue-veined English cheese made only in a very restricted area of the Vale of Belvoir in Leicestershire, England, but named after the village of Stilton, Huntingdonshire.

Matured 3-4 months; for production of blue Stilton the cheese is pricked with stainless steel wires during ripening to encourage growth of the mould Penicillium roquefortii. A rich source of vitamin B 12, a source of protein and niacin, contains 11 g of fat and supplies 125 kcal.

Stinkers Refers to fish, especially cod, that have an intrinsic unpleasant odour and flavour, often described as weedy, as a result of feeding on planktonic molluscs called pteropods.

Stiparogenic Foods that tend to cause constipation.

Stiparolytic Foods that tend to prevent or relieve constipation.

Stir fry Chinese method of cooking; sliced vegetables and meat fried for a short time in a small amount of oil, normally in a wok, over high heat with constant stirring.

Stirabout Irish name for porridge.

Stirrup cup A hot punch traditionally served to the huntsmen at the start of a hunt.

Stobb Strawberry stalk.

Stock cubes Ready made, dried, preparations of stock. Most contain relatively large amounts of salt, as well as colouring and flavouring.

Stock The juice obtained by boiling meat, fish, or vegetables, used to prepare soups, stews, and gravy. Meat and bone contain collagen, which is converted into gelatine by prolonged boiling; hence, the stock may set to a gel on cooling. The main nutritive value of stock is the mineral content.

Stockfish A traditional dried cod product in Iceland and Norway. Traditionally, it is made by hanging up headless cod in the open air, often for 6 weeks or more, until dry and hard, and the water content has fallen to about 15%, so that moulds do not grow. Larger fish are usually split and may be hung in two pieces. Some other species such as ling and blue ling may also be treated in this way. The fish can now be dried much more rapidly in a dryer which changes the characteristics of the product compared with the traditional product.

Stollen Fruit-filled loaves of yeast dough.

Stork process The process of ultra-high temperature sterilization of milk followed by sterilization again inside the bottle.

Stracciatella Soup into which beaten egg is introduced while boiling, forming curls or strands.

Strasbourg pate The true gourmet pate de fois gras, prepared from the livers of geese which have been force-fed to achieve a very high fat content.

Straw potatoes Very thin strips of potato, deep-fried. Also known as pommes allumettes.

Strawberry Fruit of Fragaria spp., a perennial herb of American origin, introduced into the UK around 1600.

Streusel Middle-European; a mixture of flour, sugar, butter, cinnamon, and chopped almonds, sprinkled over cakes and cookies.

Stringy Something that is stringy has a characteristic that is similar to string.

Stroke Also known as cerebrovascular accident; damage to brain tissue by hypoxia due to blockage of a blood vessel as a result of thrombosis, atherosclerosis, or haemorrhage. The severity and nature of the effects of the stroke depend on the region of the brain affected and the extent of damage. Hypertension and hypercholesterolaemia are major risk factors.

Strudel Sweet or savoury pastry made from paper-thin dough.

Struvite Crystals of calcium or magnesium ammonium phosphate occasionally found in canned fish and shellfish, especially in salmon, tuna, mackerel and crustacean meat, and sometimes in chilled or frozen, smoked fish. The crystals are transparent, have a glassy consistency and are harmless, but can be mistaken for shards of broken glass. Occurrence of struvite is linked to the presence of magnesium chloride, a common component of unrefined salt, in the brine used in processing, and to high, post-mortem pH in the meat. Common preventive measures include addition of either sodium hexametaphosphate or citric acid both of which sequester free magnesium and calcium ions, or lowering the pH of the product to avoid precipitation of the crystals.

Stubbs and more factor Factor for calculating the amount of fat-free meat in a product from total nitrogen content.

Stuffing Savoury mixture used to give flavour to a dish; may be placed in a body cavity, as with poultry, laid flat between two portions, or rolled into boneless joint of meat. May be a mixture of breadcrumbs, flour, chestnut, etc., with herbs and spices or may be minced meat.

Stunting Reduction in the linear growth of children, leading to lower height for age than would be expected, and generally resulting in lifelong short stature. A common effect of protein-energy malnutrition, and associated especially with inadequate protein intake.

Sublimation To change, when heated, from a solid state to a vapour without going through the liquid state.

Substrate 1. The substance on which an enzyme acts.

2. The medium on which micro-organisms grow.

Subtilin Antibiotic isolated from a strain of Bacillus subtilis grown on a medium containing asparagine. Used as a food preservative, as it reduces the thermal resistance of bacterial spores and so permits a reduction in the processing time.

Sucaryl Trade name for sodium or calcium salt of cyclohexyl sulphamate.

Succotash American; sweetcorn kernels with green or lima beans.

Sucking pig Piglet aged 4-5 weeks, usually stuffed and roasted whole.

Sucralose Trade name for synthetic chlorinated sucrose, 2000 times as sweet as sucrose, stable to heat and acid.

Sucrase The enzyme that hydrolyses sucrose to yield glucose and fructose. Also known as invertase or saccharase.

Sucron Trade name for mixture of saccharin and sucrose, four times as sweet as sucrose alone.

Sucrose Cane or beet sugar. A disaccharide composed of glucose and fructose.

Sucrose esters Di- and trilaurates and mono- and distearates of sucrose. Used as emulsifiers, wetting agents, and surface active agents, e.g. for washing fruits and vegetables, as anti-spattering agents, anti-foam agents, and anti-staling or crumb-softening agents.

Suet Solid white fat around the kidneys of oxen and sheep, used in baking and frying.

Sufu Fermented product made by inoculating soybean curd with the mould Actinomucor elegans; stored after adding salt and alcohol.

Sugar 1. Commonly table sugar or sucrose, which is extracted from the sugar beet or sugar cane, concentrated, and refined. Molasses is the residue let after the first stage of crystallization and is bitter and black. The residue from the second stage is treacle, less bitter and viscous than molasses. The first crude crystals are Muscovado or Barbados sugar, brown and sticky. The next stage is light brown, Demerara sugar. Refined white sugar is essentially 100% pure sucrose; technically described as semi white, white, and extra-white. Yields 3.9 kcal (16 kJ)/g or 24 kcal (100 kJ) per teaspoonful. Soft sugars are fine-grained and moister, white or brown.

2. Chemically a group of compounds of carbon, hydrogen, and oxygen. The simplest sugars are monosaccharides. They may contain three (triose), four (tetrose), five (pentose), six (hexose), or seven carbon atoms, with hydrogen and oxygen in the ratio C n H 2n.O n. The nutritionally important monosaccharides are hexoses: glucose, fructose, and galactose. Two pentoses are also important: ribose and deoxyribose.

Disaccharides consist of two monosaccharide units, linked by condensation. The nutritionally important disaccharides are sucrose, lactose, and maltose

Trisaccharides consist of three monosaccharide units and tetrasaccharides of four. Larger numbers of monosaccharide units make up oligosaccharides. Most tri-, tetra-, and oligosaccharides are not digested by human enzymes, but are substrates for bacterial fermentation in the large intestine.

Sugar alcohols Also called polyols, chemical derivatives of sugars that differ from the parent compounds in having an alcohol group (CH_2OH) instead of the aldehyde group (CHO); thus mannitol from mannose, xylitol from xylose, lacticol from lactulose. Several occur naturally in fruits, vegetables, and cereals. They range in sweetness from equal to sucrose to less than half. They provide bulk in foods such as confectionery, and so are called bulk sweeteners. They are slowly and incompletely metabolized so that they are tolerated by diabetics and provide less energy than sucrose: they are less cariogenic than sucrose, especially hydrogenated glucose syrup, isomalt, sorbitol, and xylitol.

The energy yields differ, but the EU has adopted a value of 2.4 kcal (10 kJ) per gram for all polyols. They are considered safe and have no specified ADI, meaning that they can be used in foods in any required amount; however, a fairly large amount, more than 20-50 g per day can cause gastro-intestinal discomfort and have a laxative effect. For labelling purposes they are included with carbohydrates, not sugars; they do not ferment and so do not damage teeth.

Sugar substitutes Also called artificial sweeteners. These compounds provide a sweet flavor but do not contain sugar. Aspartame, saccharin, and acesulfame K are three examples.

Sulphate The mineral sulphur occurs in foods and in the body in two main forms: as sulphates; and in the sulphur amino acids methionine and cysteine.

Sulphites Salts of sulphurous acid (H_2SO_3) used as sources of sulphur dioxide (SO_2).

Sulphur An element that is part of the amino acids cystine and methionine and is therefore present in all proteins. It is also part of the molecules of vitamin B 1 and biotin and occurs in foods and in the body as sulphates. Apart from these amino acids and vitamins, there appears to be no requirement for sulphur in any other form and no deficiency has ever been observed, although it is essential for plants. Not only was the old-fashioned remedy of sulphur and molasses quite unnecessary, but elemental sulphur is not used by the body.

Sulphur dioxide Preservative used in gaseous form or as salts for fruit drinks, wine, comminuted meat, and as a processing aid to control physical properties of flour; also prevents enzymic and non-enzymic browning. Protects vitamin C but destroys vitamin B 1. Prepared by ancient Egyptians and Romans by burning sulphur and used to disinfect wine.

Sulphuring Preservation by sulphur dioxide.

Sultanas Made by drying the golden sultana grapes grown in Turkey, Greece, Australia, and South Africa; the bunches are dipped in alkali, washed, sulphured, and dried. Sultanas of the European type produced in the USA are termed seedless raisins.

Summer pudding Cold sweet of stewed fruit cased in bread or sponge cake; traditionally summer soft fruits such as raspberries, blackberries, strawberries, and currants are used.

Sundae Dessert of ice cream and fruit.

Sunflower Annual plant, Helianthus annuus, residual oilseed cake is used as animal feed.

Sunlight flavour Name given to unpleasant flavours developing in foods after exposure to sunlight. In milk it is said to be due to the breakdown of methionine in the presence of vitamin B 2 ; in beer due to a change in the bitter principles from the hops.

Superchilling Reducing the temperature of fish to just below -1.5°C and maintaining this temperature under closely controlled conditions to extend the storage life of the fish foı a few days. The most beneficial temperature for white fish is -2°C but the fish becomes partially frozen which causes practical difficulties and should be avoided if the intention is to chill the fish and keep it chilled. Domestic refrigerators in some countries have been designed with a compartment intended to maintain a temperature in the region of -3°C to extend the shelf-life of chilled products. The system is claimed to have benefits for the extension of shelf life of some fatty fish. Sometimes referred to as deep-chilling.

Superglycerinated fats Neutral fats are triacylglycerols, i.e. with three molecules of fatty acid to each molecule of glycerol. Mono- and diacylglycerols are known as superglycerinated high-ratio fats or fat extenders (E471).

Glyceryl monostearate (GMS) is solid at room temperature, flexible, and non-greasy; it is used as a protective coating for foods, as a plasticizer for softening the crumb of bread, to reduce spattering in frying fats, as an emulsifier and stabilizer. Glyceryl mono-oleate (GMO) is semi-liquid at room temperature.

Superoxide dismutase An enzyme of the metal group which catalyzes the superoxide anion, a toxic form of oxygen, and converts it to hydrogen peroxide.

Superoxide One of several toxic forms of oxygen, referred to collectively as free radicals. It is produced by reduction of molecular oxygen in many biological oxidations. It is scavenged by superoxide dismutase.

Sur commande French menu term for dishes that take time to prepare and are only cooked when ordered.

Sur lattes With reference to wines, means storage 'on their side' so that the cork is kept moist and airtight.

Surfactants Surface active agents; compounds that have an affinity for fats and water and so act as emulsifiers, e.g. soaps and detergents. Used as wetting agents to assist the reconstitution of powders, including dried foods, to clean and peel fruits and vegetables, also in baked goods and comminuted meat products.

Surimi A washed, refined, and stabilised fish mince, a process originally developed in

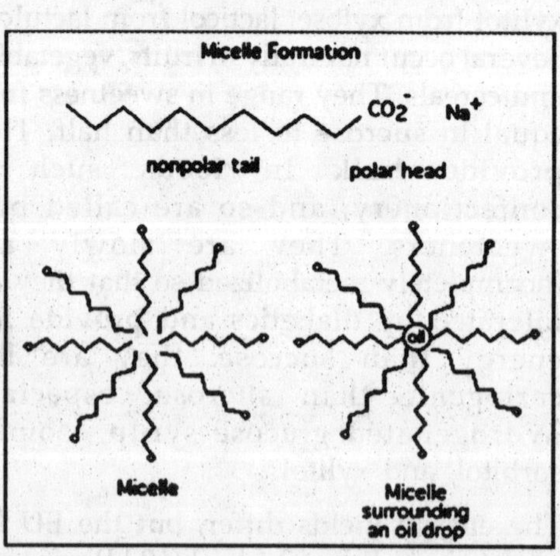

Fig. Surfactants

Japan to provide an intermediate and relatively stable frozen material suitable for the subsequent production of traditional kamaboko products. Preparation of surimi is a means of intensifying the gelling properties of fish protein and of preserving these properties during frozen storage. A typical process is outlined below. The minced fish from a bone separator is washed successively in water and dilute salt solution (0.15%) to wash out the soluble proteins, enzymes, blood, pigments, other solubles, and fat, and to concentrate the myofibrillar proteins. It is strained or refined in a refiner/strainer to remove unwanted fragments of skin and connective tissue, bones and dark muscle. Next, while keeping the temperature low, it is quickly ground finely in a silent cutter or ribbon blender, and uniformly mixed with sucrose (4%), sorbitol (4%) and polyphosphate (0.3%), or in other combinations. These act as cryoprotectants, protecting the functional gelling properties of the myofibrillar proteins during frozen storage in blocks at -30°C. Salt may also be added (2.5%) at this stage. Fish gels that have the textural characteristics termed ashi, and are light or white in colour, provide the basis for the traditional kamaboko products can be produced from the thawed surimi material by adding salt and heating at about 90°C. In this process, white fish such as Alaska pollack produce a cohesive, elastic texture (ashi) that is highly prized in Japanese kamaboko products. Consistent production of these gels and careful control of the functional properties has been used in Japan and other countries as the basis for development of a variety of novel and successful fabricated, shellfish analogue products, simulating, for example, crab, prawn, scallop and lobster meat, and these are known generically in Japan as neriseihin. A variety of white fish species may be used to successfully produce surimi but, overall, the yield is low and the process consumes large volumes of water. Surimi from fatty fish is much darker, less stable, has inferior texture, and the yield is even lower.

Surullitos Caribbean; small cylindrical biscuits made from corn meal and cheese.

Susceptors Strips of material, usually metallised polyester film/paper laminate, attached inside a microwave package to concentrate heating over the foods that need to be browned. A de-metallising process applied to the laminate is able to remove different amounts of the metal to ensure even cooking for different components or for different areas of a product.

Sushi Thinly sliced raw fish.

Sussex bread Old English loaf made from wheat and rye flour mixed with grains, given to servants; now revived.

Süssreserve Unfermented grape juice added to wines after fermentation to increase sweetness, especially in Germany, England, and New Zealand.

Sustagen Trade name for a food concentrate in powder form, also useable for tube feeding; mixture of whole and skim milk, casein, maltose, dextrins, and glucose.

Sweating Moisture exuding from the interior to the surface of a fish while cooling after smoking. If it is packed while still hot, a moist product is produced that encourages the growth of moulds.

Swede Root of Brassica rutabaga or Swedish turnip; called rutabaga in the USA.

Sweeney Osteomalacia in livestock due to phosphate deficiency.

Sweet cicely A wild plant with a smell of aniseed. The aniseed-flavoured leaves are used to flavour fruit cups, fruit salads, and cooked fruit; the main root can be boiled, sliced, and used in salads. Also known as sweet chervil.

Sweetbread Butchers' term for pancreas or thymus.

Sweeteners Four groups of compounds are used to sweeten foods: 1. The sugars, of which the commonest is sucrose. Fructose has 173% of the sweetness of sucrose; glucose, 74%; maltose, 33%; and lactose, 16%. Honey is a mixture of glucose and fructose;

2. Bulk sweeteners, including sugar alcohols;

3. Synthetic non-nutritive sweeteners, which are many times sweeter than sucrose, such as acesulphame K, aspartame, cyclamate, dulcin, P4000, saccharin, sucaryl;

4. Various other chemicals such as glycerol and glycine, and certain peptides.

Sweetness One of the five basic senses of taste.

Swells Infected cans of food swollen at the ends by gases produced by fermentation. A 'hard swell' has permanently extended ends. If the ends can be moved under pressure, but not forced back to the original position, they are 'soft swells'. 'Springers' can be forced back, but the opposite end bulges. A 'flipper' is a can of normal appearance in which the end flips out when the can is struck.

Hydrogen swells are harmless, and due to acid fruits attacking the can.

Swiss chard The spinach-like leaves and broad midrib of Beta vulgaris var. cicla, also known as leaf beet, leaf chard, sea kale beet, silver beet, white leaf beet, spinach beet. A rich source of vitamin A; a good source of vitamin C; a source of iron. The term is also used for blanched summer shoots of globe artichoke and for inner leaves of cardoon, Cynara cardunculus.

Swiss cheese American term for hard cheese such as Emmental and Gruyère: may be made in the USA or imported from Switzerland.

Swiss roll Thin rectangle of sponge cake, spread with jam or other filling and rolled into a cylinder. Known in the USA. as jelly roll.

Swordfish Oily fish, Xiphias gladius.

Syllabub Elizabethan dish made of cream curdled with wine or cider; the thickened version is used as a dessert and a thinner version as a drink.

Synapse The space between neurons where neurotransmitters are used to communicate.

Syndrome A condition defined by a cluster of related symptoms or disorders.

Synthesis The process whereby separate elements are combined to form a new complex product, synthetic chemical compound or material.

T

T3 Tri-iodothyronine, one of the thyroid hormones.

T4 Thyroxine, one of the thyroid hormones.

Tabasco A thin piquant sauce prepared by fermentation of powdered dried fruits of chilli pepper, mixed with spirit vinegar and salt. Traditionally used with Mexican and Caribbean foods.

Tabbouleh Syrian, Lebanese; salad made with bulgur, parsley, onion, mint, lemon juice, oil, and spices.

Table d'hote A meal consisting of a set number of courses at a fixed price; there is usually some choice of dishes within each course; a set menu as opposed to a la carte.

Tachycardia Rapid heartbeat, as occurs after exercise; may also occur, without undue exertion, as a result of anxiety, and in anaemia and vitamin B 1 deficiency.

Tachyphagia Rapid eating.

Taco Mexican; tortilla filled with meat, beans, and spicy sauce, and fried.

Taeniasis Infection with tapeworms of the genus Taenia.

Tafia Spirit similar to rum made from sugar cane.

Tagliatelle Pasta in ribbons 2-3 cm wide.

Tahini Middle-Eastern; paste made from sesame seeds, usually eaten as a dip; also used in preparation of hummus. A rich source of calcium, iron, and vitamin B 1, a good source of niacin; a source of protein and zinc.

Tail The abdomen or hinder part of the body of crustaceans containing the edible tail meat.

Taint An odour or flavour foreign to a product. The distinguishing feature of a taint, as distinct from off-flavour/odour, is that the chemical responsible for the taint is not produced in the product, but comes from its surroundings. Taints often have an unpleasant character, but that is not a necessary condition for taint. In the case of fish in the aquatic environment, taints are acquired from the ambient water.

Examples are the earthy/muddy taints in freshwater fish and oily taints in fish exposed to petroleum or petroleum products. Products can be tainted by strong smelling chemicals such as phenolic disinfectants in the ambient air.

Talawa Indian term for deep-fried food.

Talin Trade name for thaumatin, an extract of the berry Thaumatococcus danielli, about 3000 times as sweet as sucrose.

Tallow, rendered Beef or mutton fat prepared from parts other than the kidney, by heating with water in an autoclave. When pressed,

Separates to a liquid fraction, oleo oil, used in margarine, and a solid fraction, oleostearin, used for soap and candles.

Tamal Mexican; maize-meal pancake, similar to tortilla, but made with fat. Traditionally cooked inside the soft husks of maize.

Tamarind Leguminous tree, Tamarindus indica, with pods containing seeds embedded in brown pulp, eaten fresh, used to prepare beverages and seasonings in oriental cuisine.

Tammy To squeeze a sauce through a fine woollen cloth to strain it.

Tamper evidence Devices attached to food packages that indicate if a package has been opened or not.

Tandoor Traditional North Indian clay oven.

Tandoori Indian term for food cooked in a clay oven. The meat is marinated with aromatic herbs and spices before cooking.

Tangelo A citrus fruit, a cross between tangerine and pomelo.

Tangerine A citrus fruit, Citrus reticulata, also called mandarin; satsuma is a variety of tangerine. One medium-sized tangerine is a rich source of vitamin C; provides 1.5 g of dietary fibre; supplies 25 kcal.

Tangy Strong sharp tasting or smelling.

Tankage Residue from slaughterhouse excluding all the useful tissues; fertilizer or animal feed.

Tannia The corm. of Xanthosoma sagittifolium; known as new cocoyam or yautia in West Africa; same family as taro.

Tannins Compounds present in dark-coloured sorghum, carob bean, unripe fruits, tea, etc.; they give an astringent effect in the mouth, precipitate proteins, and are used to clarify beer and wines. Also called tannic acid and gallotannin.

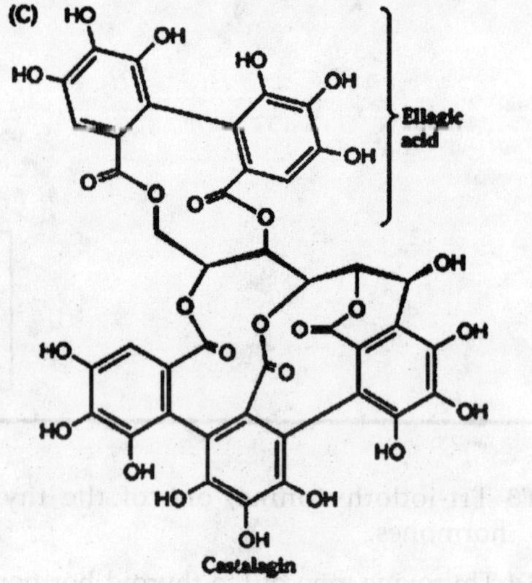

Fig. Tannins

Tanrogan Manx name for scallops.

Tansy A herb, Tanacetum vulgare. The leaves and young shoots are used for flavouring puddings and omelettes. Tansy cakes made with eggs and young leaves used to be eaten at Easter. Tansy tea made by infusing the herb was formerly used as a tonic and for intestinal worms; may be poisonous in large amounts. The root, preserved in honey or sugar, was used to treat gout.

Tapas Spanish; small savoury dishes served with wine in bars.

Tapeworm Parasitic intestinal worm; infection is acquired by eating raw or undercooked infected pork, beef, or fish

Eggs are shed in the faeces and infect the animal host. Cysticercosis is infection of human beings with the larval stage by ingestion of eggs from faecal contamination of food and water.

Tapioca Starch prepared from the root of the cassava plant; there are only traces of nutrients. The starch paste is heated to burst the granules, then dried either in globules resembling sago or in flakes. The name is also used of starch in general, as in manioc tapioca and potato flour tapioca.

Tapioca-macaroni A mixture of either 80-90 parts tapioca flour, with 10-20 parts of peanut flour; or tapioca, peanut, and semolina, in proportion 60 : 15 : 25; it is baked into shapes resembling rice grains or macaroni; developed in India. Also referred to as synthetic rice.

Tar The unwanted layer of tarry constituents of wood smoke that forms on the inner surfaces of a kiln during smoking by the continual deposition of droplets.

Taramasalata fish roe, whipped with oil, garlic, and lemon juice, then thickened with bread, to make a dip.

Taratòr Balkan; cold soup made from yoghurt, cucumber, nuts, and garlic.

Tares Traditional English name for the vetches, which are pulses.

Tarhonya Hungarian; pea-sized balls of egg and flour dough, fried in lard then simmered in water. A staple food of the Hungarian lowland plain.

Tariff A tax imposed on imported goods.

Tarka Foods seared with ghee and seasoning after cooking. Also known as chamak.

Taro Corm of Colocasia esculenta and C. antiquorum; called eddo or dasheen in the Caribbean, old cocoyam in West Africa.

Tarragon Leaves and flowering tops of the bushy perennial plant Artemisia dracunculus; the French variety is consi-dered superior to the Russian. Has a mild anise-like flavour and is used to flavour pickles; it is one of the ingredients of fines herbes. Tarragon vinegar is made by steeping the fresh herb in white wine vinegar and is used in making sauce tartare and French mustard.

Tart An open pastry case with a filling, usually of something sweet such as fruit, tasting sour or acidic.

Tartan purry Scottish; finely chopped cooked kale mixed with oatmeal. The name is probably a corruption of tarte en puree, suggesting it may originally have been served in a pastry case.

Tartar emetic Potassium antimonyl tartrate, produces inflammation of the gastrointestinal mucosa and used to be used as an emetic.

Tartar Hard gritty deposit of plaque and minerals that accumulates on and between teeth. Originally the name given by alchemists to animal and vegetable concretions, such as wine lees, stone, gravel, and deposits on teeth, since they were all attributed to the same cause.

Tartare sauce Mayonnaise flavoured with herbs, chopped capers, gherkins, etc., served with fish.

Tartaric acid A dibasic acid, chemically dihydroxysuccinic acid. Occurs in fruits, the chief source being grapes; used in preparing lemonade, added to jams when the fruit is not sufficiently acidic and in baking powder. Tartar emetic is the potassium antimonyl salt, and Rochelle salt is potassium sodium tartrate.

Tarte tatin apple pie baked with the pastry uppermost, then inverted before serving.

Taste buds Situated mostly on the tongue; about 9000 elongated cells ending in minute hairlike processes, the gustatory hairs.

Taste panel The preferred term is sensory panel.

Tatare Dish prepared from minced beef or other meat, eaten uncooked.

Taxol paclitaxel Natural product approved by the U.S. Food and Drug Administration in 1992 to treat ovarian cancer after failure of first-line chemotherapy; also used experimentally to treat breast, lung and other cancers. Manufactured by Bristol-Myers Squibb U.S. Pharmaceutical Group from bark of old-growth Pacific yew (cedar) trees. Because these forests are also home to endangered spotted owls, researchers are working to synthesize paclitaxel from twigs and needles, leaving the tree standing.

T-bone steak Cut from the thin fillet end of a sirloin of beef, containing a T-shaped section of bone.

Tea 1. A beverage prepared by infusion of the young leaves, leaf buds, and internodes of varieties of Camellia sinensis and C. assamica, originating from China. Green tea is dried without further treatment. Black tea is fermented before drying; Oolong tea is lightly fermented. Among the black teas, Flowering Pekoe is made from the top leaf buds, Orange Pekoe from first opened leaf, Pekoe from third leaves, and Souchong from next leaves.

2. An afternoon meal; may consist of a light meal or be a substantial meal as in northern Britain.

Teacake Flat round cakes made from yeast dough, normally toasted and buttered.

Teaseed oil Oil from the seed of Thea sasangua, cultivated in China; used as salad oil and for frying; similar in properties to olive oil.

Technology treadmill In rural sociology, the concept that introducing new production-enhancing technologies has not solved - and some believe has compounded - problems of farm profitability and sustainability. As farmers adopt a new technology, costs decrease and production increases, driving down prices to reflect lower per-unit costs and causing farmers to constantly seek new, cost-reducing technologies to stay ahead.

Teff A tropical millet, Eragrostis abyssinica, the dietary staple in Ethiopia; little grown elsewhere.

Teg A 2-year-old sheep.

Teiglech small biscuits made from flour, egg, and ginger, boiled in honey.

Tempeh Indonesian; soya bean fermented by a mould, Rhizopus spp.

Tempering Softening by raising the temperature of frozen fish without thawing, especially regular, cuboid fillet or mince blocks, to a level typically between -10°C and -15°C suitable for the efficient sawing, cutting, slicing or dividing of these blocks into smaller blocks, slabs, portions, fingers or sticks. Fish readily conduct heat at temperatures below -10°C, and only about 40 MJ of heat are required to soften one tonne of fish.

Templein Trade name for textured vegetable protein.

Tempranillo A grape variety widely used for wine making, although not one of the classic varieties; the wine of Rioja.

Tenderizer Usually refers to the enzyme papain, when used to tenderize meat. Similar enzymes occur in pineapple and figs. Weak acids such as vinegar and lemon juice also tenderize meat.

Tenderloin Fillet of beef or pork.

Tenderometer Instrument to measure the stage of maturity of peas to determine whether they are ready for cropping. Measures the force required to effect a shearing action.

Tenesmus Persistent ineffective spasms of bladder or rectum; intestinal tenesmus commonly occurs in irritable bowel syndrome.

Tenosynovitis Inflammation of a tendon sheath caused by strain, calcium deposits, hypercholesterolaemia, gout, rheumatoid arthritis, etc. Claimed to be relieved by high intakes of vitamin B 6, some 50 times the reference intake, but there is little evidence.

Tenuate Anorectic drug, used in the treatment of obesity.

Tepary bean Phaseolus acutifolius, also known as frijole, Mexican haricot bean, or pinto. It is able to grow during drought.

Tequila Mexican; spirit prepared by double distillation of fermented sap of the cultivated agave or maguey, Agave tequilana. Mescal and pulque are similar, made from various species of wild agave, and have a stronger flavour.

Teratogen Substance that deforms the foetus in the womb and so induces birth defects.

Terpenes Major components of the essential oils of citrus fruits; however, they are not responsible for the characteristic flavour, and, since they readily oxidize and polymerize to produce unpleasant flavours, they are removed from citrus oils by distillation or solvent extraction, leaving the so-called terpeneless oils for flavouring foods and drinks.

Terrine Oval French earthenware or china dish and also the food cooked in the dish.

Testa The fibrous layer between the pericarp and the inner aleurone layer of a cereal grain.

Tetany Over-sensitivity of motor nerves to stimuli; particularly affects face, hands, and feet. Caused by reduction in the level of ionized calcium in the bloodstream and can accompany severe rickets.

Tetracyclines A group of closely related antibiotics including tetracycline, oxytetracycline, and aureomycin. The last two are used in some countries for preserving food and as growth improvers, added to animal feed at the rate of a few mg per tonne.

Tetraenoic acid Fatty acid with four double bonds, e.g. arachidonic acid.

Tetralogy of fallot TOF is the combination of four heart defects: pulmonary stenosis, a stiffened, overdeveloped right ventricle, ventricular septal defect, and an overriding aorta that is connected to both the left and right ventricles, instead of draining just the left.

Tetramine poisoning Paralysis similar to that caused by curare, caused by a toxin in the salivary glands of the red whelk, Neptunea antiqua.

Tetrodontin poisoning Caused by a toxin, tetrodotoxin, in fish of the Tetrodontidae family and amphibia of the Salamandridae family. Occurs in Japan from Japanese puffer fish or fugu, eaten for its gustatory and tactile pleasure since traces of the poison cause a tingling sensation in the extremities.

Tetrodotoxin Tetrodotoxin is a potent, water soluble, neurotoxin associated with fish of the family Tetraodontidae, hence the name. Fish in this family are called pufferfish, or go by the Japanese name fugu. There are marked differences in toxicity among species, some species are virtually non-toxic. the toxin is present in various organs of the fish, including the muscle tissues, though concentrations are much higher in the gonads, liver, intestines and skin, than in the muscle. The illness induced by the toxin is tetrodotoxicosis. The first symptom of intoxication is a slight numbness of the lips and tongue, appearing between 20 minutes to three hours after eating toxic material, followed by paraesthesia, in the face and extremities. In more severe cases, there is increasing paralysis throughout the body, including respiratory centres. The disease is associated with a 50% mortality. The toxin is not confined to tetraodonts; other species of vertebrate and invertebrate fishery products have been reported to have the toxin - parrotfish, the blue-ringed octopus, starfish, angelfish, and xanthid crabs. The toxin is not now thought to be produced by the animals since recent studies have shown that several common marine bacterial species, including strains of the family Vibrionaceae, Pseudomonas sp., and Photobacterium phosphoreum can produce tetrodotoxin or a derivative of it.

This remains a matter of debate, and the relationship between the bacteria and the poisonous fish, and the mechanism of transfer of the toxin are not understood.

Texatrein Trade name for textured vegetable protein made by extrusion.

Texgran Trade name for textured vegetable protein.

Texture All the rheological, geometrical, and surface attributes of a product perceptible by the senses. The main sense involved in texture evaluation is an as yet poorly defined sense perceived by pressure sensors in the mouth and skin, and by position sensors in the muscles and joints, the kinaesthetic sense.

Texture profile An experimental procedure used in the sensory evaluation of foods. The assessor evaluates components of the texture of a product using descriptors from a prepared list of terms. The assessor also rates the perceived intensity of each texture attribute identified. Most scales for rating texture attributes are bipolar, that is, the end-points of the scale are described by antonyms, for example, tough/tender, firm/soft. The results are combined over members of a panel of assessors, and the resulting assemblage of terms and intensities forms the texture profile.

Textured vegetable protein Spun or extruded vegetable protein, usually made to simulate meat.

Thal platter of brass, copper, or silver on which food is served and eaten; a smaller dish is a thali.

Thaumatin The intensely sweet protein of the African fruit, Thaumatococus danielli, 1600 times as sweet as sucrose. Called katemfe in Sierra Leone and miracle fruit in the Sudan.

Thawing Melting the ice contained within frozen fish and also called defrosting. This occurs completely when the temperature throughout the fish is higher than -1°C. The time required to just melt all the ice in frozen fish is the thawing time. The heat required to thaw white fish from the temperature of cold storage of -30°C to a uniform temperature of -1°C is about 300 kJ/kg; fish with a fat content of 15% would only require about 240 kJ/kg. Fish usually lose weight on thawing. The thaw drip or drip lossmay be as high as 5% depending on the degree of deterioration which occurred during the previous period of frozen storage. Methods of thawing fall into two groups; heat can be conducted into the flesh from the surface so that thawing is limited by the rate of conduction, or heat can be generated uniformly throughout the flesh. The first group includes exposure to still air, which is very slow and not practicable commercially, or to moving warm, moist air, or to condensing water vapour, or to water thawing by immersion , by spray, or by some combination of immersion and spray. Spray thawing is difficult to control. The second group depends on controlled absorption of energy at mains, radio or microwave frequencies, and all result in much faster rates of thawing, but require careful control. Overheating must be avoided in all applications and the fish temperature should not exceed 18°C. Air blast thawing can be achieved in practice by using air at 20°C, saturated with water vapour to improve heat transfer and prevent surface drying, moving uniformly over the fish at an air speed or air velocity of 6 to 8 m/s. Typically, under these conditions, the thawing time for a 50 kg block of gutted cod, 100 mm thick, is about 4 to 5 hours - 30°C. Water thawing is acceptable for whole fish, and is typically by immersion in water at a temperature not exceeding 18°C, flowing at about 1 m/min. White fish lose their gloss and the skin becomes bleached, whilst herring will lose most of the remaining scales. The thawing time for a 100 mm thick block of whole fish is about

5 hours from -30°C under these conditions. The basis of vacuum thawing is the heat released when water vapour condenses on the fish in an evacuated chamber. Water in the evacuated chamber is heated to 18°C to produce water vapour. The water cools in contact with the frozen fish and eventually reaches a temperature below which the water begins to condense as a liquid. Since most of the air is evacuated, the rate of condensation becomes so high that it can provide heat as fast as it can be conducted to the inside of the fish, and the water vapour can easily enter the small spaces between fish in the frozen block and between blocks. The heat released at the freezing point amounts to 2400 kJ/kg of water that condenses, and approximately 125 kg of water will condense for each tonne of white fish thawed. Thawing rates are comparable with air blast and water thawing. A problem with this method is that gases normally present in the tissues are released during thawing and can spoil the appearance of soft textured fish. Sometimes the skin can be ruptured on the dorsal surfaces of white fish or on the belly walls of fatty fish when the fat content is high. In dielectric thawing the frozen fish is placed between two parallel electric plates carrying a high alternating voltage, high frequency, electric field, typically 5 kV and 80 MHz respectively. Under ideal conditions, with blocks of uniform thickness, composition and temperature, heat is produced rapidly throughout the thickness of the blocks without contact with the plates, and thawing is rapid. In practice, large blocks of whole large fish are immersed in water to create a more homogeneous electrical environment, generate heat more evenly and avoid local overheating, and the fish are conveyed through a series of separate thawing units. The thawing time for a 100 mm thick block of white fish is about 1 hour. Smaller blocks of small fish or product are not immersed and can be thawed in the packaging. Electrical resistance thawing uses the heat generated internally by the resistance of frozen fish to an electrical current passed through it. The electrical frequency causes oscillation of dipoles in the water as the direction of the electrical field changes, producing frictional energy and heating the fish. The application is a hybrid type because frozen fish is a poor conductor of electricity until its temperature rises above -10°C, at which point electrical resistance continues to decrease with increasing temperature. Fish is first immersed in cold water and warms rapidly to a temperature, still frozen, at which a high enough current is conducted to generate enough heat to complete the thawing. Alternating current at mains frequency, but low voltage, is applied to flat blocks of frozen fish in electrical contact between larger flat metal plates. The system is suitable for blocks up to about 50 mm thick and the total thawing time ranges from 30 to 50 min. Microwave power used for thawing can be at least ten times as fast as dielectric thawing but the fish is much more prone to overheating. When fish are subjected to microwaves, generated typically at about 3000 MHz, the radiation is absorbed by the flesh. Heat is generated, being greatest immediately below the surface, and progressively less as the radiation penetrates deeper into the flesh. The warmer the flesh becomes, the more readily it absorbs the energy, and the less the radiation is able to penetrate it, hence the frozen fish warms unevenly and is more likely to be overheated. Microwave heating is more suitable for cooking from frozen, especially recipe dishes and ready prepared meals, for thawing smaller items, and for tempering. Thawed fish may be referred to as slacked fish in the USA.

Theophylline An alkaloid found in tea, chemically related to caffeine, and with similar effects. Theine is 1,3-dimethyl xanthine; caffeine is trimethylxanthine.

Therapeutic diets Those formulated to treat disease or metabolic disorders.

Thermal centre The point in a product that has the highest temperature after chilling or freezing, or the lowest temperature after thawing, or has the lowest temperature in the contents of a container during heat processing. It is not necessarily the geometric centre of the product.

Thermal conductivity The quantity of heat conducted through unit area and thickness of a material in unit time, under the influence of unit temperature difference.

Thermization Heat treatment, less severe than pasteurization, e.g. heat treatment of milk for cheese-making whereby the number of organisms is diminished.

Thermoduric Bacteria that are heat resistant but not thermophilic, i.e. they survive, but do not develop at, pasteurization temperatures. Usually not pathogens but indicative of insanitary conditions.

Thermogenesis Increased heat production by the body, either to maintain body temperature or in response to food intake, diet-induced thermogenesis.

Thermogenic drugs Substances that stimulate body heat output, and thus of interest in 'slimming'.

Thermopeeling A method of peeling tough-skinned fruits in which the fruit is rapidly passed through an electric furnace at about 900°C then sprayed with water.

Thermophilic Describes micro-organisms termed thermophiles or thermophilic bacteria that grow best at elevated temperatures in the range 55 to 75°C. Bacillus stearothermophilus is an extremely heat resistant thermophilic spore forming bacteria which has been found to be responsible for the flat-sour spoilage of canned foods.

Thiaminase An enzyme present in many species of micro-organisms, plants, and fish which splits thiamin, forming products that have anti-vitamin activity. Non-enzymic cleavage of thiamin, for example by polyphenols, is also sometimes called thiaminase action. Chastek paralysis in foxes and mink fed diets rich in raw fish, and blind staggers in horses and other animals eating bracken fern, are due to acute vitamin B 1 deficiency caused by dietary thiaminase.

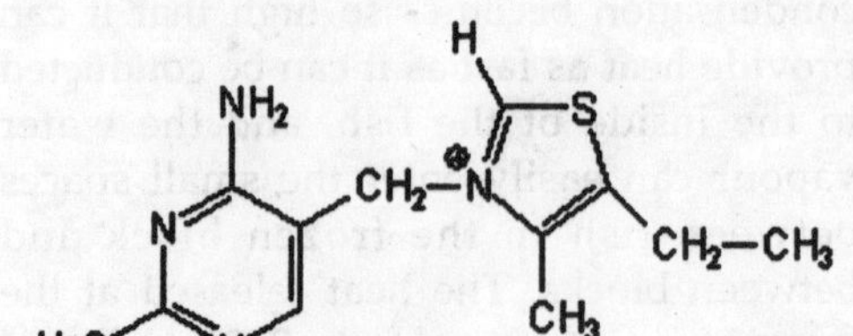

Fig. Thiamin

Thiobarbituric acid value A measure of the amount of reaction products of oxidation of lipids. It is based on measurement of the amount of malondialdehyde, a major oxidation product.

Thiobendazole Antifungal agent used for surface treatment of bananas.

Threonine An essential amino acid, chemically amino-hydroxybutyric acid. It was the last of the protein amino acids to be discovered, in 1935, in studies of nitrogen balance on subjects fed mixtures of the then known amino acids in place of proteins.

Thrombin Plasma protein involved in the coagulation of blood.

Thrombokinase Or thromboplastin. An enzyme liberated from damaged tissue and blood platelets; it converts prothrombin to thrombin in the coagulation of blood.

Thrombosis Formation of blood clots in blood vessels.

Thuricide Name given to a living culture of Bacillus thuringiensis which is harmless to man but kills insect pests. Known as a microbial insecticide. Used to treat certain foods and fodder crops to destroy pests such as corn earworm, flour moth, tomato fruit worm, cabbage looper, etc. The bacillus is mass-produced and stored like a chemical.

Thyme The aromatic leaves and flowering tops of Thymus spp. used as flavouring in soup, meat, fish, poultry dressing, and sausages.

Thymonucleic acid Obsolete name for deoxyribonucleic acid.

Thymus Chest sweetbread; a ductless gland in the chest, as distinct from gut sweetbread or pancreas. A rich source of protein, iron, and vitamin C.

Thyroglobulin The protein in the thyroid gland which is the precursor for the synthesis of the thyroid hormones. The thyroid-stimulating hormone of the pituitary gland stimulates hydrolysis of thyroglobulin and secretion of the hormones into the bloodstream.

Thyroid hormones The thyroid is an endocrine gland situated in the neck, which takes up iodine from the bloodstream and synthesizes two hormones, tri-iodothyronine (T3) and thyroxine. The active hormone is T3; thyroxine is converted to T3 in tissues by the action of a selenium-dependent enzyme. T3 controls the basal metabolic rate.

Enlargement of the thyroid gland is goitre; it may be associated with under- or over-production of the thyroid hormones. Severe iodine deficiency in children leads to cretinism.

Thyrotoxicosis Over-activity of the thyroid gland, leading to excessive secretion of thyroid hormones and resulting in increased basal metabolic rate. Hyper-thyroid subjects are lean and have tense nervous activity. May be due to over-stimulation of the thyroid gland.

Iodine-induced thyrotoxicosis affects mostly elderly people who have lived for a long time in iodine-deficient areas, have had a long-standing goitre, and then have been given extra iodine. Also known as Jodbasedow, Basedow's disease, and Graves' disease.

Thyroxine One of the thyroid hormones.

Tia maria Trade name; Jamaican rum-based coffee liqueur.

Tierce Obsolete measure of wine cask; 1/3 of a pipe, i.e. about 35 imperial gallons.

Tiffin Anglo-Indian name for a light midday meal.

Tiger nut Tuber of grass-like sedge, Cyperus esculentus, also earth or ground almond, chufa nut, rush nut, nut sedge, 5-20 mm long, usually available in partly dried condition. Mainly starch and fat; a rich source of dietary fibre.

Tikka marinated chicken threaded on skewers and grilled.

Timbale Round, fireproof china or tinned copper mould, used for moulding meat or fish mixtures; also the dishes cooked in the mould. For hot timbales the mould is lined with potato, pastry, or pasta; for cold the lining is aspic.

Tin A metal; a dietary essential for experimental animals, but so widely distributed in foods that no deficiency has been reported in man, and its function, if any, is not known. In the absence of oxygen tin is resistant to corrosion; hence widely used in tinned cans for food containers.

Tipsy cake Sponge cake soaked in wine and fruit juice, made into a trifle and reassembled into the original tall shape. The wine and fruit juice may cause the cake to topple sideways in drunken fashion.

Tiramisu dessert made of coffee-flavoured sponge or biscuit filled with sweetened cream cheese and cream, doused with syrup.

Tisane French term for an infusion made from

herbs, believed to have medicinal properties.

Tissue culture Growing plant or animal tissues or cells in test tubes or other laboratory glassware, without other contaminating organisms, for propagation, chemical production and medical research.

Toad-in-the-hole Sausages cooked in batter, also called sausage toad.

TOBEC Total body electrical conductivity, a method of measuring the proportion of fat in the body by the difference in the electrical conductivity of fat and lean tissue.

Tocino Spanish; salted pork fat.

Tocopheronic acid Water-soluble metabolite isolated from the urine of animals fed with tocopherol; it has vitamin E activity.

Toddy Warming drink made from whisky with hot water and sugar.

Toffee A sweet made from butter or other fat, milk, and sugar boiled at a higher temperature than caramels. Called candy or taffy in the USA. Variants include butterscotch and glessie. Toffee apples are apples coated with hardened syrup.

Tofu soybean curd precipitated from the aqueous extract of the soya bean.

Tokay Hungarian sweet white wine made from grapes affected by the 'noble rot'; also a variety of grape used for wine making.

Tomatillo Or ground tomato, husk-covered fruit of Physalis ixocarpa; resembles a small, green tomato.

Tomato The fruit of Lycopersicon esculentum. Tomato is a good source of vitamin C & dietary fibre.

Tonic water A sweetened, carbonated beverage flavoured with quinine, commonly used as a mixer with gin or vodka. Originally invented by the British in India as a pleasant way of taking a daily dose of quinine to prevent malaria; sometimes known as Indian tonic water.

Tonka bean Seed of the S. American tree Dipteryx odorata with a sweet, pungent smell, used like vanilla for flavouring.

Tooth-friendly sweets Name given to 'sugar' confectionery made with sugar alcohols and/or bulk sweeteners which do not ferment in the mouth and so do not damage teeth; the term originated in Switzerland. Sugar alcohols can have a laxative effect if eaten in large quantities.

Topepo American; Cross between tomato and sweet pepper.

Topside Boneless cut of beef from the top of the hind leg.

Torte Open tart or rich cake mixture baked in a pastry case, filled with fruit, nuts, chocolate, cream, etc.

Tortilla 1. Mexican; thin maize pancake. Traditionally prepared by soaking the grain in alkali and pressing it to form a dough, which is then baked on a griddle. Tortillas filled with meat, beans, and spicy sauce are tacos.

2. In Spain, an omelette made by frying potatoes and onions with eggs; may be served hot or cold; also used for a variety of filled omelettes.

Total communication The use of additional ways of communicating, such as sign language or communication boards, gestures, electronic devices, etc., to expand the conversational abilities of the child with severe speech and language disabilities.

Total viable count Also called Aerobic plate count, Standard plate count and other tems. A frequently performed microbiological count. It is the number of bacteria growing on a non-specific solid bacteriological growth medium under the specified conditions. A sample of product is blended in an appropriate solution and aliquots of the suspension, after dilution as necessary, are applied to the medium. The inoculated plate is incubated in air and, after a specified time, the number of visible colonies are counted. The results are typically expressed as colony forming units /g. An important variable in the experimental procedure is the temperature of incubation. An incubation temperature around 35°C favours mesophiles, and provides a general index of sanitary or hygienic conditions of the handling and processing of the product. Lower temperatures, around 20°C, favour psychrophiles and are more indicative of spoilage bacteria on chilled fish.

Total Volatile Bases Equivalent terms are Total Volatile Basic Nitrogen and Total Volatile

Toughness A term referring to an attribute of texture, an aspect of quality, the attribute of high resistance to deformation and disintegration.

Tournedos Thick steak cut from the 'eye' of the fillet or undercut of beef.

Tourte shallow round tart or flan made from pastry, filled with sweet or savoury mixture.

Toxin A natural substance produced by an organism capable of causing harm to another organism, including man. The meaning is sometimes broadened to include any chemical causing harm to an organism, but it would be better to use 'poison' to refer to the broader class with toxin as a subset. In nature, toxins are taken into animals by ingestion. The word is frequently compounded to show the origin of the toxin. For example phytotoxin for a toxin present in plants; phycotoxin is frequently used for a toxin present in algae; mycotoxins for a toxin present in fungi. The term biotoxin has the same meaning as toxin as defined above, but it is often used in writings on food poisoning to emphasis the difference from poisons of non-biological origin. Sometimes the compound word indicates the target of the toxin, for example neurotoxins attack the nervous system and hepatotoxins affect liver function. Many of the toxins found in fish and shellfish are neurotoxins and some are hepatotoxins. Others are compounded to reflect the initial source of identification of the toxin, such as pectenotoxin or brevetoxin. There is some confusion in the nomenclature in this area.

Transaminase activation test An enzyme activation test for vitamin B 6 nutritional status, based on the fact that transaminases require the vitamin for activity.

Transaminase Any enzyme that catalyses the reaction of transamination.

Transamination The transfer of the amino group (NH_2) from an amino acid to an acceptor. Pyridoxal phosphate, the metabolically active form of vitamin B 6, acts as the intermediate carrier of the amino group. The enzymes catalysing the reaction are known as transaminases or aminotransferases.

Transgenic Carrying one or more genes introduced using recombinant DNA technology.

Transition A coordinated set of activities for a student with a disability designed to promote the transition from school to post-school activities, including post-secondary education, vocational training, integrated employment, continuing and adult education, adult services, independent living, or community participation.

Transketolase test An enzyme activation test for vitamin B 1 nutritional status, based on the fact that the enzyme transketolase in red blood cells requires the vitamin for activity.

Translocation Breaks in two separate chromosomes may be followed by unification of the fragment of one chromosome with the wrong fragment of another chromosome. Translocations in Down syndrome always have a 21st chromosome as one of the two translocated chromosomes. The second involved chromosome is usually either another 21st chromosome or a 14th chromosome.

Translucent Almost transparent, allowing some light through so that objects can be seen through it but not very clearly.

Transparent Able to be seen through; that allows light through, so that objects on the other side can be seen clearly.

Trapistine herb-flavoured liqueur made by Trappist monks at the Abbey de la Grace de Dieu.

Treacle First product of refining of molasses from beet or sugar cane extract is black treacle, slightly less bitter; will not crystallize.

Trebbiano A grape variety widely used for wine making, although not one of the classic varieties; makes the relatively thin wines of central Italy if not blended with other varieties.

Trehalose Mushroom sugar, also called mycose, a disaccharide of glucose. Found in some fungi, manna, and some insects; hydrolysed to glucose.

Trematodes Trematodes or flukes are members of the zoological Class Trematoda of the Phylum Platyhelminthes. They are small, but not microscopic, flattened, and generally leaf-shaped animals. Within the Class there are two Orders Monogenea, and Digenea. The latter includes members that are parasitic in freshwater fish and crustacean shellfish and can be transmitted to man causing disease. Examples of genera of Digenea posing hazards to humans include, but are not confined to, Clonorchis, Opisthorcis, Paragonimus, Metagonimus and Nanophyetus. Digenetic trematodes have animals, including man, as definitive hosts, and two intermediate hosts. The first is an aquatic snail; the second is a freshwater vertebrate fish or a freshwater crustacean shellfish, depending on genus.

Triacylglycerols Sometimes called triglycerides, simple fats or lipids consisting of glycerol esterified to three fatty acids. The major component of dietary and tissue fat. Also known as saponifiable fats, since on reaction with sodium hydroxide they yield glycerol and the sodium salts (or soaps) of the fatty acids.

Trichinosis Disease that can arise from eating under-cooked pork or pork sausage meat; due to Trichinella spiralis, a worm that is a parasite in pork muscle; destroyed by heat and by freezing.

Trichosanthin Eight ribosome-inactivating proteins derived from Chinese cucumber, also named chosen-karasu-uri. Seeds produce triterpene, the first naturally occurring anti-inflammatory drug used in traditional Chinese medicine. Synthesized protein activates normal human serum. The plant's roots can be genetically engineered to produce potential anti-tumor and anti-AIDS virus drugs, as well as biological nematicides and fungicides for crop protection.

Trifle Cold sweet made from sponge cake soaked in fruit juice or sweet wine, covered with custard sauce and whipped cream, and decorated.

Trifyba Trade name for processed wheat bran from husk of Testa triticum tricum, containing 80 g of dietary fibre/100 g with reduced content of phytate.

Triglycerides Esters of fatty acids with glycerol, more correctly called triacylglycerols.

Trigonelline A metabolite of niacin, chemically N-methylnicotinic acid, excreted in the urine in small amounts after consumption of relatively large amounts of nicotinic acid. It also occurs in some foods; it has no vitamin activity, but a considerable amount is converted to active niacin during the roasting of coffee.

Tri-iodothyronine One of the thyroid hormones.

Trimethylamine An amine of formula N. Formed in the muscle tissue of stored unfrozen fish by the bacterial metabolism of trimethylamine oxide. It has a strong ammoniacal smell and the ammoniacal smell of spoiled fish is likely to be due to TMA rather than ammonia. The muscle tissue of living fish contains only traces of TMA, typically less than 1mg TNA nitrogen/100 g. Not all species in the bacterial flora of fish when caught metabolise TMAO and TMA formation is slow until the psychrophiles, which include important TMA-producing species, begin to increase in numbers. TMA content then increases approximately exponentially with storage time, reflecting the increase in bacterial count. TMA content, like TVB, can be used as an index of spoilage, but it is more precise than TVB because there is not the problem of dependence of results on the experimental procedure. TMA is most frequently determined by a colorimetric procedure,or by gas liquid chromatography (GLC). A variety of other procedures based on different principles from those just listed have been described, but have not achieved popularity or application. The procedures for measurement of TMA are not fast enough or simple enough for use in routine quality control.

Trimethylamine oxide A compound of formula NO of wide occurrence in animals and in bacteria. It is present in the flesh of marine species of fish and crustacean shellfish, but generally not in the muscle of freshwater fish. It is metabolised by some species of spoilage bacteria of fish, and is split enzymatically into formaldehyde and dimethylamine in the muscle of some species of fish, including during frozen storage.

Tripa Corsican; pudding made from beetroot, spinach, herbs, and sheep's blood, stuffed into the stomach of a sheep, and boiled.

Tripe Lining of the first three stomachs of ruminants, usually calf or ox. Sold 'dressed', i.e. cleaned and treated with lime. According to the part of the stomach there are various kinds, such as blanket, honeycomb, book, monk's hood, and reed tripe. Contains a large amount of connective tissue which is changed to gelatine on boiling.

Triple sec Sweet orange-flavoured liqueurs such as cointreau and curacao. **triticale** Cross between wheat and rye which combines the winter hardiness of the rye with the special baking properties of wheat.

Trisomy 21 From the Greek tri = "three". About 90% of individuals with Down syndrome this karyotype: a total of 47 chromosomes, instead of the usual 46. Also see: Down syndrome.

Tropical oils Suggested term for vegetable oils that contain little polyunsaturated fatty acid, such as coconut and palm oils.

Trout Freshwater oily fish; brown trout is Salmo trutta, rainbow trout is S. gairdneri. A rich source of protein, vitamin B 12, and selenium; a good source of vitamin B 6.

Truffles 1. Edible fungi growing underground, associated with roots of oak trees; very highly prized for their aroma and flavour. Most highly prized is the French black or Perigord truffle, Tuber melanosporum, added to pate de foie gras. Others include: white Piedmontese truffle, T. magnatum; summer truffle T. aestivum, and violet truffle T. brumale.

2. Chocolate truffles, a mixture of chocolate, sugar, cream, and often rum, covered with chocolate strands or cocoa powder.

Trusoy Trade name for full-fat soya flour, heat-treated.

Trypsin A proteolytic enzyme of the pancreatic juice, an endopeptidase. Active at alkaline pH. Secreted as the inactive precursor, trypsinogen, which is activated by enteropeptidase.

Trypsin inhibitors Proteins found in soya and other beans which inhibit trypsin; they are denatured and hence inactivated by heating.

Trypsinogen The inactive precursor of trypsin, secreted in the pancreatic juice and activated by enteropeptidase.

Tryptophan An essential amino acid. In addition to its role in protein synthesis, it is the precursor of the neurotransmitter 5-hydroxytryptamine and of niacin. Average intakes of tryptophan are more than adequate to meet niacin requirements without the need for any preformed niacin in the diet.

It is destroyed by acid, and therefore not measured when proteins are hydrolysed by acid before analysis; determination of tryptophan requires alkaline or enzymic hydrolysis of the protein.

TSP Trade name for textured soya protein, prepared by extrusion through fine pores to give a fibrous, meat-like texture to the final product.

Tuber Botanical term for underground storage organ of some plants, e.g. potato, Jerusalem artichoke, sweet potato, yam.

Tuberin The major protein of potato, a globulin.

Tumbet Spanish; baked vegetables and egg.

Tun Obsolete measure; a large cask holding 216 imperial gallons of ale; 252 gallons of wine.

Tuna fish Species of Thunnus and Neothunnus, oily fish. Albacore tuna is specifically Thunnus alalunga; bonita tuna and skipjack tuna are different species.

Turkey A poultry bird, Meleagris gallopavo.

Turkish delight Sweet made from gelatine and concentrated grape juice, flavoured with rose water. Also sometimes made with marshmallow.

Turmeric Dried rhizome of Curcuma longa, grown in India and South Asia. It is deep yellow and used both as condiment and food colour; used in curry powder and in prepared mustard. Its pigment is used as a dye under the name curcumin or Indian saffron.

Turnip Root of Brassica campestris eaten as a cooked vegetable.

Turnover Small pie or pasty; the filling is placed on one half of a piece of rolled-out pastry and the other half is folded over to make a semicircular envelope.

Turrón Spanish nougat.

Turtle Marine reptile; the main species for food is the green turtle, Chelonia mydas, so called because of the greenish tinge of its fat. It is farmed to a small extent, but mainly caught in the wild.

Tutti-frutti Mixture of soft fruits bottled in brandy. Also ice cream containing mixed preserved fruit.

Tweed kettle Scottish; salmon cut into small cubes before poaching. Also called salmon hash.

Tyramine The amine formed by decarboxylation of the amino acid tyrosine; chemically p-hydroxyphenylethylamine.

Tyropita cheese pie prepared with feta and cottage cheeses, eggs, and phyllo pastry.

Tyrosinase Enzyme that oxidizes the amino acid tyrosine and other phenolic compounds to form brown and black pigments. Absent from albinos and the white areas of piebald animals. It is present in some fruits and vegetables, e.g. potato and apple, and is responsible for the dark colour produced when cut raw food or the juice is exposed to air.

Tyrosine A non-essential amino acid that can be formed in the body from the essential amino acid phenylalanine, hence it has some sparing action on phenylalanine. In addition to its role in proteins, tyrosine is the precursor for the synthesis of melanin, and for adrenaline and noradrenaline.

U

Ubiquinones Coenzymes in the respiratory chain in mitochondria, also known as coenzyme Q or mitoquinones; widely distributed in nature. Chemically derivatives of benzoquinone with isoprene side-chains. There is no evidence that they are dietary essentials; they may have antioxidant activity.

UHT Ultra high temperature.

UI Upper Intake Level. The maximum, tolerable, level of daily nutrient intake that is unlikely to pose risks of adverse health effects to almost all of the individuals in the group for whom it is designed. UI is one of the guidelines encompassed in DRI established by the National Academy of Sciences.

Uitsmijter Dutch; sandwich of cooked meat or ham and fried egg.

Ulcer A crater-like lesion of the skin or a mucous membrane resulting from tissue death associated with inflammatory disease, infection, or cancer. Peptic ulcers affect regions of the gastro-intestinal tract exposed to gastric juices containing acid and pepsin: gastric in the stomach and duodenal in the duodenum. Treatment was formerly conservative, with a bland diet, followed if necessary by surgery; specific antagonists of histamine receptors have improved treatment enormously. May be caused or exacerbated by infection with Helicobacter pyloris.

Ullage Liquid left in cask or bottle after some has been removed.

Ultra-high-temperature sterilization Sterilization at higher temperatures, and for shorter times, than high-temperature short-time sterilization.

Ultrasonic homogenizer High-speed vibrator giving a cavitation force of 60 tons per square inch in the liquid. Used to cream soups, disperse dried milk, disperse essential oils in soft drinks, stabilize tomato puree, prepare peanut butter, etc.

Ultraviolet irradiation Light of wavelength below the visible range. Wavelength for maximal germicidal action is 260 nm; it has poor penetrating power and is only of value for surface sterilization or sterilizing air and water. Also used for tenderizing and ageing of meat, curing cheese, and prevention of mould growth on the surface of bakery products.

Ultraviolet from sunlight is responsible for skin tanning, and the formation of vitamin D from 7-dehydrocholesterol in the skin.

Umami Name given to the special taste of monosodium glutamate, protein, certain amino acids, and the ribonucleotides. The Japanese name for a savoury flavour, now considered one of the five basic senses of taste.

Umbles Edible entrails of any animal which used to be made into pie, umble pie or humble pie.

Uperization A method of sterilizing milk by injecting steam under pressure to raise the temperature to 150°C. The added water is evaporated off.

Upside-down cake A cake or sponge pudding which is baked with a layer of fruit under the sponge mixture, then served upside down, with the fruit uppermost.

Urea A chemical of formula 2 CO.. It is the end-point of nitrogen metabolism in animals and some fishes. It is a feature of elasmobranchs, that they accumulate urea in their tissues. Urea has a bitter taste and contributes to the characteristic flavour of this group of fish. It is degraded to ammonia by enzymes in muscle tissue and by spoilage organisms of fish, leading to the very strong, pungent, odour of spoiling elasmobranchs.

Urethane Ethyl carbamate, used as intermediate in organic syntheses, as a solubilizer, and as the precursor for polyurethane foam. Found in small amounts in liqueurs made from stone fruits, wines, and some distilled spirits where it is formed by reaction between alcohol and nitrogenous compounds; a cause for concern since it is genotoxic. uric acid The end-product of purine metabolism in man and other apes; other mammals have the enzyme uricase, which converts the uric acid to allantoin, which is more soluble in water. Gout is the result of excessive formation of uric acid, and/or impaired excretion; it is only slightly soluble in water, and in excess it crystallizes in joints as gouty nodules under the skin, and sometimes in the kidney.

V

Vacherin Circular cakes of meringue and cream.

Vac-ice process Alternative name for freeze-drying.

Vacreation Deodorization of cream by steam distillation under reduced pressure; developed in New Zealand.

Vacuum A place or region containing no solid, liquid or gas.

Vacuum dried salt Salt made by evaporating brine under a vacuum.

Valine An essential amino acid, rarely, if ever, limiting in foods. Chemically, amino-isovaleric acid.

Value-added The processing of products so that their selling price is higher than that of the raw materials from which they were made.

Vanadium A mineral known to be essential to experimental animals, although sufficiently widespread for human dietary deficiency to be unknown. Its precise function is unknown, although it acts as an activator of a number of enzymes.

Vanaspati Purified hydrogenated vegetable oil; similar to margarine and usually fortified with vitamins A and D. Also used to prepare ghee.

Vanilla Extract of the vanilla bean, fruit of the tropical orchid Aracus aromaticus and related species. It was discovered in Mexico in 1571 and could not be grown elsewhere, because pollination could be effected only by a small Mexican bee, until artificial pollination was introduced in 1820. The main growing regions are now Madagascar and Tahiti. The major flavouring principle is vanillin, but other substances present aid the flavour. Ethyl vanillin is a synthetic substance which does not occur in the vanilla bean; 31/2 times as strong in flavour, and more stable to storage than vanillin, but does not have the true flavour.

Vanilla sugar Sugar flavoured with vanilla by storing the bean and sugar together.

Vareniki Rounds of noodle dough stuffed with meat, vegetables, curd cheese, etc. Similar to ravioli.

Vareschaga Pork casseroled with beetroot juice and rye breadcrumbs.

Variety meat American name for offal.

Vasoconstriction Constriction of the blood vessels; the reverse of vasodilatation.

Vasodilatation Dilation of the blood vessels; the reverse is vasoconstriction. Caused by a rise in body temperature; serves to lose heat from the body.

V-cutting The pin bones in the anterior portion of the fillet can be removed in many commercially important species of roundfish by making a cut through the skinned fillet along both sides of the line of pin bones from the neck end towards the tail of the fillet, such that both cuts meet just behind the position of the last pin bone. Thus, an acute angled, v-shaped cut is made, the V-cut, which contains the pin bones and can be removed to leave a V-cut fillet. The proportion of fillet removed by V-cutting varies according to the species and degree of skill and experience exercised in cutting. It ranges from 4 to 15% for roundfish such as cod, haddock and whiting but usually lies in the range 8 to 12 % when cut by hand. Some heading machines are described as V-cut heading machines which refers to the type of angled head cut directly behind the collar bone as opposed to heading machines that either make a so-called round-cut or a straight-cut to remove the head.

Veganś Those who consume no foods of animal origin.

Vegemite Australian; trade name for yeast extract.

Vegetable butters Naturally occurring fats that melt rather sharply because they contain a preponderance of a single triacylglycerol. Cocoa butter from the cocoa bean, used in chocolate; Borneo tallow or green butter from the Malaysian and Indonesian plant, Shorea stenopiera, resembles cocoa butter; shea butter from the African plant, Butyrospermum parkii, softer than cocoa butter. Mowrah fat or illipe butter from the Indian plant, Bassia longifolia, used for soap and candles.

Vegetable protein products General term to include textured soya and other bean products often made to simulate meat. Basic material is termed flour when the protein content is not less than 50%; Concentrate, not less than 65%; isolate, not less than 90% protein.

Vegetable spaghetti Spaghetti squash, a gourd with fibrous strands of flesh resembling spaghetti after cooking.

Vegetables Plants or parts of plants cultivated for food. Some foods that are botanically fruits, such as tomatoes and cucumbers, and seeds, such as peas and beans, are included with the vegetables; some plants, such as rhubarb, are classed as fruit, although they are not botanically fruits. The distinction in popular usage depends on whether they are eaten as savoury or sweet dishes.

As a source of nutrients most of the vegetables are useful sources of vitamin C and minerals, the root vegetables supply carbohydrate, but only the seeds are an important source of protein. Green and yellow or orange vegetables and fruits are sources of vitamin A as carotene.

Vegetarian A person who excludes some or all animal food products from his diet. Lacto-ovo vegetarians eat eggs and dairy but no meat; lacto-vegetarians eat dairy but no eggs or meat; vegans eat no animal products at all.

Vent Anal opening at the rear of the belly cavity of a fish, through which are discharged wastes from the intestine and kidney, and also eggs or sperm.

Verbascose A non-digestible tetrasaccharide, galactose-galactose-glucosefructose, found in legumes; it is fermented by intestinal bacteria and causes flatulence.

Verjuice Literally 'green juice'; sour juice of crab apples formerly used in cooking meat, fish, and game dishes. Now normally replaced by lemon juice.

Vermouth Fortified wine flavoured with herbs and quinine. French vermouth is dry and colourless; Italian may be red or white and is sweet. Drunk as an aperitif, either with soda or with gin or vodka. Name originally derived from German Wermut for wormwood, a toxic ingredient that was included in early vermouths. Sweet or Italian vermouth, 15-17% alcohol, 12-20% sugar. Dry or French type, 18-20% alcohol, 3-5% sugar.

Veronique Garnish of white grapes.

Vert-pre, au A dish garnished with watercress and straw potatoes, often served with maitre d'hotel butter; also a dish served with green mayonnaise.

Verveine du velay herb liqueur, strongly flavoured with verbena; there are two types, yellow and green.

Vestibular Individuals who tend to have gravitational insecurity, use a wide-based gait and may have decreased sense of balance and protection reactions.

Vetch Old term applied generally to legumes; originally Vicia spp., also called tares.

Vichy carrots Young carrots, lightly scraped, cooked in water and butter until all the liquid has disappeared and the carrots are glazed with butter.

Vichy water Mineral water from springs at Vichy in southern France.

Vichy, a la A dish prepared or garnished with carrots.

Vichyssoise Cold leek and potato cream soup.

Vicine One of the toxins in broad beans, responsible for the acute haemolytic anaemia of favism.

Vienna bread Loaf with a very crisp, thin, highly glazed crust, with cuts on the upper surface, coarser than ordinary bread and with gas holes. It is baked in an oven which retains the steam.

Vienna flour Specially fine flour used to make strudel pastry, Vienna bread, and cakes.

Viennese coffee Ground coffee containing dried figs.

Villi, intestinal Small, finger-like processes covering the surface of the small intestine in large numbers, projecting some 0.5-1 mm into the lumen. They provide an enormous surface area for the absorption of digested food from the small intestine.

Vindaloo Indian curry, marinated and cooked in vinegar; highly spiced with chilli pepper and hence highly pungent.

Vinegar Not less than 4% solution of acetic acid; the product of two fermentations, first with yeast to convert sugars into alcohol, then this liquor, called gyle, is fermented with Acetobacter spp. to form acetic acid.

In most countries vinegar is made from grape juice

Malt vinegar is made from malted barley and may be distilled to a colourless liquid with the same acetic acid content but a more mellow flavour. Cider vinegar is made from apple juice; vinegars may be flavoured with a variety of herbs. Non-brewed condiment is a solution of acetic acid, 4-8%, coloured with caramel. Balsamic vinegar is made from grape juice that has been concentrated over a low flame and fermented slowly in a series of wooden barrels, made only around Modena, Italy.

Vintage The year of production of a wine, from the French vendage, for grape harvest. Used mainly for superior quality wines, when each year's production is matured separately. Ordinary and table wines are not generally dated, and indeed the production from more than one year may be blended.

Viogner A grape variety widely used for wine making, although not one of the classic varieties.

Violet The sweetly scented flowers of the wild violet are candied or crystallized and used as decorations in confectionery, or to make a sweet souffle. The flowers can be used to flavour syrups, and both flowers and leaves can be used in salads.

Virol Trade name for a vitamin preparation based on malt extract.

Virus Microscopic particle that contains genetic information, but must invade a cell to reproduce.

Viscometer instrument for measuring the viscosity of liquids.

Viscosity Term used of liquids to define their resistance to flow.

Vitamers Chemical compounds structurally related to a vitamin, and converted to the same overall active metabolites in the body. They thus possess the same kind of biological activity, although sometimes with lower potency.

When there are several vitamers, the group of compounds exhibiting the biological activity of the vitamin is given a generic descriptor.

Vitamin A A nutrient that promotes the growth and health of cells throughout the body and works as an antioxidant. Dark orange and green, leafy plants are good sources of beta-carotene, which is converted to Vitamin A in the body.

Vitamin A toxicity Retinol in excess of requirements is stored in the liver, bound to proteins, and is a cumulative poison. When the storage capacity is exceeded, free retinol causes damage to cell membranes. Carotene is not toxic in excess, since there is only a limited capacity to form retinol from carotene. The recommended upper limits of habitual daily intake of retinol are about 12.5 x reference intake for adults, but only 2.5 x reference intake for infants. Retinol is also teratogenic in excess, and for pregnant women the recommended upper limit of daily intake is 3300 μg.

Vitamin B 1 Thiamin; essential for the liberation of energy from foods, especially carbohydrates, and for nerve conduction. Deficiency, especially when associated with a carbohydrate-rich diet, results in the disease beriberi – degeneration of the sensory nerves in the hands and feet fast spreading through the limbs, with fluid retention and heart failure. Relatively acute deficiency, especially associated with alcohol abuse, results in central nervous system damage, the Wernicke-Korsakoff syndrome.

Good sources are whole grain and enriched bread and cereals, meat, yeast, potatoes, and peas; cooking losses can be as much as 50%.

Vitamin B 2 Riboflavin. Involved in a wide range of oxidation reactions, of fats, carbohydrates, and amino acids. Deficiency impairs cell oxidation and results clinically in a set of symptoms known as ariboflavinosis. These include cracking of the skin at the corners of the mouth, fissuring of the lips, and tongue changes; seborrhoeic accumulations appear round the nose and eyes.

Occurs mainly in yeast, liver, milk, eggs, cheese, and pulses; milk and milk products are probably the most important source in the average diet.

Vitamin B 3 Non-existent; term once used for pantothenic acid and sometimes, quite wrongly, used for niacin.

Vitamin B 4 Name given to what was later identified as a mixture of the amino acids arginine, glycine, and cystine.

Vitamin B 5 Name given to a substance later presumed to be identical with vitamin B 6 or possibly nicotinic acid; also used for pantothenic acid.

Vitamin B 6 Generic descriptor for three compounds: 1. The hydroxyl (alcohol) compound, pyridoxine;

2. The aldehyde, pyridoxal; and

3. The amine, pyridoxamine; and their phosphates. All are equally active biologically.

Deficiency causes abnormalities of the metabolism of the amino acids tryptophan and methionine; in rats deficiency causes convulsions and skin lesions and in dairy cows and dogs, anaemia with abnormal red blood cells. Dietary deficiency leading to clinical signs is not known in human beings, apart from a single outbreak in babies fed a severely over-heated preparation of formula milk in the 1950s; they showed abnormalities of amino acid metabolism and convulsions resembling epileptic seizures, which responded to supplements of the vitamin.

Rich sources include nuts, meat, fish, wholegrain cereals, and beans.

Vitamin C Also known as ascorbic acid. A nutrient that aids the body in iron absorption; heals cuts and wounds; and works as an antioxidant. Found in citrus fruits, bell peppers, broccoli, strawberries, and tomatoes.

Vitamin D Vitamin D 3 is calciol or cholecalciferol; formed in the skin by the action of ultraviolet light on 7-dehydrocholesterol, and hence not strictly a vitamin. However, in northern latitudes sunlight exposure may not be adequate to meet requirements, and a dietary source may become essential. Vitamin D 2 is a synthetic vitamer produced by irradiation of ergosterol. The name vitamin D 1 was given originally to. an impure mixture and is not used now.

The metabolic function of the vitamin is to control the body's calcium metabolism. It stimulates the absorption of dietary calcium from the intestine and calcium turnover in bone. Deficiency causes rickets in young children, osteomalacia in adults.

It is not widely distributed in foods, but is found in egg yolk, butter, fatty fish, and enriched margarine. There are no reference intakes for adults, since it is assumed that normal sunlight exposure will meet requirements; for the house-bound elderly the reference intake is 10 μg. The obsolete international unit of vitamin D = 25 ng calciol; 1 mg calciol = 40 iu.

Vitamin E The generic term for all tocopherols and their derivatives having the biological activity of RRR-alpha-tocopherol.

Vitamin F Sometimes used for the essential fatty acids.

Vitamin K Fat-soluble vitamin essential for the production of prothrombin and several other proteins involved in the blood clotting system, and the bone protein osteocalcin. Deficiency causes impaired blood coagulation and haemorrhage; sometimes called the antihaemorrhagic vitamin. Two groups of compounds have vitamin K activity: phylloquinones, found in all green plants, and a variety of menaquinones synthesized by intestinal bacteria. Dietary deficiency is unknown, except when associated with general malabsorption diseases. However, some new-born infants are at risk of developing haemorrhagic disease as a result of low vitamin K status, and it is general practice to give a single, relatively large dose of the vitamin by injection.

Vitamins, fat-soluble Vitamins A, D, E, and K, which are all soluble in lipids, but not in water.

Vitamins, water-soluble Vitamin C and the B vitamins, which are all soluble in water, but not in lipids.

Vita-wheat Trade name for a crispbread.

Vodka Made from neutral spirit, i.e. alcohol distillate mainly from potatoes, with little or no acid, so that there is no ester formation and hence no flavour. Polish vodka is flavoured with a variety of herbs and fruits.

Vodkatini A martini cocktail prepared with vodka rather than gin.

Volume The amount of space that is enclosed within an object or solid shape.

High-volume bread means a larger loaf.

Vongole small clams or venus mussels.

Votator Machine used for the continuous manufacture of margarine; the fat and water are emulsified, and the subsequent conditioning process carried out in the same machine.

W

Wafer Wafer can refer to a crisp, sweet, very thin flat dry cake which is often used to decorate ice creams. Wafers can also be made into cookies with cream flavoring sandwiched between them. They frequently have a waffle surface pattern. The word also refers to the special wafers made for Anglican Holy Communion services; the word host is used in the Roman Catholic liturgy. These holy wafers often have an image of the crucified Christ imprinted on them.

Walnuts The rough-shelled English walnut, black walnut, hickory nut, and butternut are all botanically walnuts. Common English walnut is Juglans regia.

Warfarin A chemical which antagonizes the action of vitamin K; used to reduce blood clotting in people at risk of thrombosis and, at much higher doses, as a rat and mouse poison.

Washing This term may refer to a number of aspects of fish handling and processing. Cleaning of the gut cavity of gutted fish by hand with jets of water or by means of a mechanical washer prior to boxing. The bleeding and washing steps may be combined in a mechanical system. It may refer to the cleaning of iced fish or thawed fish in water to remove mucous, slime or loose scales prior to filleting, or to preparation for some other process such as smoking or canning, or to the cleaning of shellfish prior to shucking. It may refer to the removal of excess salt from dry salted fish prior to smoking. It may refer to the process of treating fish mince or minced fish with successive water and brine washes to remove soluble proteins, pigments and lipids during the production of surimi.

Wassail 1. Spiced ale, drunk especially on festive occasions.

2. Salutation or toast drunk to a person's health.

Water activity The microbial and chemical stability of dry food and feed products depends on the a w of the product. It is a measure of the availability of water in the product, to react chemically, or to support the growth and metabolism of microorganisms, such as bacteria and moulds, during spoilage. It is expressed as the ratio of the water vapour pressure of the product, or solution, to that of pure water at the same temperature. The a w of pure water is 1.0. The a w of a saturated solution of sodium chloride at 25°C is 0.765. Reduction of the a w in a product to low levels, i.e. less than 0.75, by direct removal of water by drying, or by the addition of solutes such as salt or sugar, or by some combination of these factors, can restrict the growth of many microorganisms. Most bacteria cannot grow below a w 0.9, but Staphylococcus aureus can grow down to approximately a w 0.85, red obligate halophiles will grow at 0.75, and some yeasts and moulds can grow as low as a w 0.6 - 0.7. Dried foods with an a w less than 0.6 are microbiologically stable, such that the shelf life is not limited by microbial spoilage, provided that they remain dry. The microbial and chemical stability of dried fish products during processing and storage is dependent on their a w. In freshly caught fish, a w is above 0.95, and can be decreased by drying and salting processes to reduce the rate of microbial growth, as well as the range of microorganisms that can grow. The water activity of a dried fish product can be calculated from the water content and salt content.

Water balance The balance between intake and excretion of fluids. Average daily intakes are: as drinks, 1-1.5 L; as aqueous part of food, 0.5 L; and formed in the body by oxidation of foodstuffs, 300-500 mL; total 2-3 L. Losses from the lungs, 400-500 mL; through the skin 400-500 mL; in faeces 80-100 mL; in urine 1-1.8 L.

Total body water is 40-44 L, as: blood plasma, 2-3 L; extracellular fluid, 10 L; and intracellular fluid, 27-30 L. The kidneys control the volume of extracellular water by excreting water. Ingestion of sodium chloride (salt) raises the osmotic pressure of the extracellular water, causing thirst.

Water chestnut Seeds of Trapa natans and T. bicornis.

Water content The amount of water in a product. It is commonly determined by the loss in weight on heating a sample of the product at 105°C for 18-24 hours.

Water hardness Soap-precipitating power of water due to the formation of insoluble calcium and magnesium salts of the soap. Temporary hardness is removed by boiling, permanent hardness is not. May be measured in degrees Clarke; one degree = 1 part of calcium carbonate per 100 000 parts of water.

WatercressLeaves of Nasturtium officinale and N. microphyllum x officinale; eaten raw in salads.

Water-glass Sodium silicate; used at one time to preserve eggs, by forming a layer of insoluble calcium silicate around the shell, and so sealing the pores.

Waterless cooking Cooking in a heavy pan with tightly fitting lid, with a steam vent; only a minimal amount of cooking liquid is needed, but the food is not cooked under pressure.

Waterlow classification A system for classifying protein-energy malnutrition in children based on wasting and the degree of stunting.

Waterzooi Belgian; chicken or fish simmered in white wine with vegetables.

Waxes Esters of fatty acids with long-chain monohydric alcohols, e.g. beeswax, an ester of palmitic acid with myricyl alcohol; spermaceti, palmitic acid with cetyl alcohol. Animal waxes are often esters of the steroid alcohol cholesterol.

Waxing Coating fruits and vegetables with a thin layer of edible wax. In the case of apples and oranges this replaces the natural wax that is removed when the crop is washed; in the case of vegetables it is an addition; in both instances the waxing prevents loss of moisture, prolongs storage life, and improves the appearance.

Weaning foods Foods specially formulated for infants aged between 3 and 9 months for the transition between breast- or bottle-feeding and normal intake of solid foods.

Weenie American name for small sausages, abbreviation of wienerwurst.

Weetabix Trade name for a breakfast cereal prepared from wheat flakes. Two biscuits is a rich source of vitamin B 2 ; a good source of vitamin B 1, niacin, iron, ad copper; a source of folate; provides 4.8 g of dietary fibre; contains 1.4 g of fat of which 11% is saturated; supplies 140 kcal.

Weight, desirable Standardized tables of desirable weight for height for adults are based on life expectancy; both undernutrition and obesity are associated with increased risk of premature death.

Weight-for-height For children, can be used as an alternative to weight-for-age as an index of nutritional adequacy; for adults it is the only acceptable way of expressing weight relative to ideal or desirable weight.

Wellcome classification A system for classifying protein-energy malnutrition in children based on percentage of expected weight for age and the presence or absence of oedema. Between 60 and 80% of expected weight is underweight in the absence of oedema, and kwashiorkor if oedema is present; under 60% of expected weight is marasmus in the absence of oedema, and marasmic kwashiorkor if oedema is present.

Welschriesling A grape variety widely used for wine making, although not one of the classic varieties. Also known as Italian riesling; the wines cannot legally be labelled simply riesling.

Welsh rarebit Melted cheese, mixed with mustard powder, pepper, and brown ale, served on toast. Buck rarebit is Welsh rarebit topped with a poached egg.

Wernicke-korsakoff syndrome The result of damage to the brain as a result of vitamin B 1 deficiency, commonly associated with alcohol abuse. Affected subjects show clear signs of neurological damage with psychiatric changes characterized by loss of recent memory and confabulation.

Wet fish A term sometimes used to describe unfrozen fish, especially chilled fish, that are also called, sometimes incorrectly, fresh fish.

Wetzel grid Children are grouped by physique into five groups, ranging from tall and thin to short and thick-set. A healthy child will grow, as measured by height and weight, along one of these channels at a standard rate; if he/she deviates from the channel, malnutrition is suspected.

Wey The measure of 48 bushels of oats or 40 bushels of salt or 'corn'.

Whale Meat of Baleanoptera spp. A rich source of protein, iron, and niacin; vitamin B 2 & fat.

Wheat The most important of the cereals and one of the most widely grown crops. Many thousand varieties are known but there are three main types: Triticum vulgare, used mainly for bread; Triticum durum, largely used for pasta; and Triticum compactum, too soft for ordinary bread. The berry is composed of the outer, branny husk, 13% of the grain; the germ or embryo, 2%; and the central endosperm, 85%.

Wheatfeed Also called millers' offal and wheat offals; by-product from milling of wheat, other than the germ; bran of various particle sizes and varying amounts of attached endosperm.

Whelks Shellfish; several types of spiral-shelled marine molluscs, Buccinum undatum, Fusus antiquus. A rich source of protein, niacin, vitamin B12, and iron.

Whey The residue from milk after removal of the casein and most of the fat; also known as lacto-serum. Contains about 1% protein together with all the lactose and water-soluble vitamins and minerals, and therefore has some food value, although it is 92% water.

Whey cheese can be made by heat coagulation of the protein and whey butter from the small amount of fat. Dried whey is added to processed cheese; most whey is fed in liquid form to pigs.

Whim-wham Scottish; similar to syllabub.

Whiskey, whisky A grain spirit distilled from barley, rye, maize, or other cereal which has first been malted and then fermented. Most brands of whisky are a blend of pure malt whisky with spirit distilled from grain. The distilled spirit is diluted to about 62% alcohol and matured in wooden casks; Irish and Scotch whisky, made from malted barley, are matured for at least three years. Bourbon, made from malted maize, for at least one year. Sour mash bourbon is made from mash that has yeast left in it from a previous fermentation. Other American and Canadian whiskies are made from rye. Diluted after maturation and generally around 40% alcohol by volume, 220 kcal per 100 mL.

Both spellings permitted but generally whisky is the Scotch variety and whiskey the Irish and American varieties. Name derived from the Gaelic uisge beatha, water of life.

White fish Fish species in which the main reserves of fat are in the liver, e.g. cod. Fish with less than 2% fat in the flesh. Also referred to as lean fish which has other meanings. White fish, as the description of fish content in a product, means that the product may contain material from more than one white fish species. Not to be confused with the local or trivial name, whitefish, for the Japanese icefish.

White foots Fine, white precipitate of calcium and other salts deposited in jars of meat cured with rock salt.

White pudding Sausage made from white meat cereal, and spices. The French version, boudin blanc, includes eggs and onions. Irish white pudding is made from flake or leaf lard and oatmeal, spiced; served sliced and fried.

White spirits Distilled spirits from fermented fruit; eau de vie or alcool blanc in French, schnapps in German.

White spot Blemish on dried salted white fish, caused by formation of crystals of disodium hydrogen phosphate, usually on stale raw material stored in very dry conditions. Not to be confused with infections of the organism Ichthyophonus hoferi in the condition known as greasy haddock.

Whitebait Traditionally a mixture of young herrings and young sprats; they are caught together and impossible to separate on a commercial scale.

Whiting White fish, Merlangius merlangus.

Whole grain The entire edible part of any grain, including the germ, endosperm, and bran. Refined grain products typically have the bran removed.

Wholefoods Foods that have been minimally refined or processed, and are eaten in their natural state. In general nothing is removed from, or added to, the foodstuffs in preparation. Wholegrain cereal products are made by milling the complete grain.

Wholesome Description applied to food that is fit for human consumption as far as hygiene is concerned.

Wine Fermented juice from grapes, also made with other fruits and even vegetables with the addition of sugar. Red wines are made by fermenting the juice together with the skins at 21-29°C; white wines normally from white grapes by fermenting the juice alone at 15-17°C; rose by removing the skins after 12-36 hours, or by mixing red and white wines.

Beverages made by fermenting other fruit juices and sugar in the presence of vegetables, leaves, or roots are also called wines, although the legal definition may be restricted to the fermented grape. White wines are graded as dry, to sweet, on a scale of 1 to 9. Red wines are graded from A to E. Wines generally contain 9-14% alcohol, dry wines 70 kcal, sweet wines 120 kcal, and about 1 mg of iron per 100 mL; there are only traces of vitamins.

Wood chips Fragments of wood larger than sawdust that produce more flames and heat, and less smoke than sawdust when burned. See smoking and smoking kiln.

Woodruff A wild plant of woodlands on chalk or limestone; the leaves have the smell and flavour of new mown hay, and are used to flavour alcoholic or fruit drinks, or to make a herb tea.

Worcestershire sauce Worcester sauce is a widely used fermented liquid condiment. It is currently made with vinegar, molasses, corn syrup, water, chilli peppers, soy sauce, pepper, tamarinds, anchovies, onions, shallots, cloves and garlic. It is a flavouring used in many dishes, both cooked and uncooked, and particularly with beef. Filipino cooking uses it frequently as a marinade, especially with pork. It is an important ingredient in Caesar salad and in a Bloody Mary. Lea & Perrins ships it in concentrate form to be bottled abroad. Though a fermented anchovy sauce called *garum* was a staple of Greco-Roman cuisine and of the Mediterranean economy of the Roman Empire, "Worcester sauce" is one of the many legacies of British contact with India. While some sources trace comparable fermented anchovy sauces in Europe to the 17th century, this one became popular in the 1830s.

Wuntun Small dough parcels containing meat, boiled or deep-fried, eaten as an accompaniment to a meal or in a soup. Also known as chiao-tzu or wraplings. Sweet wuntun is deepfried, served with sugar or syrup, and is stuffed with sweetened bean paste.

X

Xanthan gum Xanthan gum is a natural gum polysaccharide used as a food additive and rheology modifier. It is produced by a biotechnological process involving fermentation of glucose or sucrose by the *Xanthomonas campestris* bacterium. It was discovered by an extensive research effort by the United States Department of Agriculture, which involved the screening of a large number of biopolymers for their potential uses.

One of the most remarkable properties of xanthan gum is its capability of producing a large increase in the viscosity of a liquid by adding a very small quantity of gum of the order of one percent. In most foods it is used at 0.5% or as low as 0.05%. The viscosity of xanthan gum solutions decreases with higher shear rates, this is called pseudoplasticity. Foods need high viscosity at low shear rates to be stable. But when consumed they cannot seem to be thick and heavy in the mouth. Due to the pseudoplastic properties of xanthan gum it can seem thin in the mouth but still have good stabilization properties. Unlike other gums it is very stable under a wide range of temperatures and pH, and is accepted as a safe food additive in USA and Europe, with E number E415.

In foods, xanthan gum is most often found in salad dressings and sauces. It helps to stabilize the colloidal oil and solid materials found in these products. It is also used in frozen foods and beverages. Toothpaste often contains xanthan where it serves as a binder to keep the product uniform. In the oil industry xanthan gum is used in large quantities. In most cases, the xanthan gum is used to thicken drilling fluids. These fluids serve to carry the solids cut by the drilling bit back to the surface. The widespread use of what are called horizontal wells and their demand for good control of drilled solids has lead to the expanded use of xanthan gum. Xanthan has also been used to bind fine tobacco solids together so that they can be formed into a strip that is used to extend the normal cut tobacco leaves.

Xanthia Cocktail made with equal amounts of yellow, Chartreuse, cherry brandy, and gin.

Xanthine A purine, interediate in the metabolism of adenine and guanine to uric acid. Caffeine is 1,3,7trimethylxanthine; theophylline is 1,3-dimethylxanthine; theobromine is 3,7-dimethylxanthine.

Xanthophylls Yellow-orange hydroxylated carotene derivatives; occur in all green leaves together with the chlorophyll and carotene, also present in egg yolk, Cape gooseberry, rose hips, etc. Most have no vitamin A activity. Include flavoxanthin (E-161(a)), lutein (161(b)), cryptoxanthin (E-161(c), which is converted into vitamin A), rubixanthin (161(e)), rhodoxanthin (161(f)), canthaxanthin (161(g)).

Xenobiotic Substances foreign to the body, including drugs and some food additives.

Xerophthalmia Advanced vitamin A deficiency in which the epithelium of the cornea and conjunctiva of the eye deteriorates because of impairment of the tear glands, resulting in dryness then ulceration, leading to blindness.

Xylitol A five-carbon sugar alcohol found in raspberries, endive, lettuce, and formed in the body as an intermediate in glucose metabolism; 80-100% of the sweetness of sucrose; used in sugar-free hard sweets and gelatine gums. Apart from being of low cariogenicity, xylitol is said to have an effect in suppressing the growth of some of the bacteria associated with dental caries.

Xyloascorbic acid Term used to distinguish vitamin C from isoascorbic acid, which is araboascorbic acid and has only slight vitamin C activity.

Xylose Xylose is an aldopentose – a monosaccharide containing five carbon atoms and including an aldehyde functional group. It has chemical formula $C_5H_{10}O_5$. Xylose is found in the embryos of most edible plants. It is said to be one of eight sugars which are essential for human nutrition, the others being galactose, glucose, mannose, N-acetylglucosamine, N-acetylgalactosamine, fucose, and sialic acid.

Y

Yakhni Indian bouillon or stock made from meat and bones, sometimes with the addition of vegetables. Garhi yakhni is concentrated and sets to a jelly.

Yam Yam is the common name for members of the genus *Dioscorea*. There are more than 150 different varieties of yam. Some species are cultivated for the consumption of their starchy tubers in Africa, Asia, Latin America and Oceania. The word *yam* comes from Portuguese *inhame* or Spanish *ñame*, which both ultimately derive from the Wolof word *nyami*, meaning "to eat". They are used in a similar fashion to potatoes, and are sometimes called a potato substitute, though in areas where yams are grown, potatoes are the newer vegetable.

Yam tubers can grow up to seven feet in length and weigh up to 150 pounds. The yam has a rough skin which is difficult to peel, but which softens after heating. Yam skins vary in color from dark brown to light pink. The majority of the yam is composed of a much softer substance known as the "meat". This substance ranges in color from white to bright orange in ripe yams. Yams are a primary agricultural commodity in West Africa and New Guinea. They were first cultivated in Africa and Asia about 8000 B.C., probably because they were able to flourish in environments where growing leafy vegetables and keeping livestock was difficult. To this day, the yam tuber remains crucial to survival in the region; it can be stored for four to six months without refrigeration, which makes it a valuable resource for the yearly period of food scarcity at the beginning of the wet season. Yams are used as food plants by the larvae of some Lepidoptera species including *Palpifer sordida*.

Yard-long bean Caribbean name for asparagus bean, Vigna sesquipedalis.

Yeast One-celled fungi that reproduce by budding. Some ferment carbohydrates, important for baking and brewing. Many biochemical properties of yeast are similar to those of higher organisms.

Yerba dulce The leaves of the Paraguayan shrub, Stevia rebaudiana, the source of stevioside and rebaudioside.

Yestamin Trade name for a variety of preparations of dried Saccharomyces yeast used to enrich foods.

Yield Generally, the amount of product that is obtained from the raw material. Of most interest in fish technology is the yield of edible material. Typical yields of edible meat from some major groups of fish are vertebrate fish, as skinless fillets40%shrimp and prawn 45% crab 30% bivalve shellfish

15% cephalopod 70%. Within these groups, yields vary with species, and within species with size, season, and condition.

Yoghurt Milk coagulated and fermented with two types of bacteria, Streptococcus thermophilus and Lactobacillus bulgaricus; may be stirred or set. May be pasteurized, when most of the bacteria are destroyed, otherwise termed live yoghurt. Bioyoghurts also contain Lactobacillus acidophilus and Bifidobacterium bifidum, which are claimed to enhance the growth of beneficial bacteria in the intestine.

Z

Zabaglione Zabaglione is an Italian dessert made with egg yolks, sugar, a sweet liquor, and sometimes cream or whole eggs. It is a very light custard, which has been whipped to incorporate a large amount of air. Zabaglione is traditionally served with fresh figs and is sometimes also spelled sabayon, zabaione, or zabajone. It is also popular in Argentina, where it is known as sambayón.

Zabaglione originated in Venice when this city ruled the Adriatic. Originally, sweet Cyprus wine was used but with the retreat of the Venetian Republic, Marsala wine began to be used instead. Another change in the recipe is the use of sugar instead of honey, the original ingredient. Classical zabaglione uses raw egg yolks but today many may prefer to prepare it in a bain-marie. However, it is mainly recommended to use a simple double boiler with a heat resistant bowl not reaching the water and only simmer to avoid scrambling the eggs. Beaten egg white is also widely replaced by whipped cream.

Zeaxanthin One of the carotenoid pigments in maize, egg yolk, and Physalis; has no vitamin A activity; used as a colouring.

Zebrine Variety of aubergine with purple and white stripes. zedoary root Root of the Indian plant Curcuma zedoaria, a member of the ginger family. Used in the manufacture of flavours and bitters.

Zein The major protein of maize, very poor in lysine and tryptophan.

Z-enzyme Enzyme found associated with amylases, that attacks the few ß-1,3-links present in amylose. Pure, crystalline ß-amylase will convert only 70% of amylose to maltose; it requires the presence of the Z-enzyme for complete conversion.

Zeocarb An ion exchange resin.

Zinc An essential mineral which forms the prosthetic group of a large number of enzymes, and also the receptor proteins for steroid and thyroid hormones and vitamin A and vitamin D. Deficiency results in hypogonadism and delayed puberty, small stature, and mild anaemia; it occurs mainly in subtropical regions where a great deal of zinc is lost in sweat, and the diet is largely based on unleavened wholemeal bread, in which much of the zinc is unavailable because of the high content of phytate.

Meat, fish, legumes, and wholegrain cereals are rich sources.

Zinfandel Red grape variety especially grown in California for wine making. Zinfandel wines may be blush or red. White zinfandel wines are really rose. The wines have a characteristically fruity flavour.

Zooplankton Wide variety of very small marine crustaceans and other invertebrates, mixed with the young of larger fish, which live upon the phytoplankton and serve, in turn, as a food supply of small fish and other marine life.

Zubrowka vodka flavoured with a single sprig of bison grass in each bottle.

Zucchini Italian variety of marrow developed to be harvested when small. American and Australian name for courgette.

Zucker rat A genetically obese strain of rat used in research.

Zuppa inglese rich cake or trifle.

Zwieback German name for twice-baked bread or rusk. Ordinary dough plus eggs and butter, baked, sliced, baked again to a rusk, and sometimes sugar-coated.

Zymase Zymase is an enzyme complex that catalyze glycolysis, the fermentation of sugar into ethanol and carbon dioxide. As the conversion takes place, the reaction will gradually slow down. They occur naturally in yeasts. Zymase was first isolated from the yeast cell in 1897 by a German chemist named Eduard Buchner who fermented sugar in the laboratory. A British chemist by the name of Sir Arthur Harden divided zymase into two varieties dialyzable and nondialyzable in 1905.

Zymogens A zymogen or a proenzyme, is an inactive enzyme precursor. A zymogen requires a biochemical change for it to become an active enzyme. The biochemical change usually occurs in a lysosome where a specific part to the precursor enzyme is cleaved in order to activate it. The amino acid chain that is released on activation is called the activation peptide. Very often the suffix "-ogen" can be added to the name of the enzyme to indicate that it is in the precursor form. Examples of zymogens are trypsinogen, chymotrypsinogen, pepsinogen, most of the proteins of the coagulation system, some of the proteins of the complement system, the caspases. Zymogen is another term for a pro-enzyme. Some enzymes have a pro- prefix to denote their inactive status eg. Proalbumin which is converted into albumin.